T0314538

Spatial Analysis of Coastal Environments

At the convergence of the land and sea, coastal environments are some of the most dynamic and populated places on Earth. This book explains how the many varied forms of spatial analysis, including mapping, monitoring and modelling, can be applied to a range of coastal environments, such as estuaries, mangroves, seagrass beds and coral reefs. Presenting approaches to modelling, which draw on recent developments in remote sensing technology, geographical information science and spatial statistics, it provides the analytical tools to map, monitor and explain or predict coastal features. With detailed case studies and accompanying online practical exercises, it is an ideal resource for undergraduate courses in spatial science. Taking a broad view of spatial analysis and covering basic and advanced analytical areas such as spatial data and geostatistics, it is also a useful reference for ecologists, geomorphologists, geographers and modellers interested in understanding coastal environments.

Sarah M. Hamylton is a Senior Lecturer at the University of Wollongong (Australia), where she specialises in the spatial analysis of coastal environments. Her research applies geospatial technology to tropical coastal environments. Her maps have helped establish marine protected areas, and her models of how climate change impacts coral reefs have informed national coastal policies.

'I wish this book had been around when I was a student! It ticks all the boxes: the primary focus on spatial analysis and interrogation of geospatial data is essential for sound, sustainable and evidence-based decision-making, and will give invaluable practical skills to students and practitioners alike; while the adoption of landscape ecology as the underpinning conceptual framework emphasises the need for joined-up, holistic and ultimately spatially-determined thinking in coastal science and management. The author shows a deep understanding of her subject matter, and her enthusiasm for, and love of, the coast stands out. Even the more complex ideas and methods are explained clearly and in an easily accessible, student-friendly manner. Although written for students of the coast, many of the concepts and methods introduced here will be readily transferrable to other areas of Earth Science specialism where geospatial expertise is needed.'
Darius Bartlett, *University College Cork, Ireland*

'*Spatial Analysis of Coastal Environments* is a rare and overdue resource that provides a comprehensive overview as well as an introduction to an array of important spatial analytical techniques and issues. Students and professionals new to coastal GIS will find the introductory coverage of data sources, mapping principles and analysis techniques easily accessible. Experienced researchers, coastal managers and planners, and instructors will take value from the coverage of advanced techniques such as geostatistics, modelling and characterising uncertainty. The book is richly and usefully illustrated with both conceptual and case study maps and graphics. I expect this book to fill an important void and, through its readers, further expand the scientific and practical application of GIS to coastal environments.'
Thomas R. Allen, *Old Dominion University, USA*

'It should become essential reading for students of coastal environments, demonstrating how spatial analysis methods, together with GIS, can enrich and bring new insights to the study of this important field.'
Robert Haining, *University of Cambridge, UK*

Spatial Analysis of Coastal Environments

SARAH M. HAMYLTON
University of Wollongong, New South Wales

CAMBRIDGE
UNIVERSITY PRESS

University Printing House, Cambridge CB2 8BS, United Kingdom
One Liberty Plaza, 20th Floor, New York, NY 10006, USA
477 Williamstown Road, Port Melbourne, VIC 3207, Australia
4843/24, 2nd Floor, Ansari Road, Daryaganj, Delhi – 110002, India
79 Anson Road, #06-04/06, Singapore 079906

Cambridge University Press is part of the University of Cambridge.

It furthers the University's mission by disseminating knowledge in the pursuit of education, learning, and research at the highest international levels of excellence.

www.cambridge.org
Information on this title: www.cambridge.org/9781107070479
10.1017/9781107707412

First published 2017

Printed in the United Kingdom by TJ International Ltd. Padstow Cornwall

A catalogue record for this publication is available from the British Library.

ISBN 978-1-107-07047-9 Hardback

Additional resources for this publication at www.cambridge.org/hamylton

For Gran, who was always interested.

CONTENTS

Colour plate section can be found between pages 128 and 129

FOREWORD

As the noted econometrician and geographer Luc Anselin once said, 'Spatial is special'. Many of us without realising have probably first learned about spatial analysis at a very young age, perhaps as soon as we began to walk and to become fully aware of our geographic surroundings. More often than not, the first step in spatial analysis is understanding *where*. As a small child, you became aware of where you were, in the bedroom or in the kitchen. Your skills then started to expand into a second development phase, that of navigation. How do you go from one room to the next, and then how do you go from home to school? You didn't need to take a formal course in spatial analysis, but it was still becoming ingrained in your everyday life. Spatial analysis is a diverse and comprehensive capability that includes the simple visual analysis of maps and imagery, the determination of how places are related, the finding of the best locations and paths, the selection of the most appropriate site, the detection and quantification of patterns, and even advanced predictive modelling. The process of spatial analysis occurs every day in the human brain. However, over the last four decades, our ability to solve complex spatial problems has grown exponentially with technologies that include the global positioning system, real-time sensors, navigation algorithms, and the technology that brings them all together, the geographic information system or GIS. Spatial analysis has always been a hallmark of GIS, the 'numerical recipes' within, which set GIS apart from other forms of computerised visualisation and information management. With GIS we pose questions and derive results using a wide array of analytical tools to help us understand and compare places. One special place is the coast.

The coast is critical. Nearly 50% of the world's population lives within 60 km of a coast, and the United Nations estimates that by 2020 this will rise to 75%. This is because the coastal zone not only is a beautiful place to be, but also is critical for human life itself, with its provision of food and energy and its modulation of the weather. At the same time, the news is replete with stories of the hazards of sea level rise, hurricanes, tsunamis, rogue waves, coastal flooding, shark attacks, toxic spills, oxygen-poor 'dead zones', and even modern-day pirate attacks. With 60% of global coral reefs under immediate threat, 35% of global mangroves in dangerous decline, 80% of native oyster stocks in danger, and a bevy of other threats, the coast has become not only a desired destination but also a focus of intense restoration. Here spatial analysis is particularly needed for restoration site selection, long-term monitoring, and designing of 'greener' infrastructures, preferably with large doses of community involvement and political will.

Spatial is special. The coast is critical. When you combine both into a *textbook* aimed at second and third year undergraduates (rather than the customary edited compilation of applied research papers), you have a unique book for raising awareness and inspiring action. This is because it is the next generation of coastal planners, conservationists, policy-makers, and well-informed 'citizens of the planet' that must be taught the basics of thinking spatially and critically about the coast, in rigorously manipulating and analysing spatial data, and in rapidly prototyping and delivering repeatable GIS solutions for a range of coastal applications. It is via spatial analysis that the next generation will see innovations in the problems they will be able to solve, in the stories that they will be able to tell, and the decisions that will help make their organisations and governments more successful. And it is within the classroom that students learn languages of many kinds, spatial analysis being one of those languages. This language consists of a core set of questions that we ask, a taxonomy that organises and expands our understanding, and a set of fundamental steps embodying how we solve spatial problems. Languages also have dialects, and *coastal* spatial analysis is one such important dialect.

This book is a great exemplar of not only *how* to conduct spatial analysis in coastal environments, but also why this is critically important. With this book, instructors and students will see that this is not just about eyeballs on a map, but about the invisible rubber bands of mathematical manipulation of different layers of data. With this book, you also possess an effective guide to coupling the appropriate data with the analysis to effectively communicate scientific results, and to transforming those results into actionable information, information that is useful in decision cycles for coastal restoration and resilience. The arrival of this text couldn't be more timely, as the time is *now* for governments, communities, NGOs, and, yes, universities to go beyond just an exploration and discussion of ideas. We need to teach our students to effectively use spatial analysis to help guide the planet towards a more resilient future ... before time runs out.

Dawn J. Wright
Chief Scientist, Environmental Systems Research Institute (Esri)
Courtesy Professor of Geography and Oceanography, Oregon State University

PREFACE

Learning to think in spatial terms has changed my outlook considerably. Many research challenges, management problems and policy issues are geographical at heart. The application of spatial analysis has the potential to contribute insight through mapping, monitoring and modelling activities that express and develop our understanding of a range of phenomena. My primary motivation in writing this book is to raise awareness of the many and varied forms of spatial analysis that exist, and how these can be profitably applied to coastal environments.

Over the last 15 years I have conducted many different types of geospatial analysis working as a coastal environmental consultant or researcher. This includes work in Fiji, Thailand, the Philippines, Saudi Arabia, the Seychelles, Belize, the British Indian Ocean Territory (Chagos Islands), New Caledonia and Australia. Over this time I have experienced first-hand the profound ways in which maps and analytical results influence discussions around coastal policy or management decision-making. This book represents a timely and accessible consolidation of the learning outcomes from the projects I have worked on, intended for both students and people working in coastal environments.

While the applications described in this book have a coastal focus, the techniques are applicable across a wide range of different phenomena. For example, the tests described for the detection of point clusters of shark attacks in Chapter 3 were originally taught to me in the context of detecting hotspots of crime in cities or disease across countries. My aim is to provide a comprehensive outline of the *theoretical principles* underpinning the conduct of spatial analysis in coastal environments. This incorporates widely applicable basic spatial theory, skills and approaches that are helpful for any spatial analysis. While different software and statistical packages for conducting analysis may come and go, the theoretical and fundamental knowledge base upon which they are founded endures. Such a knowledge base is valuable and necessary for thinking creatively and critically to analyse phenomena in coastal environments. My intent is also to capture the state of the art in spatial analysis of coastal environments through carefully selected case studies that emphasise recently developed geospatial technology. For example, the use of unmanned aerial vehicles (drones) to map coastlines and the construction of interactive Web-based geographic information system (GIS) interfaces to deliver the mapped results of spatial analysis are both developments that have emerged relatively recently in the geospatial industry.

Another motivation for this book is to provide a text for the increasing number of undergraduate courses in the UK, USA and Australia that teach spatial sciences within a coastal context. In spite of a substantial growth in the application of geospatial technologies to coastal environments and a growing body of Higher Education courses on the topic, no textbook exists that comprehensively covers the application of GIS to coastal environments. I hope that both students and practitioners wanting to learn more about how to collect and analyse spatial information on coastal environments will find this book useful. Although some of the topics covered are complex, they are introduced at a basic level and elaborated on in a user-friendly manner. A variety of pedagogical features have been included to facilitate comprehension and to guide learning, including real-life case studies demonstrating the application of analytical theory, a series of associated practical exercises offering the reader the opportunity to manipulate and analyse companion datasets, summaries of key points at the end of each chapter and a comprehensive glossary of important terminology.

This book covers all aspects of spatial analysis. It enables the reader to break down, interrogate and summarise patterns in coastal environments. Analytical methods draw from a range of disciplines, including ecology, geomorphology, geography and statistical modelling. The book builds in complexity, beginning with the collection of spatial data and basic geographical analysis of raster and vector data. Mapping and monitoring exercises are presented that draw on recent developments in remote sensing technology, geographical information science and spatial statistics to represent coastal features and assess how they have changed over time. The development of empirical geographical models to explain and predict coastal phenomena is described. Many processes in coastal environments produce patterns that are spatially autocorrelated, and a range of geostatistical techniques are presented for quantifying and characterising the spatial structure that underpins this autocorrelation. Methods for evaluating the uncertainty of spatial analysis are reviewed, along with techniques for the effective and responsible presentation and communication of spatial analysis results.

Chapter 1 defines the scope and structure of the book, providing introductory guidance on spatial analysis to the coastal researcher while also introducing coastal environments to the spatial analyst. Chapters 1, 2 and 3 provide an introductory foundation for carrying out spatial analysis, including an overview of different types of analysis, data considerations and basic analytical operations. A tripartite framework of mapping, monitoring and modelling activities is adopted for the application of increasingly complex levels of spatial analysis to coastal environments. This forms an organisational principle for Chapters 4, 5 and 7. Technical and analytical considerations are set out relating to each of these activities, including the use of remote sensing technology and the nature of datasets commonly employed for coastal mapping in both two and three dimensions, tools for observing coastal change over time, repositories of baseline data against which change can be measured and the development of both predictive and explanatory models. Chapter 6 introduces key spatial statistical principles that are often invoked in the process of model development. Any form of spatial analysis should be accompanied by an evaluation of the uncertainty associated with that analysis, and Chapter 8 presents a range of methods for achieving this. Finally, Chapter 9 provides advice on how to present and communicate the results of spatial analysis effectively, including through online Web viewers and factors to be considered in the organisational uptake of spatial analysis.

ACKNOWLEDGEMENTS

This book has taken shape over the last two years at the University of Wollongong, but has its roots in research that goes back around 14 years to Fiji. This is where I first dived on a coral reef and fell in love with coastal environments. At the time I was volunteering for Coral Cay Conservation, which set me (like many UK-based students of environmental science) on a long and enjoyable path of coastal research. Since then, I have had the good fortune to discover all the exciting prospects that spatial analysis offers for better understanding coastal environments. Particularly formative in this regard was Bob Haining's introduction to analysing and modelling with spatial data, principally the issue of spatial autocorrelation. Around this time, I spent an invaluable week at the University of Exeter, where Alastair Harborne taught me how to make maps from remote sensing images, a skill that has been central to my research since.

Several research fellowships have enabled me to undertake field-trips that have contributed to the examples described in the book. These include a Domestic Research Studentship (University of Cambridge), a Khaled bin Sultan Living Oceans Foundation Fellowship (for which I am particularly grateful to Prince Khaled bin Sultan and Captain Phil Renaud), an Australian Bicentennial Fellowship (King's College, London) and a Lizard Island Research Station Fellowship.

I have had the benefit of working with many excellent colleagues and research collaborators over the years. In particular, the examples set by Tom Spencer and David Stoddart, whom I am proud to call my academic father and grandfather, respectively, have inspired me to do research that makes a difference. The opportunity to publish this book with Cambridge University Press was created by Colin Murray-Wallace, whose interest and encouragement has been much appreciated. Similarly, Colin Woodroffe has provided invaluable help through intellectual discussions, proofreading and an unfailing generosity in supporting my growth as a researcher. Marji Puotinen, whose GIS course I inherited, provided me with excellent guidance as to the skills necessary for a well-rounded education in GIS, alongside the *'Geographic Information Science & Technology: Body of Knowledge'*. I am also indebted to Professor Noel Cressie, who provided general feedback on the manuscript, including the need for a more thorough discussion of directional spatial dependences in coastal environments.

I am grateful to the University of Wollongong for allowing me the time and intellectual space to see this book project through to completion. Many thanks are due to two anonymous reviewers for positive and productive comments on the initial book proposal. Parts of this book have been tested on undergraduate and postgraduate students at the Universities of Cambridge, Wollongong and Sydney, whom I thank for the opportunity to refine exercises and chapters. I am grateful to John Hedley, and PhD students Tom Doyle and Stephanie Duce for feedback. Their sharp eyes and useful suggestions improved early chapter drafts. Susan Francis and Zoe Pruce at Cambridge University Press provided unwaveringly positive support and help throughout the two years it has taken to put this book together.

A number of individuals and organisations have given permission for me to use diagrams, and they are acknowledged in the captions of the appropriate figures. In particular, Matt Smith has generously allowed me to use his photographs, both as a front cover and throughout the book. Thanks are due to the following people for allowing me to use their data in case studies and examples throughout the book: Tom Spencer and Annelise Hagan (University of Cambridge) for the Aldabra vegetation change detection case study; Holly East (University of Cambridge) for the analysis of lagoon volumetric and shoreline planimetric change at Diego Garcia Atoll; Richard Barnes (University of Cambridge) for the invertebrates datasets in Moreton Bay; Eglantine Chapuis (Centre d'Estudis Avanşats de Blanes) for the population data collected along the intertidal rocky shorelines of Catalonia; Rafael Carvalho (University of Wollongong) for the bathymetric soundings from the Shoalhaven Estuary; Javier Leon (University of the Sunshine Coast) for the idea of modelling uncertainty of tsunami inundation; John Hedley for the confidence interval map of Sykes Reef; and Rob Thieler (US Geological Survey) for the Nantucket shoreline change example.

Finally, thanks to my mother, father and brother, Kathryn, Gordon and Michael, who have given me such support over the years. And, of course, to my family – Valerie, Michael, Henry and Alexander – whose enthusiasm, tolerance and encouragement have helped immeasurably.

1 The Application of Spatial Analysis to Coastal Environments

1.1 Geographical Foundations of Spatial Analysis

The use of quantitative techniques in geographical enquiry stems from a simple belief that, in many situations, numerical data analysis or quantitative theoretical reasoning provides an efficient and reliable means of obtaining knowledge about the world. Spatial analysis is often associated with quantitative geography, which evolved during the 1950s and 1960s as the increased availability of numerical data provided a basis for geographical reasoning. The quantitative revolution marked a major turning point in investigative methods, with a transition from a descriptive to an empirical geography (Livingstone, 1993; Inkpen and Wilson, 2013). Quantitative geography can broadly be defined as the development of spatial theory through analysis of numerical spatial data and the construction and testing of mathematical models of spatial processes (Fotheringham et al., 2000). From a philosophical viewpoint, it is often aligned with positivist perspectives that use empirical approaches to define laws pertaining to spatial processes. These often have the objective of explaining observations made about the natural world, or predicting its characteristics in space and time. Yet quantitative techniques do not provide a flawless argument. Indeed, it is precisely because of the need to account for error and uncertainty in both data collection and model formulation that statistical methods have come to dominate quantitative geography (Fotheringham et al., 2000). A common objective of many quantitative geographers is therefore to conduct analysis that maximises knowledge on spatial processes while simultaneously minimising error. Spatial analysis has advanced progress towards this objective over the last few decades, assisted by developments in technology such as remote sensing instruments and the global positioning system (GPS) and allied computational and statistical methods for analysing large spatially referenced datasets. The aim of this book is to explore the exciting foundation that such developments provide for analysing, and better understanding, coastal environments.

1.2 Definition of Spatial Analysis: Academic and Analytical Origins

Spatial analysis is defined as a collection of techniques and statistical models that explicitly use the spatial referencing associated with each data value or object under study (Upton and Fingleton, 1985; Fotheringham and Rogerson, 2013). The spatial sciences are largely based on empirical methods that involve making observations and collecting data about the natural world (Upton and Fingleton, 1985; Fotheringham et al., 2000). Spatial analysis is therefore clearly differentiated from non-spatial analysis.

The latter does not engage with the spatial referencing of data values but interrogates data points and associated trends regardless of their geographical distribution. Within this broad definition, several families of spatial analysis techniques can be defined, including exploratory and confirmatory spatial data analysis, geostatistics (or spatial statistics) and map algebra (Figure 1.1). It is also important to distinguish between spatial analyses that incorporate aspects of uncertainty, such as locational uncertainty, and those that do not, as these can influence the results of spatial statistical analyses (Cressie, 2002).

Spatial analysis has a role to play in supporting the search for scientific explanation. To this end, two families of spatial analytical techniques can be further differentiated as exploratory and confirmatory spatial data analysis. Exploratory spatial data analysis (ESDA) builds on methods of exploratory data analysis to investigate observable patterns in a given dataset (Tukey, 1977). This may involve identifying broad-scale trends, spatial clusters or outliers. ESDA is often a precursor to confirmatory spatial data analysis, which further evaluates datasets using traditional statistical tools of inference, significance and confidence limits to quantify the extent to which data observations could be expected to occur by chance alone, or because of some underlying process (Gelman, 2004). Confirmatory or inferential statistical methods therefore provide a basis for making statements (or inferring) about large populations of data samples on the basis of smaller ones. Where significant relationships are found, inference can then be used in a predictive manner to simulate values either into the future or across unsampled geographic areas. Spatial models can therefore be used as important tools for furthering our understanding of natural phenomena through explanation and prediction.

Spatial analysis has a more general role to play in problem solving because of spatial dependence. This is the phenomenon whereby observations close together in geographic space are more alike than those that are further apart. This generic property of many phenomena can be exploited in problem solving, often forming an important underpinning principle to spatial statistical or geostatistical analysis. This structured component of spatial dependence can be usefully employed as a statistical predictor in various geostatistical techniques, for example, interpolation methods such as kriging (see Chapter 6). Yet it also raises issues that make it difficult to apply classical (i.e. non-spatial) statistical techniques to a problem. This is because the spatial dependence introduces data redundancy and a departure from fundamental statistical assumptions, such as the independence of observations.

The techniques of map algebra, or cartographic modelling, provide another clearly distinguishable class of spatial data analysis that treats maps as datasets in their own right. Map-based operations are used to generate new maps, often through the application of simple algebraic manipulations. By overlaying multiple maps on top of each other, the analyst is able to implement logical conditions (AND, OR, XOR) through arithmetic operations (+, −, ×) to identify areas on the map that meet a range of user-specified criteria. For example, the condition 'AND' is implemented with the + operand. It is achieved by adding two maps together to identify the areas on a map that simultaneously satisfy a set of conditions on two or more variables (Arbia et al., 1998). By structuring a sequence of multiple arithmetical operations, it is possible to successively apply multiple criteria to interrogate the attributes associated with a given area as a basis for routines such as multi-criteria decision analysis (see Chapter 3) (Kitsiou et al., 2002).

Many frameworks of enquiry combine several different components of spatial analysis into a powerful and practical methodological framework. For example, interpolation may be used to convert discrete point samples of data to a continuous raster grid. Then each individual grid cell might have equations applied to it that were derived from confirmatory methods (e.g. autoregression) using a cellular automata approach before the resultant layer is manipulated alongside another layer using map algebra techniques. In this way, several different components of spatial analysis are brought together into a framework for investigating the natural world.

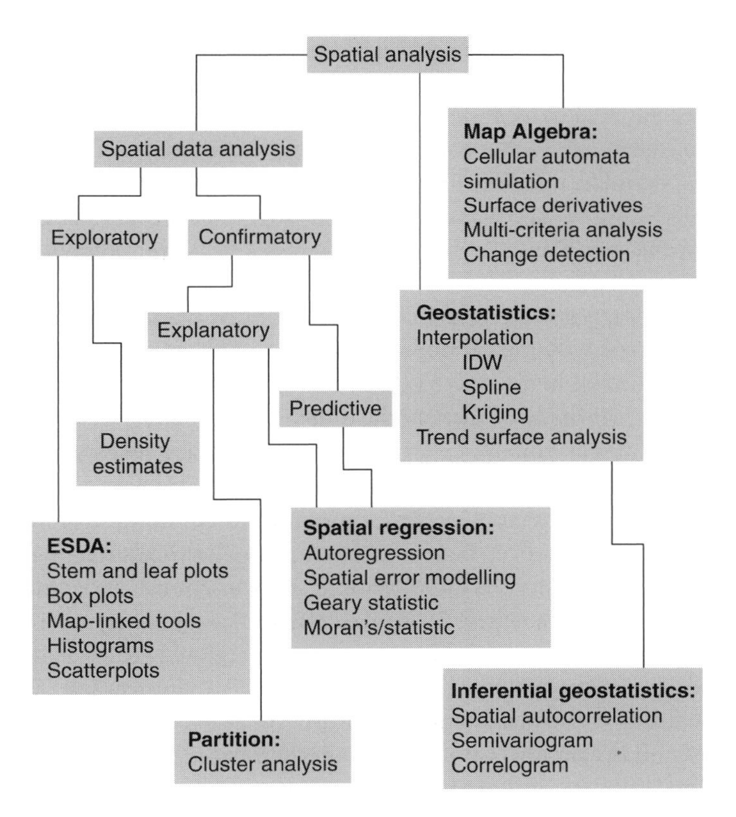

Figure 1.1 Taxonomy of the different types of spatial analysis described in the analytical chapters (Chapters 3, 5, 6 and 7) of this book. Boxes with bold headings represent established techniques; other boxes further up the taxonomy group these techniques by their analytical objectives.

1.3 Location Matters in Coastal Environments: Place and Space

A spatial analyst understands places by constructing explanations that draw on the ideas and methods of allied disciplines, including geology, biology, environmental science, oceanography, biogeography, landscape ecology or geomorphology (or some combination of these). One distinguishing feature of the environmental sciences, as opposed to many areas of experimental science, is that they are observational sciences. Outcomes have to be taken as they are found. The analyst is not usually in a position to test their ideas by experimenting with processes that they think may be important for a particular explanation. In terms of data, they cannot manipulate levels of explanatory variables (a variable being any characteristic, number or quantity that can be counted) or replicate outcomes. Nevertheless, an observational scientist is able to use information on location to induce variability in an explanatory variable as a form of 'natural laboratory'. Location can act in two distinct ways in this regard: location as *place* (and, by extension, context) and location as *space* (Haining, 2003).

Coastal environments are made up of a diverse array of places. Each of these places is subject to a unique set of local environmental conditions. In the context of spatial modelling, the treatment of location as 'place' draws on the diversity of places to vary the value of associated variables that may be of interest to the analyst. Consider the case of trace metals found in the mussels of the northwestern Mediterranean. A regional coastal survey measured levels of trace metals in mussels across stations at 15 different shoreline locations adjacent to urban areas with different levels of industrialisation (Fowler and Oregioni, 1976). The varying levels of industrial activity gave rise to measurable differences in levels of heavy-metal pollution into neighbouring coastal waters. This information can be combined with mussel population data to analyse the relationship between levels of water pollution in the direct vicinity of the mussel beds and the occurrence of trace metals in the soft tissue of these populations. In such an analysis, it is necessary to account for the effects of possible external confounding processes or variables that may correlate with the variables explicitly under consideration (e.g. mussel age, hydrodynamic mixing of the coastal waters, etc.). Highest trace-metal values were found in mussel tissue samples that were extracted from beds adjacent to port cities, and areas in the vicinity of river discharge. This indicates the importance of *place* as context for mussel beds along the coastline (Fowler and Oregioni, 1976). The effect of place can operate at a hierarchy of scales, from the immediate neighbourhood up to regional scales, and may include both contextual and compositional effects. For example, variation in trace-metal concentration might also be explained in terms of mussel age or the specific types of industrial development on the adjacent coastline (the compositional effect).

Gradients of different localised environmental conditions are discernible within coastal environments. The alternative 'space' view of location enters into explanation in the observational sciences by emphasising how objects are positioned *with respect to one another* along such gradients and how this relative positioning may enter explicitly into explanations of observable environmental variability. Such variability might arise from interactions between the places that are a function of those spatial relationships, and can be expressed through measurements such as distance, gradient and neighbourhood context. A gradient is a local property of a space that arises from the relative magnitude of a variable in two neighbouring areas (Burrough and McDonnell, 2011). For example, the levels of wave energy in a semi-exposed bay during a storm reflect the water depths and degree of exposure of a single location. But they also reflect the seafloor composition of neighbouring locations that determine wave energy through their capacity for frictional attenuation. Thus, the gradient between these two areas induces an effect in both locations that is not purely a consequence of the inherent properties of the two respective neighbourhoods. Rather, the instantaneous level of wave energy for a given location reflects the multiple paths along which the wave may have travelled to reach that location. Returning to the example of trace metals in the mussel beds of the NW Mediterranean, a *space* treatment of location might investigate the concentration of trace metals in mussel tissue as a function of the distance of individual mussels from the point of river discharge at an individual shoreline station. In a meta-analysis of the influence of local environmental conditions on the characteristics of 103 low-lying reef islands along the Great Barrier Reef (GBR), distance from the mainland coastline emerged as a strong predictor of island character. This reflected the variable sea-level history of each island across the sloping GBR shelf. A gradient of island age was inferred that corresponded to the length of time available for island evolution, with more evolved islands built up on higher platforms nearer the coastline than their younger seaward counterparts that had to build up from lower antecedent platforms (Hamylton and Puotinen, 2015). Finally, analyses that draw on the contextual surroundings of a given data point represent another family of contextual analyses that treat location as *space*. These include hydrological flow models that use slope information to map river flow through catchments, or the analysis of viewsheds that express visible areas accounting for the elevation of the neighbourhood surrounding a given location.

1.3.1 Examples of the Role of Place and Space Within Coastal Environments

Early work on coastal environments examined why different areas of the intertidal zone gave rise to distinct benthic community assemblages (Stephenson and Stephenson, 1949; Chapman and Trevarthen, 1953). More recently, spatially referenced datasets covering large geographic areas (e.g. 100 km × 100 km) provide consistent and synoptic information on physical and environmental characteristics of the coastal zone. These include digital elevation models extending from the terrestrial to the marine domain, and hydrodynamic models of wave power or current strength. They represent important sources of information on potential environmental explanatory variables. Confirmatory data analysis techniques draw on *location* to link data values that correspond to dependent and independent variables in overlapping datasets.

Changes to the way that humans use space within coastal zones have drawn attention towards the geographical variation of coastal development, including land-use transitions associated with expanding urban areas, and industries such as agriculture and aquaculture reclaiming riparian swamplands (Davies, 1972). More recently, coastal spatial datasets have provided a foundation for coastal vulnerability assessments that combine information from multiple sources to form an index of vulnerability to the various impacts that climate change can have on coastal environments. These include sea-level rise and the increased frequency and intensity of storms and cyclones. Vulnerability indices are often calculated from information on physical variables such as geomorphology, coastal slope, relative sea-level rise, shoreline erosion or accretion, tidal range, wave height, exposure and infrastructure (Gornitz, 1991; Gornitz et al., 1994; Abuodha and Woodroffe, 2010). They may also incorporate socioeconomic information, such as population, cultural heritage, roads, railways, land use and conservation status (McLaughlin et al., 2002). Thus, the development of vulnerability assessment tools, such as DIVA (Hinkel and Klein, 2009) and Smartline (Sharples et al., 2009), has provided a framework for integrating multiple datasets for an assessment of contextual vulnerability. This draws explicitly on a treatment of location as *place*.

Species distribution models (SDMs) correlate species occurrence with environmental data in geographic *space* to explain and predict the distribution of a given species of interest. Although SDMs were first applied in the terrestrial domain, they are increasingly being used in marine environments, where their success at constructing reliable explanations and predictions depends on unique physical properties of marine habitats and biological characteristics of marine organisms being accounted for. These properties include enhanced dispersal ability, species interactions and shifting environmental requirements throughout life-history stages (Robinson et al. 2011). Owing to both their commercial value and data availability, fish are often the focus of such models. These have practical value for planning marine protected areas and designating essential fish habitats, alongside permissible fishing zones (Leathwick et al., 2008; Valavanis et al., 2008; Maxwell et al., 2009). Maxwell et al. (2009) modelled the distributions of commercially harvested plaice, sole and ray in UK waters to inform conservation planning. Motivated by population declines due to habitat degradation and fisheries by-catch mortalities, models for marine mammals are also relatively common (Redfern et al., 2006; Panigada et al., 2008), particularly for informing habitat conservation (Bailey and Thompson, 2009). Ecological patterning of benthic organisms (those that live on the seafloor) has also been a common research focus, largely due to the distinctive distribution of features such as seagrass beds, coral patches and delta channels that are observable from aerial imagery.

1.4 Spatial Processes in Coastal Environments

Four generic spatial processes that are of particular significance in coastal environments are diffusion, dispersal, marine species interactions and environmental controls. These operate across a range of geographical and temporal scales.

1.4.1 Diffusion

Diffusion occurs when a substance undergoes a net movement along a concentration gradient (i.e. from high to low concentration). At any point in time, it is possible to specify in which areas the given substance is present, and in which it is not. Along the coastline, the presence of a water body often fundamentally influences the mechanism responsible for the spread (e.g. through tidally or wind-drive currents and waves). Such hydrodynamic influences will determine how the substance, such as an oil spill across the sea surface, diffuses and its rate of diffusion. The quantification of diffusion coefficients in coastal sediments has received substantive research focus in relation to contamination by pollutants and the associated practical implications of compromised groundwater quality, as also has the downward mixing of gases commonly found in the atmosphere and how this effects the distribution and population structure of sediment benthic invertebrate communities (Hyland et al., 1999).

1.4.2 Dispersal

Dispersal refers to the tendency for marine organisms to spread away from an existing population or site of origin (Kinlan and Gaines, 2003). The potential for dispersal is determined by the structure of the coastal environment and the method by which an organism disperses. Many marine organisms, including plants and benthic animals, either shed their reproductive propagules or have a pelagic larval stage during which their relatively small size allows them to disperse through a comparatively large and complex fluid environment, providing a potential for wide transport by currents (Kinlan and Gaines, 2003). In the marine environment, there are comparatively fewer physical barriers to dispersal than in the terrestrial environment (Steele, 1991). Physical barriers in the sea include continental and island land masses (Begon et al., 2006), ocean frontal systems and currents (Kinlan and Gaines, 2003), vertical stratification of the water column (Longhurst, 2010), changes in substrate and bathymetry (Gaines et al., 2007) and heavily polluted regions. Nevertheless, marine dispersal often occurs across large distances, sometimes at the scale of ocean basins; taxonomic studies suggest that the corals colonising the reefs of the west Indian Ocean were seeded by larvae originating from the Indo-Pacific region (Sheppard, 1998). Similarly, rates of range expansion in marine macroalgae far exceed those of terrestrial plants, and the distances that insects on land disperse are also generally lower on average than marine invertebrate and fish species (Kinlan and Gaines, 2003). Indeed, many marine organisms have been observed to disperse farther and faster than their terrestrial counterparts, which reflects the increased connectivity that results from a combination of the presence of a fluid medium and fewer barriers in marine systems than on land (Carr et al., 2003; Cowen and Sponaugle, 2009).

1.4.3 Marine Species Interactions

There are many different types of ecological interactions that could influence the spatial distribution of a given species, including competition, facilitation, parasitism, feeding (herbivory and predation), mutualism, symbiosis and disease (Robinson et al., 2011). These present both positive and negative influences on a given population. Competitive interactions are likely to be easier to include in models of benthic distributions as opposed to pelagic fish species because of the paucity of data on competitive exclusion in pelagic habitats (Bilio and Niermann, 2004). Path analysis and correlative species distribution models have been used to explore the relative influence of interspecific competition and the environmental correlate of temperature on barnacle distributions (Poloczanska et al., 2008). Multivariate population data have been used to test whether the presence of a dominant competitor mediated the climatic influence on the subordinate species. The influence of temperature has been shown to be significant for the dominant competitor, and only indirectly affected the abundance and distribution of the inferior competitor.

The relative importance of interactions and environmental conditions to marine species distributions is highly dependent upon the feeding behaviour of different organisms. In marine systems, the availability and density of prey may be an important factor in determining the distribution many apex predators (mammals, tuna and some sharks) that are less constrained by environmental conditions due to their ability to maintain body temperatures independently of surrounding environmental conditions (Torres et al., 2003; Redfern et al., 2006; Wirsing et al., 2007). By contrast, many benthic invertebrates cannot thermoregulate and tend to be opportunistic generalist feeders that consume a wide variety of prey suspended in the water column. Consequently, they are restricted more by ambient environmental conditions than by specific dietary requirements. In many cases, species interactions *per se* are difficult to represent as variables. Nevertheless, their spatial structure can be captured in models as interaction parameters that draw on statistical methods of characterising autocorrelation.

Organisms can aggregate because of biological processes that are separate from environmental ones, such food availability, predator or competitor avoidance, mating behaviour, limited dispersal and advection, that cause spatial autocorrelation (Ritz, 1994). Spatial autocorrelation refers to the correlation of a single characteristic as a function of its position in geographic space, such that characteristics at proximate locations tend to be more closely related than their distant counterparts. This fundamental property of most datasets arises because of processes that abide by Tobler's first law of geography: 'everything is related to everything else, but near things are more related than distant things' (Tobler, 1970). Positive spatial autocorrelation means that geographically nearby coastal features, such as the ecological composition of a littoral zone, tend to be similar because of spatially structured processes and neighbourhood interactions. If a spatially dependent species distribution model has greater explanatory success than its non-spatial counterpart, the theoretical implication that follows is that neighbourhood interactions play an important role. This invites greater consideration of interaction between sites, which can be captured by spatially explicit models.

1.4.4 Environmental Controls

The arrangement of individuals of a given species will be fundamentally dependent on the presence of environmental controls. For example, sea surface temperature and the presence of rocky reef habitat were found to be a major predictor of the spatial abundance of two abalone species in southern Australia (Mellin et al., 2012). Shorelines typically traverse a vertical series of littoral sub-environments that represent critical gradients in environmental conditions. Factors such as degree of aerial exposure, water submergence, salinity, wave exposure, temperature and light availability all vary along these vertical gradients (Harley and Helmuth, 2003). Across broader spatio-temporal scales, the configuration of geological foundations provides a first-order control on presence or absence of marine habitat.

1.5 When Is It Desirable to Conduct Spatial Analysis for Coastal Management and Research?

The presence of spatial autocorrelation in datasets introduces several deviations from the assumptions of classical statistics. These warrant attention in studies of coastal environments. Indeed, it is only through the application of spatial (as opposed to classical or ordinary) statistics that these deviations can be addressed (Haggett and Chorley, 1967). For example, where classic statistics are used, stationarity is assumed across a dataset. This corresponds to constant statistical characteristics (such as mean and covariance values of a measured property) across a study area. In coastal environments that are characterised by high levels of environmental variability and associated heterogeneity, such assumptions often fail to be met because of contextual processes that operate at a hierarchy of scales. Spatially dependent influences can only be detected using techniques that draw explicitly on the location of attributes in space to investigate structures within the data. For example, when attempting to explain or predict the variable communities that inhabit adjacent intertidal littoral zones, the presence of interactive (or neighbourhood-context) effects in ecological communities suggests a need for a spatial (as opposed to classic or non-spatial) model (Cliff and Ord, 1981). It is therefore often desirable to adopt a spatial approach to the analysis of data. While the presence of spatial autocorrelation presents an analytical challenge, there are a number of practical benefits that arise from its incorporation into the analysis of coastal environments. Discussed further in Chapter 6, these include:

- quantifying the extent and pattern of spatial dependence across coastal zones,
- signifying the presence of redundant information in field datasets,
- indexing the nature and degree to which fundamental assumptions of classic (i.e. non-spatial) statistical techniques are violated,
- indicating the nature (spatial vs non-spatial) of an observable pattern to be modelled, and
- offering an opportunity to partition out and utilise spatially structured components of model error as a surrogate for a missing variable.

1.6 Spatial and Temporal Scales of Analysis

Most coastal environments are subject to numerous interrelated processes that define the physical space within which patterns are observed and organisms or landforms such as beach spits interact. Process scales on coastlines can be expressed in terms of their operational spatio-temporal ranges (Figure 1.2). These scale ranges have been identified for numerous purposes relating to coastal environments, including assessment of climate change impact (Slaymaker and Embleton-Hamann, 2009), studying coastal morphodynamics (Cowell and Thom, 1994; Woodroffe, 2002) and landscape ecology (Forman, 1995). Within these spatio-temporal scales, physical characteristics such as rocks and sediments may be examined by geologists and geophysicists, living organisms may be described by biologists and ecologists, environmental processes may be explored by oceanographers, and human use of the coast may be analysed by economists, human geographers, planners or social scientists. The appropriate scale for conducting spatial analysis depends on the questions that are to be addressed, particularly the dimensions of a phenomenon or process under consideration and the delineation of corresponding process ranges. This can often be subjective. For example, if 'water movement' is considered to be an important process for the evolution of sandy islands, then relevant timescales encompass a wide range of variability. They might include glacially driven fluctuations in sea level (10^5 a), which fundamentally determine the time periods during which islands are able to evolve, or the passage of individual waves (1–10 s), which are significant determinants of modern sediment deposition (Gourlay, 1988). From a process-led perspective, it is therefore possible to delineate spatio-temporal scales at which associated features of interest can be studied. It is also important to align these scales of analysis with the operational scales of instruments used to gather information (e.g. mapping technology such as airborne or spaceborne remote sensing).

Figure 1.2 illustrates the overlapping ranges of spatio-temporal scales at which environmental processes operate (and are analysed) in coastal environments. A continuum exists between the small and large extremes, with process scales that overlap in extent, but within which distinct disciplinary focus ranges can often be delineated. On the one hand, a biologist may use a microscope to observe instantaneous physiological processes occurring within the cells of a marine organism at smaller spatio-temporal scales (in the lower left of the diagram). On the other hand, a geologist may collect fossil or rock evidence of past sea levels where shorelines were submerged in a previous interglacial period (in the upper right of the diagram). The dimensions corresponding to the scales at which the biological and geological ranges meet in the middle also coincide with the scales at which information is acquired from coastal environments. This is often referred to as the 'landscape' or 'seascape' scale. It represents a common scale at which map-based coastal spatial analysis is conducted. This is noteworthy because, in a disciplinary sense, coastal studies have hitherto tended to

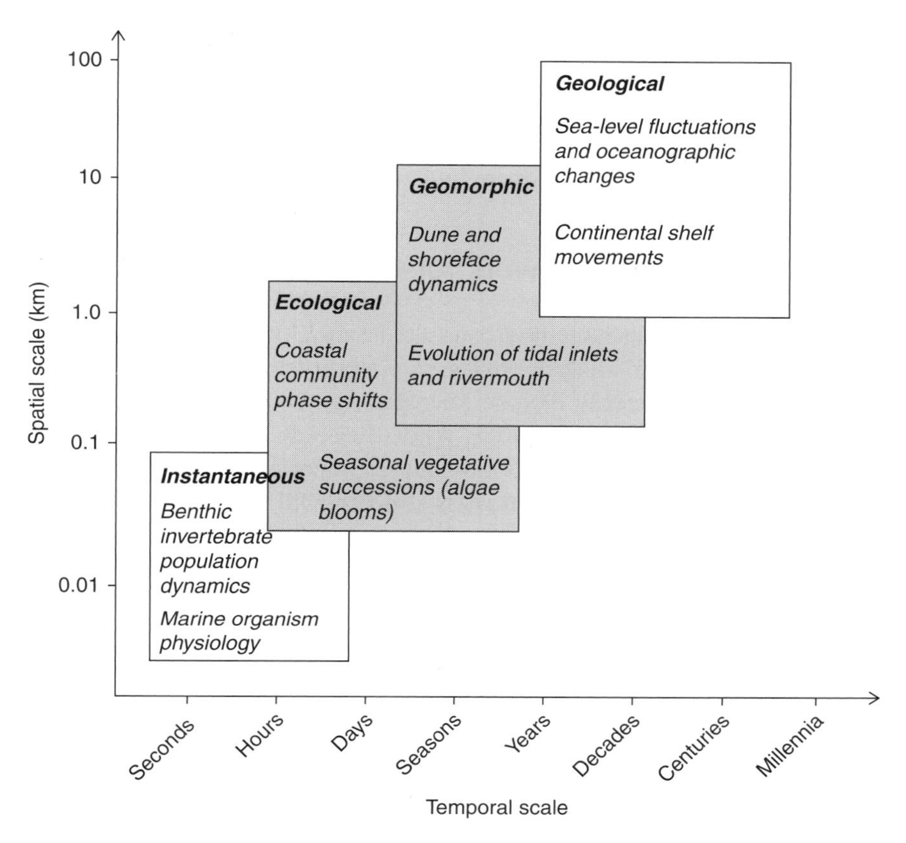

Figure 1.2 Hierarchy of processes operating within coastal environments at different temporal (instantaneous, up to days or weeks; ecological, up to a hundred years; geomorphic, hundreds to thousands of years; and geological, up to tens of thousands of years) and spatial scales.

diverge at these spatio-temporal scales, adopting either an ecological or geomorphic stance and rarely combining the two. Spatial analysis therefore emerges as a versatile technique with the potential to bridge this divide.

Although spatial analysis may be undertaken for a particular snapshot in time, it focuses on structures that are the time-integrated result of a range of temporally sequenced processes. Different coastal disciplines rarely fall neatly into discrete range categories that correspond to the specific temporal and spatial scales of the processes they focus on. Rather, they represent a continuous spectrum of overlapping and sometimes disjointed scales, such that a marine biogeographer might compare the dynamics of benthic invertebrate populations sampled in cores that are a few centimetres in diameter across different continents separated by ocean basins (Barnes, 2010). Likewise, a microbiologist studying the physiological traits of unicellular organisms in a laboratory environment might use techniques such as remote sensing to upscale their findings to broader geographic units. For example, Doo et al. (2012) used satellite remote sensing to scale up reductions in foraminifera calcification that were measured under ocean warming and acidification experiments in growth tanks to ascertain the wider implications for an entire coral reef. Thus, although the physical and biological studies of natural environments have traditionally been independent of each other, the coast has been a place where integration of these approaches has been necessary.

1.7 A Disciplinary Framework for Coastal Spatial Analysis

Figure 1.3 illustrates the interconnected nature of tools and techniques that can be applied for spatial analysis within an operational framework that spans a range of disciplines. These may include microbiology, ecology, biogeography, ecomorphology, geomorphology or geology. While they may be employed within a specific disciplinary context, these tools have a widely applicable and distinctive role to play at different stages of the spatial analysis process. Mapping enables features of interest to be observed because a remotely sensed image provides an accurate representation of structural features within coastal landscapes in the form of a digital map. This can then be deconstructed using landscape ecology techniques into patch assemblages for which landscape metrics can be calculated to yield a dataset on a structural feature. This is often viewed as a response variable in a model. Because corresponding spatial information has been retained in this dataset, it can be built

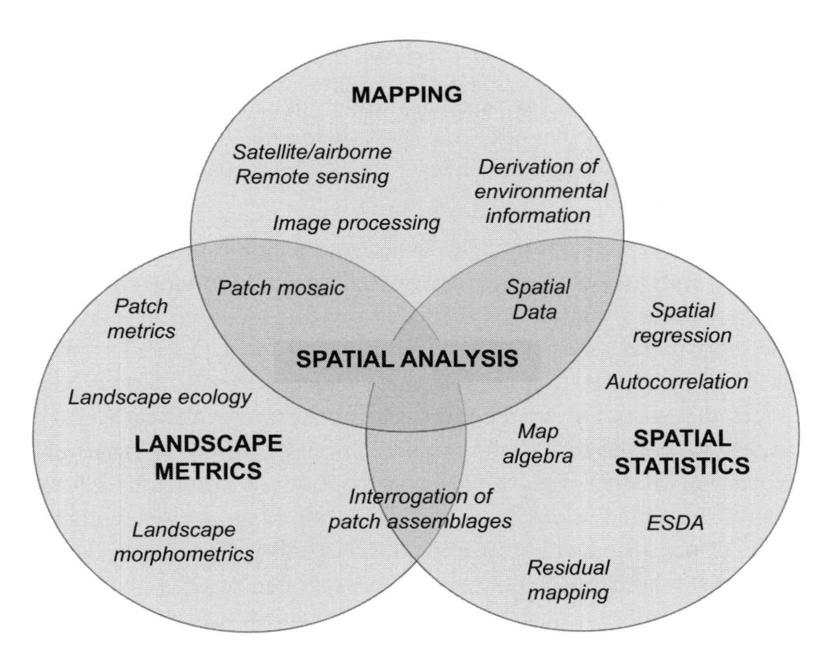

Figure 1.3 The overlapping roles of mapping, landscape metrics and spatial statistics within the framework for developing conducting spatial analysis in coastal environments.

explicitly into the data analysis using spatial statistical techniques. Together, these activities enable the characteristics of a coastal environment to be conceptualised and represented in digital form, while spatial statistics provide an empirical basis for experimentation and associated theoretical development. In this way, these tools act together to form the basis of the analysis, where independently this would not have been possible.

1.8 The Coastal Environment and Geospatial Analysis

The coastal zone is defined as the area between the tidal limits, including the continental shelf and coastal plain (Viles and Spencer, 2014). The shoreline specifically represents the actual margin of the land and the sea (Woodroffe, 2002). The intertidal zone might support many sub-environments, including beaches, dunefields, cliffs, tidal and brackish water wetlands, coral reefs, estuaries, embayments, deltas, coastal marshes, mangroves, seagrass and algae beds and developed urban coastal settlements (Figure 1.4). Global coastal environments have been classified on the basis of their tectonic setting (Inman and Nordstrom, 1971), morphological features (Summerfield, 2014) and wave environments (Davies, 1972). Global variation in these conditions has given rise to a range of different coastal types, including mountainous shorelines, narrow-shelf hills and plains, and deltaic, reef and glaciated coastlines (Inman and Nordstrom, 1971).

Coasts are situated at the confluence of the atmosphere, land and sea. They are therefore highly dynamic interface zones that are subject to variable weather and movements of sediments, debris and nutrients powered by waves, tides and currents. Such movements drive the erosional and depositional processes that shape the coastal profile, forming features such as barriers, longshore bars, troughs and cusps. Adjacent land masses deliver sediments and freshwater through rivers from terrestrial catchments (Figure 1.5). Fifty per cent of the world's population lives within the coastal zone (Neumann et al., 2015). This zone is also home to some of the world's most productive and diverse non-human ecosystems, such as coral reefs, mangrove forests

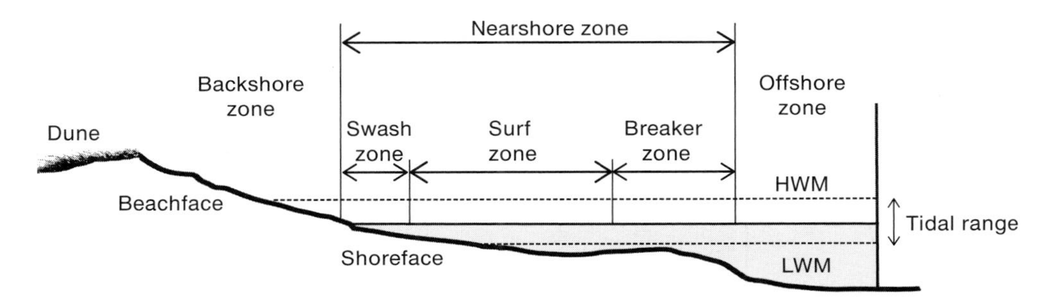

Figure 1.4 A cross-sectional profile of the key features of the coastal zone.

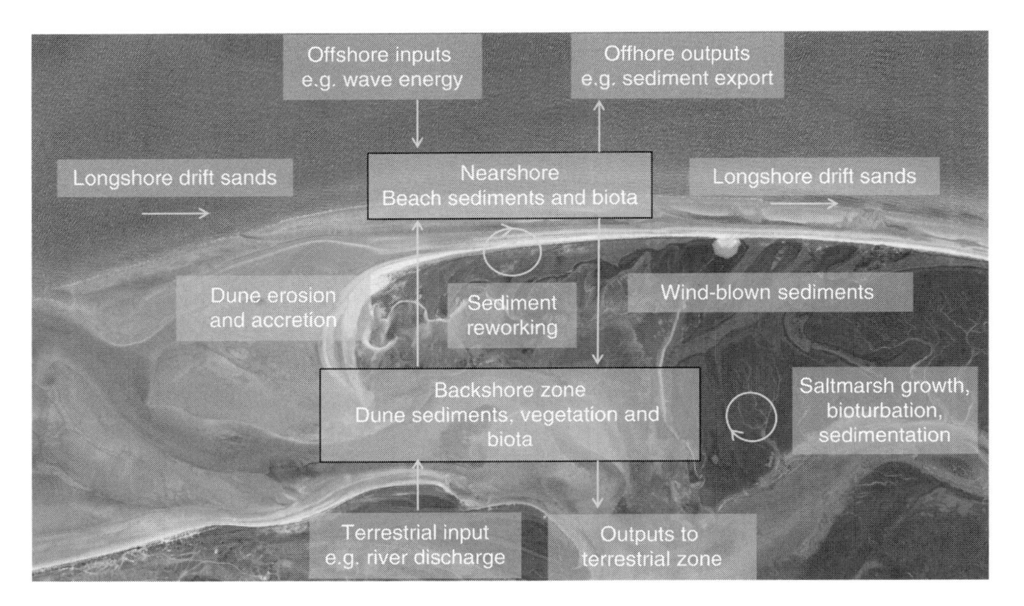

Figure 1.5 A simple coastal biogeomorphological system. (A black and white version of this figure will appear in some formats. For the colour version, please refer to the plate section.)

and saltmarshes. These ecosystems aid sedimentation, providing an important buffer zone and filtration system along the shoreline itself. The mutual co-adjustment of coastal form and process is termed *coastal morphodynamics* (Cowell and Thom, 1994; Woodroffe, 2002), into which can be incorporated the workings of coastal ecological and morphological systems. Such a biogeomorphological perspective treats the coast as a series of interconnected and interacting systems consisting of sediment and water movement, as well as biological ecosystems that are subject to both spatial and trophic linkages (Viles, 1988).

Many components of coastal environments are in a balance controlled by physical and biological processes, which can easily be upset by natural or human-induced disturbances (Viles and Spencer, 2014). In this context, spatial analysis offers an opportunity to explore, infer and predict coastal trends in both space and time. These trends can often be predicted at broad geographical scales that enable scientists or managers to make meaningful observations and statements about the behaviour of landforms and biological communities. Table 1.1 provides some examples of management issues across different types of coastal environment, with corresponding spatial analyses conducted to address these issues.

Table 1.1 Some examples of management issues by coastal environment type and associated spatial analysis.

Coastal environment type	Spatial analysis examples
Sandy coastlines	• Simulation of sea-level rise impacts on coastal erosion and sand dune plant succession (Bruun, 1988; Feagin et al., 2005) • Predicting spatial and temporal variation in longshore drift sand volumes (Cooper and Pilkey, 2004) • Assessing the downstream hydrological and geomorphic effects of large dams in American rivers (Graf, 2006)
Rocky coasts (cliffs and platforms)	• Mapping spatial and temporal variability of factors driving cliff failure along the Oregon coastline, including tectonic uplift, subsidence, wave attack, groundwater seepage, beach and cliff geology, El Niño events, severe storms and high water levels (Shih and Komar, 1994) • Use of terrestrial laser scanning to monitor hard rock coastal cliff erosion in north Yorkshire, UK, including undercutting, detachment of blocks, debris, soils, falls, slides and flows (Rosser et al., 2005)
Coastal wetlands	• Quantifying area and width reductions to coastal barrier features in a delta complex wetland in south Louisiana from historical maps and airborne imagery (Williams et al., 1992) • Tracking the conversion of marshes, mangroves and tidal flats to fish and shrimp farms and new housing in Mai Po, Hong Kong (Nelson, 1993) • Modelling surface elevation dynamics in estuarine wetlands (Rogers et al., 2012)
Coral reefs	• Identifying sources of riverine sediment to fringing coral reefs (McKergow et al., 2005) • Modelling the route and timing of lionfish invasion in the western Atlantic and Caribbean (Johnston and Purkis, 2011) • Modelling the risk of cyclone damage to coral reefs on the GBR from cyclone tracks and wind conditions (Puotinen, 2007)
Cold coasts: permafrost, sea ice, glaciers and fjords	• GIS simulations of rock fall events from paths of steepest descent, flow paths and changes in direction of permafrost progression (Noetzli et al., 2006) • Simulating ice-sheet response to climate change (Hanna et al., 2013) • Modelling riverine nitrogen delivery to Vejle Fjord, Denmark, from leaching, point source emissions, soils, terrain and groundwater head elevation and nitrate decay (Skop and Sørensen, 1998)
Urbanised coasts	• Bathymetric and volumetric change analysis of dredged pits in Lower Bay complex, New York Harbour (Wilber and Iocco, 2003)

1.9 Landscape Ecology and Coastal Morphodynamics

1.9.1 Landscape Ecology: a Brief Background

Ecology is broadly concerned with understanding the framework of organisms related by energy at different scales of organisation into a functional ecosystem (Odum et al., 1971). Population characteristics such as the number and distribution of species can be linked to functional drivers, such as the production and transfer of energy to better understand ecological dynamics.

Carl Troll first defined landscape ecology as 'the study of the entire complex cause–effect network between the living communities and their environmental conditions which prevails in a specific section of the landscape and becomes apparent in a specific landscape pattern or in a natural space classification of different orders of size' (Troll, 1939). The strong influence of landscape configuration on local community structure has meant that landscapes are increasingly viewed as patches of habitat with edges and boundaries, linked by corridors and barriers, in a matrix of 'non-habitat' (Forman, 1995; Dauber et al., 2003). The scale of ecological studies was therefore extended through subdisciplines such as metapopulation ecology and autecology. This latter disciplinary focus has found a practical outworking in the more contemporary fields of predictive habitat mapping (Brown et al., 2011) and species distribution modelling (Franklin, 2010).

Predictive habitat mapping and species distribution modelling offer an alternative approach to the traditional mapping of habitats and species from remotely sensed imagery. They statistically model species–environment relationships to simulate the geographical distribution of a given species (Miller, 2010). Statistical relationships are established between spatially continuous information on physical environmental drivers (e.g. bathymetry, rugosity) that is usually derived from remote sensing datasets and point samples of benthic habitat or species cover. This statistical relationship is subsequently used to extrapolate predictions of benthic habitat or species cover over larger spatial areas (Guisan and Zimmermann, 2000; Brown et al., 2011). The result is a map that predicts the cover of individual seafloor habitats or species, such as live coral or seagrass. Similarly, statistical procedures such as spatially lagged autoregressive modelling (Chapter 7) and spatial error modelling (Chapter 8) can generate maps that account for spatially structured processes and neighbourhood interactions underpinning the distribution of seafloor habitats and species. Such approaches demonstrate the value of digital maps as datasets that can help us better understand coastal environments through the specification of models.

1.9.2 Decomposing, Measuring and Interrogating Landscapes Through Patches

Landscape ecologists typically divide landscapes into patches to provide a representation of reality that can be readily understood. Patches are surface areas differing in appearance from their surroundings. Patches embody the idea that the distribution of organisms in space is an ecological response to non-uniform distribution of resources, which in many cases can be taken as a proxy for habitat suitability, or a range of related ecological functions, a principle that is central to landscape ecology (Farina, 2008). They may vary widely in size, shape, type, heterogeneity and boundary characteristics (Forman and Godron, 1986). By quantifying these characteristics of patches, it is possible to express the structure of an observed landscape in digital form within a computer.

For a landscape ecologist, structure is ultimately of importance because it relates to ecological function. Table 1.2 and Figure 1.6 provide examples of some patch characteristics that have been related to ecological properties.

Before the effects of landscape patterns on ecological processes can be investigated, those patterns must be quantified. Metrics allow patch characteristics to be quantified so that their emergent assemblage properties can be compared across a landscape. Examples of metrics might include:

- tone, colour and location on a satellite image (Farina, 2008),
- diversity, dominance, contagion, fragmentation and patch shape complexity, e.g. perimeter-to-area ratio (Frohn, 1997), and
- number of classes, dominance, contagion, fractal dimension, average patch perimeter/area ratio (Riitters et al., 1995).

Studying patch dynamics brings insight into the underlying process causing a patch. However, patterns are difficult to discern from information on a single patch alone. Analysis is therefore more commonly carried out on assemblages. In particular, patch assemblages can be interrogated along a temporal or spatial axis. Many physical and ecological processes transcend individual patches. Landscapes are therefore commonly treated as a mosaic of patches. Heterogeneous landscapes can be represented as mosaics composed of many patches of different composition that are product of scaling up from the individual patch, through patch assemblages, to a point where broader trends are recognisable. A large area in equilibrium that contains many patches can be viewed as a *shifting mosaic* (Forman, 1995). The geographical boundaries of this mosaic are defined by the dimensions of the processes that underpin their formation. Assemblages originating from different areas of the landscape subject to different influences can be compared statistically through their assemblage properties.

1.9.3 Landscape Ecology Models in Coastal Environments

A variety of landscape ecology techniques can be used to explore the relationship between spatial patterns observable in marine and coastal landscapes and associated ecological processes (Paine and Levin, 1981; Hinchey et al., 2008). Landscape structural features such as coral reef habitat patch size, complexity, diversity and richness have been related to reef fish assemblages in the US Virgin Islands (Grober-Dunsmore et al., 2008). Similarly, reductions in the density of damselfish on patches of fringing reefs around Lizard Island (Great Barrier Reef) have been related to increasing

Table 1.2 A summary of some key characteristics of landscape patches and the associated ecological functional role.

Structure	Function	Reference
Patch matrix heterogeneity, isolation and boundary discreteness	Species and habitat diversity, disturbance, age	Forman and Godron (1981)
Configuration	Interaction through wind carriage of moisture and dust, fire, species movement and transport through networks (e.g. sediment transport in rivers)	Forman (1995)
Patch size	Productivity, water flux and species dynamics	MacArthur and Wilson (1967)
Patch shape	Dispersal, home range suitability and the edge effect	Patton (1975)

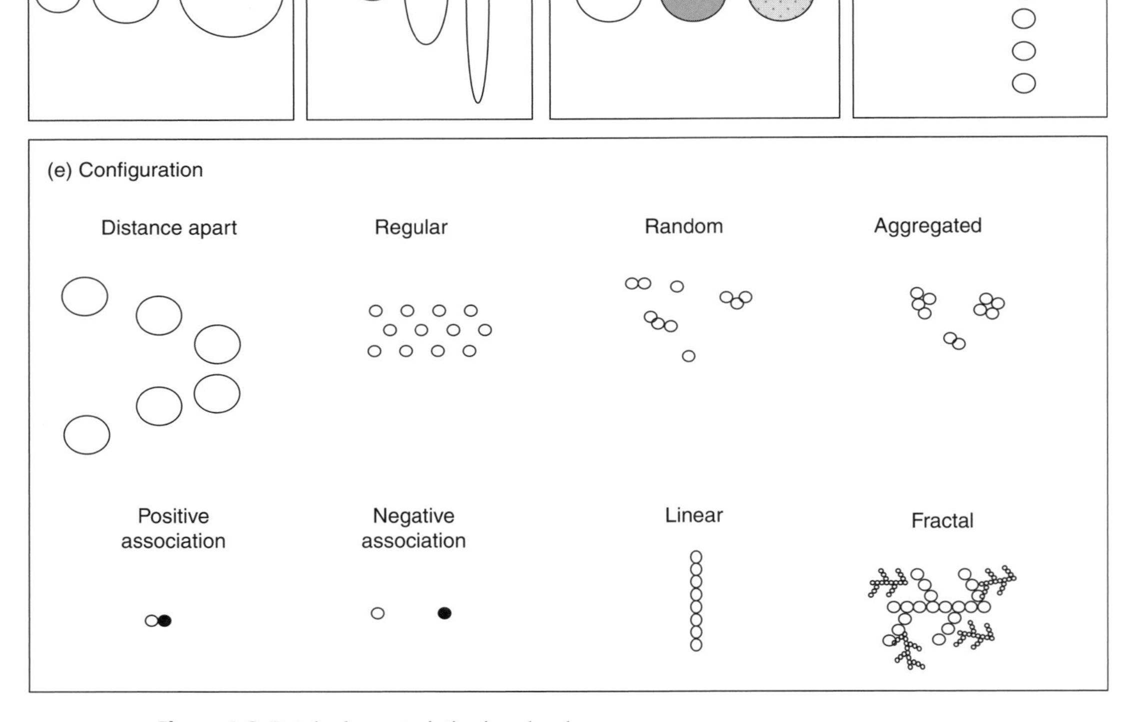

Figure 1.6 Patch characteristics in a landscape.

lengths of boundaries intersecting reef habitat (Meekan et al., 1995; Bartholomew et al., 2008).

Seagrasses are well suited to landscape ecology analysis because they exist as distinct, dark patches of biota in shallow subtidal zones within a matrix of lighter soft sediments or hard substratum (Robbins and Bell, 1994). Causes of patchiness include physical disturbance, changes in substrate, resource availability, diseases and competition with algae (Duarte, 1991; Ceccherelli et al., 2000; Ralph and Short, 2002; Farina, 2008). Seagrass beds form patterns across a gradient of increasing hydrodynamic activity. These range from continuous beds in low-energy zones to widely dispersed, discrete patches in high-energy zones where waves break (Fonseca and Bell, 1998). Patch formation has been related to the size and frequency of periodical disturbance, with coverage characteristics such as bare sediment, patchy and continuous meadow typical of high-, medium- and low-frequency disturbance, respectively (Figure 1.7). Landscape ecology metrics such as percentage cover of seagrasses, bed perimeter-to-area ratio, sediment organic content and percentage silt–clay have been found to have an inverse relationship to wave exposure and current speed (Fonseca and Bell,

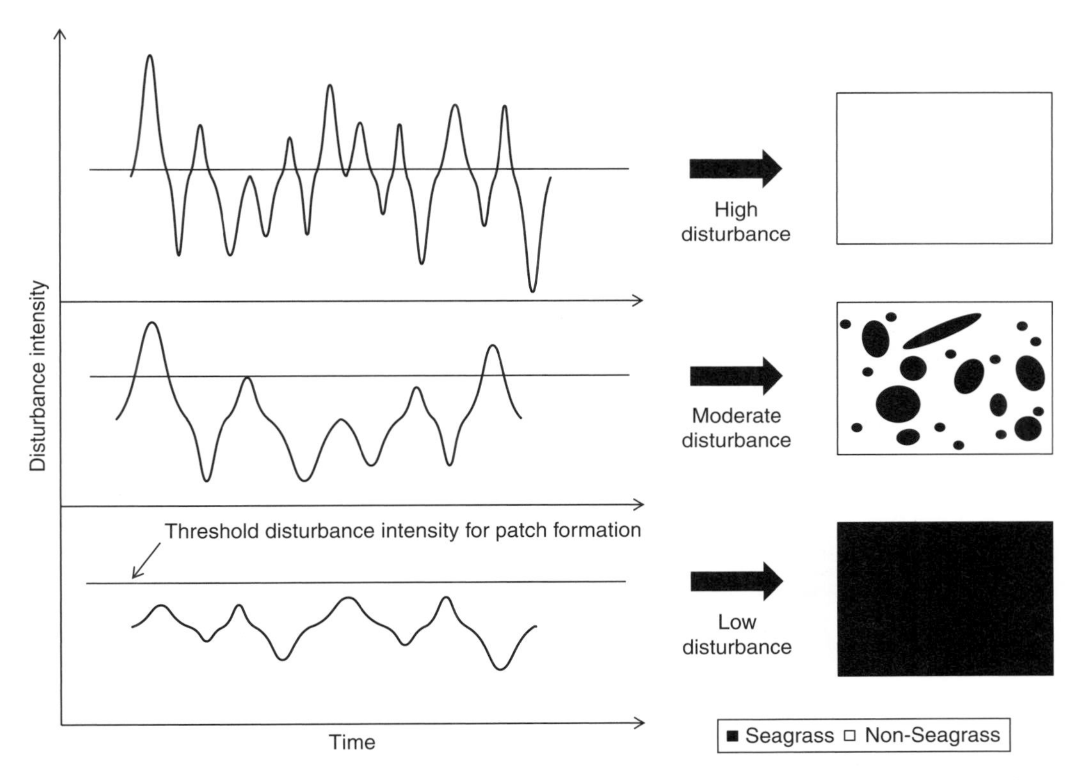

Figure 1.7 The effects of disturbance amplitude and frequency on seagrass coverage, showing the absence of seagrass, patchy seagrass and continuous, unbroken seagrass beds associated with high, moderate and low disturbance regimes, respectively.

1998). The physical setting of seagrass beds at Beaufort, North Carolina, USA, was linked to wave exposure, tidal current speed and water depth, which collectively governed the spatial heterogeneity of seagrass distribution and abundance. Large-scale disturbances such as hurricanes have been found to influence the spatial dynamics of *Halophila* sp. patches and generate new landscape patches by redistributing the sediment and associated seed banks locally (Bell et al., 2008).

In each of these examples, it can be seen how observable features of the landscape (e.g. patches) are linked with biophysical processes towards a better understanding of the spatial patterns that often characterise coastal environments. Such studies illustrate the value of landscape-scale analysis for enhancing our understanding of the population dynamics of marine ecological communities.

1.9.4 Coastal Morphodynamics: a Brief Background

Geomorphology is the study of landforms and the processes that create them (Summerfield, 2014). The mutual co-adjustment of process and form is of central interest to coastal scientists in the context of morphodynamics (the scientific study of landforms). This often draws on statistical techniques, which have been increasingly applied to geomorphological problems since the 1960s and 1970s. This was when a greater knowledge of surface processes saw an increased focus on the development of predictive models of short-term landform change. For example, many morphodynamic models of beach response to wave energy were based on an increased understanding of sediment entrainment and transport that came from laboratory studies. It was not until later that rigorous spatial techniques were adopted by geomorphologists, notably to the analysis of sediment parameters (Haggett and Chorley, 1967). Chorley and Kennedy (1971) further classify the application of spatial statistical techniques to geomorphology under the headings of trend surface analysis (polynomial, spectral, Fourier), regional taxonomy (factor, discriminant or cluster analysis), automated cartography (interpolation of contours and elevations), simulation (of hydrological networks and spits), slope, terrain, shape, volume and point pattern analysis of features such as sinkholes. Technological developments have further refined some of these techniques. As with landscape ecology, digital maps representing landforms can be used as a physical realisation of the functional context in which coastal processes operate (Spencer et al., 2009). Within this functional context, the morphological properties of landform elements interact with process-driven movements of mass and flows of energy through the landscape.

It is necessary to define the boundaries of a system before it can be analysed. This is achieved by outlining a conceptual model (Figure 1.8). A conceptual model identifies whether the system is open to the transmission of energy and mass, or closed. The elements of the system will need to be defined (i.e. the content of the boxes in Figure 1.8) and the relationships between these (i.e. the processes driving their interrelationships, depicted by the arrows in Figure 1.8). Variables representing the structural form of the landscape, as well as the geomorphic and environmental processes acting upon it, will need to be specified. As well as the identity of these variables, the way in which they relate to each other will also need to be specified, i.e. whether they

Figure 1.8 A conceptual model of sediment accretion on Bora Bora (French Polynesia), identifying the different components (sources, sinks) of the sedimentary system and processes linking them (production, breakdown, export, transport) around different sub-environments of the atoll lagoon (terrestrial, seagrass, coral, mangrove). (A black and white version of this figure will appear in some formats. For the colour version, please refer to the plate section.)

are independent (i.e. causal) or dependent variables. Coastal processes may be further subdivided into exogenic processes, i.e. those resulting from the action of water, ice and wind, and often resulting in erosion (or in the case of sand dunes, aeolian build-up of dune structure over time), or internal endogenic processes.

In the case of the conceptual model of sediment dynamics at Bora Bora Atoll, French Polynesia (Figure 1.8), the components of importance to the system include the sources and sinks of different sediment types (e.g. the calcified sediments from the surrounding coral reef and the terrigenous sediments from the central volcanic island peaks). The processes that form relationships between these components include the erosion and transport of sediments from drainage basins on the island to the adjacent lagoon sink. Carrying out a visual inspection, or statistical exploration, of patterns observable on a map of the study area may prove to be a practical and useful exercise for constructing a conceptual model. Exploratory spatial data analysis (ESDA) techniques (see Section 3.10) may yield useful statistical clues as to the identity of meaningful independent (predictor) variables.

Coastal processes may be subject to positive or negative feedbacks. Positive feedbacks respond to changes in inputs in such a way that they reinforce the original input. This may cause a 'snowball effect' that moves the system towards a new state once a threshold has been exceeded. A threshold is a given level or magnitude of a process above or below which a response may suddenly and rapidly change, leading to a bigger landform response. For example, the threshold of sediment motion for a grain of sand determines whether it remains settled on the seafloor, or becomes entrained in the water column and gets transported elsewhere. A negative feedback has the opposite effect of adjusting the system in such a way as to minimise the effect of externally generated changes. This is a common response of geomorphic systems that leads to self-regulation of their form. It helps a system to maintain an equilibrium in which the structure of a landform remains essentially constant due a balance between form and process over time (e.g. the delivery, deposition and removal of sediment to a beach over time (Woodroffe, 2002)).

1.9.5 Morphometric Models in Coastal Environments

Morphometric studies quantitatively analyse the geometry and topographic features of landforms. Examples of morphometric features in coastal environments include beach width, length and height. Morphometric models characterise coastal features in a standardised manner so that they can easily be compared to other features across a landscape (e.g. the comparison of beach lengths). Such comparisons may provide a better understanding of the mechanisms underlying their formation. For example, the spacing of sand ridges among extensive sand ridge fields on the inner shelf of the Atlantic Bight, North America, has been related to stability models that propose the evolution of an irregular initial topography, with development towards an ordered array of bedforms in response to hydrodynamic flows (Figueiredo et al., 1981). Similarly, chaos theory, self-organisation and fractals have been used to model the organisation of beach cusps by simulating patterns in sand ripples and dunes (Baas, 2002). The utility of maps for quantifying features has long been recognised (Clarke, 1966). This utility draws on the increased availability of quantitative datasets that characterise terrain, such as digital elevations models, which can easily be interrogated in geographic information system (GIS) software. Landscape changes can also be monitored by taking repeat morphometric measurements of features over time. For example, along the Baltic Sea coastline of Lithuania, beach width, length and height were measured repeatedly at 70 stations from 1993 to 2008 to model dynamics of beach growth and spit advancement (Jarmalavičius et al., 2012).

1.10 Unique Challenges to the Conduct of Spatial Analysis in Coastal Environments

Coastal environments present some unique challenges to the conduct of spatial analysis. These commonly arise because of the presence of the sea. They include the variable range of tidal datums particularly for referencing vertical heights (e.g.

tidal mean sea level vs chart datum) and the uncertainty that this variability introduces to the definition of coastal boundaries, the rapid speed of change in shoreline features (Ellis, 1978) and the practical limitations imposed on conducting fieldwork underwater.

Tidal datums are base elevations defined by phases of the tide. They are used for establishing coastal boundaries, limits of the territorial sea, water depths, critical shorelines on charts and the limits of various coastal management responsibilities (Ellis, 1978). Hirst and Todd (2003) categorise the problems associated with defining coastal boundaries into describing, visualising and realising problems. Descriptive confusion arises from vague or ambiguous descriptions in legislation and policy documents. References to a datum, or coordinates may often be omitted entirely. To visualise or map a coastal boundary, it is necessary to represent it as a digital feature within a GIS, which raises questions of how a coastline should be defined and how accurately it can be mapped. The *Tidal Interface Working Group* of the *Intergovernmental Committee on Surveying and Mapping* defined some commonly used tidal interface boundaries, including highest astronomical tide, mean high water springs, mean high water, mean sea level, mean low water springs, mean low water and lowest astronomical tide (McCutcheon and McCutcheon, 2003). The practical realisation of an accurate and unambiguous definition of a coastline also presents a challenge in relation to height datums, with offsets being apparent between national reference grids (e.g. the Australian Height Datum) and datums defined by gauge measurements of tidal fluctuations (Hirst and Todd, 2003). Given that mapping often forms a foundation for much geospatial analysis (Figure 1.3), it is therefore important for the analyst to be aware of the complexities associated with describing, visualising and creating digital representations of the coastline.

Spatial referencing not only refers to the longitude and latitude (i.e. x and y) measurements for a given coordinate, but also incorporates the height above a common datum (the z coordinate). As mentioned previously, there are a number of potential tidal and land datums that could serve as a reference point for the definition of height measurements. When information is collected from a survey vessel on the water surface, tidal fluctuations will influence the height of that survey vessel above the seabed. This is the case when conducting a bathymetric survey of water depths using an echosounder attached to the hull of a boat. It is therefore necessary to apply corrections to any datasets collected to account for any offsets due to the state of the tide, particularly as tidal ranges in some locations can be as dramatic as 11 m.

Most spatial analysis involves the collection of some form of *in situ* information, such as community census surveys or sediment samples. On shorelines with low topographic gradients, the difference between low and high tides can represent the submergence or exposure of substantial areas of intertidal shoreline. Moreover, the accessibility of estuarine environments depends on the ability of boats to access field sites through channels. The navigability of channels depends on the depth of water. Planning and implementing fieldwork therefore requires an accurate knowledge of how tidal fluctuations interact with local coastal topography. Because weather in coastal zones is often highly changeable, the feasibility of visiting field sites further depends on the coincidence of appropriate weather conditions with a favourable tidal window. This limits the amount of field sampling that can be done, restricting the spatio-temporal range of samples collected to seasons and geographical areas that coincide with favourable fieldwork conditions. Consequently, areas that coincide with difficult conditions (i.e. the breaker zone) are notoriously under-represented in coastal fieldwork campaigns. In some situations, a feature of interest (e.g. a kelp bed or coral reef) may require a detailed population survey. The analyst may need to employ self-contained underwater breathing apparatus (SCUBA) to achieve the necessary time and proximity to the field site to collect data for analysis. At a regional scale, this may result in a clustered distribution of samples, that is governed largely by the number of dives a fieldworker can do at a given site and how far they can swim underwater before surfacing because of an empty air cylinder.

Another more general challenge to the conduct of spatial analysis is the lack of spatial referencing associated with datasets. Location information is necessary for all truly spatial analytical techniques and, in some situations, it is beyond the control of the analyst. For example, to detect shoreline change, it is necessary to compare the location of a recently mapped shoreline against a historical map or aerial photograph of the same shoreline. Implicit in such a comparison is the fact that the historical map of the same shoreline also has spatial referencing information associated with it (i.e. coordinates corresponding to a given datum and projection). This standardised referencing information allows the two shoreline depictions to be brought into a frame of reference so that a comparison can be made. Often a common frame of reference does not exist, and a map will need to be georeferenced prior to analysis. Where the results of analysis are fundamentally dependent on the geographical location of features represented (as is the case with change detection analysis), care will need to be taken to georeference the information in a way that does not introduce bias into results.

1.11 Directional Dependence and the Coastline

Directional dependence occurs if a phenomenon exhibits different values when measured in different directions. Coastlines are special environments because they exhibit unique empirical properties that arise because of directional dependence. For example, the properties listed below that arise from directional dependence have implications for the evaluation of processes and structures that occur in coastal environments.

- Located at the border between the land and sea, coasts act as boundaries or natural barriers to land- or sea-based processes.
- Environmental gradients occur in perpendicular and parallel directions along and across coastlines, meaning that many coastal features are 'anisotropic'.
- Coastal environments are subject to both ecological and statistical 'edge effects'.

At a fundamental level, the coast acts as a barrier or boundary to the landward movement of marine processes and organisms and, similarly, to the seaward movement of terrestrial phenomena. This means that marine or land-based observations are generally characterised by a sharply delineated border. Obviously, the point datasets representing observations of sharks around the coastlines of Florida and California do not extend onto the land (see Figure 3.6). This neatly delineates a geographic area necessary for sampling a given phenomenon. It also often gives rise to a distinctive spatial distribution of samples that requires appropriate analysis because a lot of spatial statistical exploratory and inferential procedures rely on the computation of locally derived values across space. The presence of a boundary means that neighbouring values will be lost around the edges of a geographical area. Some localised values will therefore be calculated on a reduced number of data samples, perhaps because a certain process no longer occurs in a given direction (e.g. tidal inundation ceases in both landward and upward directions at the line of the highest astronomical tide). This needs to be accounted for in inferential or exploratory statistics using tools such as weights matrices (see Section 6.6).

Because of the dramatic environmental gradients found along coastlines, physical and biological processes and features tend to be aligned both along and across the shore. For example, sediments are transported along the shore as a result of waves that have been refracted along the coastline carrying suspended sediment particles that are deposited as their energy is attenuated as they travel along beaches (Komar, 1998). Features such as seagrass beds and spur and groove formations on coral reefs grow and are eroded in parallel alignment with the direction of incident waves (Hamylton and Spencer, 2011; Duce et al., 2016). Intertidal organisms are also structured across an environmental gradient that represents the degree of exposure and submergence throughout the tidal cycle that is experienced by a variety of coastal habitats such as cliff faces, saltmarshes, seagrasses or sands and muds. Directional dependence, also known in a statistical sense as 'anisotropy', presents a challenge to the empirical evaluation of coastlines. It can be explored by calculating statistical functions such as the semivariogram or covariance function (see Chapter 6) along different directional axes (i.e. 0° to 359°). Features such as the statistical range (the distance of separation, discernible from the levelling out of plotted points on a semivariogram, at which paired point samples are no longer autocorrelated) can then be compared along different axes. For example, the major (longest) range and minor (shortest) range may be calculated for different directions, alongside an anisotropic factor (the ratio of the major to minor range) to indicate the strength of the directional influence.

The presence of a border between two distinct environments may give rise to an ecological 'edge effect' at the coast. An ecological edge effect occurs because of the increased structural mixing and fragmentation of habitat that characterises borders where different habitats meet. These are known as ecotones and they commonly contain habitats populated by different communities. In turn, these give rise to geographically focused ecological interactions (e.g. predator–prey relationships) that support greater levels of biodiversity. For example, edge effects have been observed in the diversity of intertidal benthic macroinvertebrate communities traversing the interface between seagrass beds and sand flats in Moreton Bay, Queensland, Australia (Barnes and Hamylton, 2016). Thus, while seagrass beds are favourable habitat for benthic macrofauna by virtue of their provision of greater refuge from predation, greater availability of food, enhanced larval entrainment and more habitat structure (Tanner, 2005; Smith et al., 2011), adjacent sand flats may be preferable for bioturbators such as sand prawns and soldier crabs that turn over sediment to feed (Branch and Pringle, 1987; Siebert and Branch, 2007). Each of these two habitats is dynamic, interchangeable and engineered by the organisms that occupy them, which in turn are capable of expanding into adjacent territory of neighbouring communities. This results in distinct trends of macrofaunal biodiversity and abundance across habitat borders that are common in coastal environments.

1.12 Coastal Management Questions for Which Geospatial Analysis Offers Insight

Table 1.3 outlines a range of coastal management questions for which different types of spatial analysis can offer insight. Exploratory techniques draw on the geographical referencing of data points to answer questions such as *'where are the outliers?'*, *'where is my model performing well?'* and *'where are the residuals indicating uncertainty?'* The answers may provide important clues as to the identity of potential missing variables from a model constructed to explain an observed distribution in a coastal environment. For example, a coastal manager may be interested in selecting a series of marine reserves using habitats as surrogates for biological diversity (Ward et al., 1999). Such an approach is only possible if a reliable relationship can be established between habitats and species assemblages and biological diversity. This relationship may be influenced by a range of both environmental and ecological influences, which can be identified using a stepwise model-building approach that maps outliers and residuals for clues as to model performance and potential missing covariates.

Confirmatory and inferential methods of analysis use statistical probability to discern the extent to which an observed distribution of coastal features (benthic organisms, geomorphic landforms or urban development) can be explained on the basis of processes that are likely to influence them. Many approaches to both explanatory and predictive modelling of coastal landscapes, ecosystems and geomorphological landforms adopt a perspective that views features of interest in terms of structure and function (see Section 7.3). *Structure* commonly refers to the observable characteristics of systems, e.g. the sizes, shapes, numbers and configurations of components such as seagrass or kelp beds, and rock, reef or island formations. *Function* refers to the external environmental conditions shaping these characteristics and internal interactions between these elements, i.e. the flow of energy, materials or organisms among the components. Coastal models use either point- or raster-based approaches to draw

Table 1.3 Some spatial analysis approaches and corresponding questions that can be addressed for coastal environments.

Spatial analysis approach	Analytical question
Exploratory spatial data analysis	• Where are the outliers? • Where is my model performing well? • Where do the residuals indicate uncertainty?
Explanatory spatial data analysis	• Is it possible to explain the distributions of coastal features based on our understanding of the factors influencing their distribution?
Predictive spatial data analysis	• Is it possible to reliably predict the distributions of coastal features based on our understanding of the factors influencing their distribution?
Geostatistics	• Can I use sampled values at known locations to predict values at unsampled locations? • Can an observed geographical distribution be captured in a mathematical model?
Inferential geostatistics	• Is there a spatial pattern to an observed distribution? • Can spatial dependence be modelled to reveal clues about the processes underpinning an observed geographical distribution?
Map algebra	• Where do multiple maps adhere to Boolean criteria (NOT, AND, OR, XOR)? • Where is the best or worst location to select on the basis of a range of criteria? • How much has the landscape changed (e.g. image differencing)?

empirical statistical links between structural and functional (i.e. process-based) features of coastal environments. Examples include modelling the presence of *Montastrea* spp. coral as a function of wave exposure along the Belize Barrier Reef (Chollett and Mumby, 2012), drawing morphometric links between the features of coastal streams and rates of uplift in northern California (Merritts and Vincent, 1989), and linking beach morphological features including cusps, ripples and dune patterns to vegetation dynamics (Baas, 2002). Defining empirical links between observable structural features and underpinning processes in coastal environments also raises the possibility of simulating or predicting the future form of these features by adjusting the related processes in line with our anticipations of future environmental scenarios. In doing so, spatial analysis enables reliable predictions to be made about the potential future distributions of coastal features, based on our understanding of the contemporary factors influencing their distribution.

Empirical relationships can be used to extrapolate measurements across geographical and temporal dimensions to predict the nature of an unsampled feature. Such an exercise helps to overcome the practical challenges that constrain fieldwork in coastal environments (e.g. the presence of a water body and variable weather or sea conditions). For example, maximum likelihood automated classification techniques were used to extrapolate relationships between seafloor complexity variables derived from multi-beam echosounder bathymetry (slope, aspect, rugosity, backscatter, etc.) and eight different benthic biological communities covering 54 km^2 of seafloor across the central coast of Victoria, Australia (Ierodiaconou et al., 2011).

Confirmatory and inferential models are commonly applied within the context of a data availability scenario in which continuous coverage of information is available on a variety of environmental, physical or process-based variables (which are often derived from remote sensing datasets), accompanied by a localised set of *in situ* data reported by discrete points. These may constitute biological community surveys or geological sediment samples. This is a common data availability scenario that can be utilised for a wide variety of spatial analysis applications in coastal environments. These include mapping ecological characteristics of the seafloor from acoustic sonar (sound navigation and ranging) data (for a review see Brown et al., 2011), species distribution modelling (Miller, 2010) and mapping depositional facies across shallow carbonate banks (Hariss et al., 2013).

Interpolation is the procedure of predicting the value of attributes or variables at unsampled sites from measurements made at point locations within the same area (Burrough and McDonnell, 2011). Geostatistical techniques can be used to interpolate a series of localised point samples to a continuous surface across a geographical area. Interpolation techniques include inverse distance weighting, nearest-neighbour, kriging or spline techniques. The distinction between interpolation methods and predictive confirmatory methods is that an interpolator models the spatially structured component of variation that is observable from a given sample population, without drawing on additional information from environmental correlates.

Map algebra methods perform cartographic analysis that brings together two or more maps (usually in the form of raster layers) to produce a new raster layer (or 'map') using algebraic operations (e.g. addition or subtraction) (Tomlin, 1990). These techniques can be used for a range of raster-based data query objectives (such as multi-criteria decision analysis), cellular automata or simulation models that predict the composition of a cell in the future, or assess the historical rates and distribution of changes in coastal environments.

1.13 A Framework for Applying Spatial Analysis to Coastal Environments: Mapping, Monitoring and Modelling

Figure 1.9 illustrates the analytical framework adopted for Chapters 4, 5 and 7. Together, these chapters build through a hierarchy of analytical depth from mapping (making a map), to monitoring (making multiple maps and comparing them), to modelling (using a map as a dataset in its own right). These offer a collective, compelling view of the utility of maps in spatial analysis. Mapping lies at the heart of spatial analysis in the coastal zone. It is an activity that provides a digital representation of

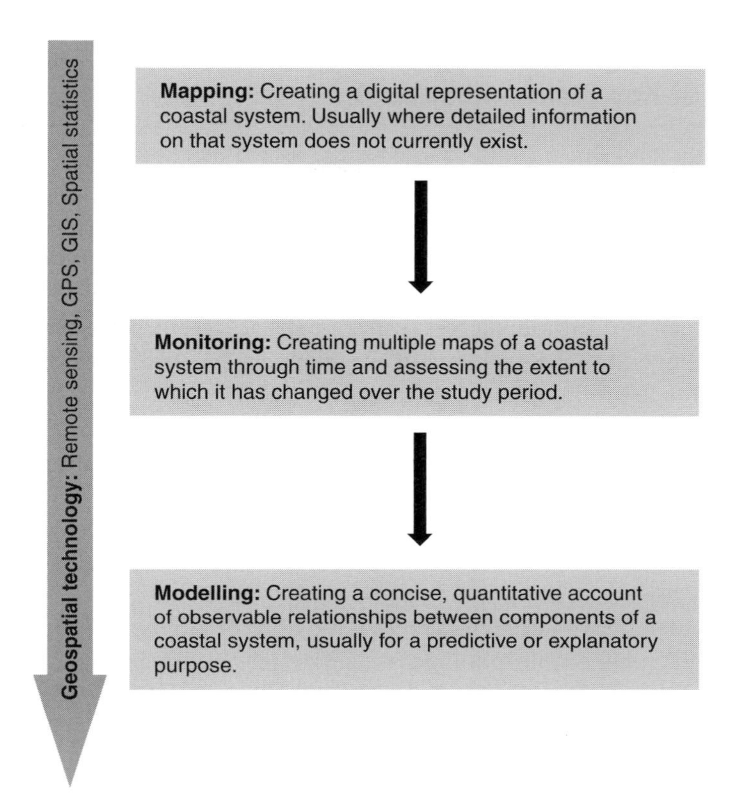

Figure 1.9 Schematic representation of the hierarchical framework adopted for Chapters 4, 5 and 7 of increased analytical depth from mapping to monitoring to modelling.

a coastline, which itself is a rich dataset that can be further interrogated to enhance our understanding of the features represented. Indeed, a primary motivation for writing this book was to demonstrate that digital maps are useful datasets. They provide important opportunities for further spatial analysis. Digital maps can represent the beginning of a line of enquiry, as opposed to the end of a mapping exercise.

A powerful and practical methodological framework for the conduct of spatial analysis can be developed by combining different types of spatial analysis (Figure 1.1). It is the aim of this book to present and review these different components of spatial analysis within different coastal environmental settings. In doing so, an overview of the analytical tools available for the investigation of coastal environments is provided that spans a range of enquiry objectives. It is hoped that, by compiling an overview of the analytical tools available into a single volume, the reader is able to see the potential links between the different techniques available to the spatial analyst and is afforded a critical perspective of how they can be brought together for analysing coastal environments.

1.14 Summary of Key Points: Chapter 1

- Spatial analysis is a collection of techniques and statistical models that explicitly use the spatial referencing associated with each data value or object under study.
- An observational scientist uses information on location to induce variability in an explanatory variable as a form of 'natural laboratory'. Location can act as *place* (and, by extension, context) and location as *space* for this purpose.
- Processes such as diffusion, dispersal, marine species interactions and environmental controls operate across a range of geographical and temporal scales in coastal environments to produce spatial patterns.
- Some processes in coastal environments are spatially autocorrelated (their characteristics at proximate locations are more similar than their distant counterparts). This makes it desirable to analyse them using spatial methods.
- Coastal environments are subject to processes that act across an interlinked range of temporal and spatial scales. These processes define the scales at which spatial analysis takes place.
- Coastal spatial analysis operates across a range of disciplines (microbiology, ecology, biogeography, ecomorphology, oceanography, geomorphology or geology). Each of these contributes distinct analytical tools and methods for analysis, e.g. landscape ecology and coastal morphodynamics.
- Coastal environments present some unique challenges to their analysis, particularly regarding the effects of tides and weather.
- Coastal environments exhibit directional dependence because they act as natural barriers between the land and sea, they host environmental gradients in perpendicular and parallel directions along and across coastlines, and they are subject to ecological and statistical 'edge effects'.

2 The Nature of Spatial Data

2.1 The Spatial Data Matrix as a Structure for Storing Geographic Information

A 'spatial perspective' relies on knowledge of the distribution and organisation of phenomena over geographical space. It therefore follows that spatial analysis cannot be separated from location, i.e. the coordinates associated with information. Indeed, spatial analysis was defined in Section 1.2 as a collection of techniques and statistical models that explicitly use the spatial referencing associated with each data value or object that is specified within the system under study. A clear association must therefore be maintained between quantitative data and the spatial coordinates that locate them. If this is not the case, any analysis undertaken with the dataset will not be spatial in nature.

Most environmental science datasets consist of a set of observations, each of which refers to a single characteristic at a single location. Individual data records may sit within a broader matrix that incorporates multiple records of the same characteristic for different locations, or multiple records of different characteristics for the same location. The nature of the former dataset enables an investigation of spatial variation, such that the different points can be mapped to display spatial patterning. The latter dataset enables a location-specific inventory of integrated phenomena at a given place to be investigated. A dataset that contains the whole series of characteristics for the whole series of different places provides a wealth of opportunities for geographical enquiry. This is most efficiently organised in the form of a matrix, referred to as the spatial data or geographic matrix (Berry, 1964). The spatial data matrix organises information in the form of multiple records of different characteristics as columns, with each row representing a unique location. This can be represented generically as

$$\begin{array}{c} \text{Data on } k \text{ variables} \quad \text{Location} \\ \begin{bmatrix} Z_1(1) & Z_2(1) & \cdots & Z_k(1) & S(1) \\ Z_1(2) & Z_2(2) & \cdots & Z_k(2) & S(2) \\ \vdots & \vdots & \vdots & \vdots & \vdots \\ Z_1(n) & Z_2(n) & \cdots & Z_k(n) & S(n) \end{bmatrix} \end{array} \quad \begin{array}{c} \text{Case 1} \\ \text{Case 2} \\ \vdots \\ \text{Case } n \end{array}$$

although it is more commonly referred to by the shorter form

$$\{z_1(i), z_2(i), \ldots, z_k(i) \mid s(i)\}_{i=1,\ldots,n} \tag{2.1}$$

where $z_1, z_2, \ldots, z_k$ refer to k variables and s to location, and all observations relate to the same time period, with multiple matrices employed to record multi-temporal information (Haining, 2003).

Figure 2.1 illustrates how locational information and spatial objects can be linked within a dataset. This guiding organisational principle for geographical information provides a structure for most databases stored using contemporary geographic

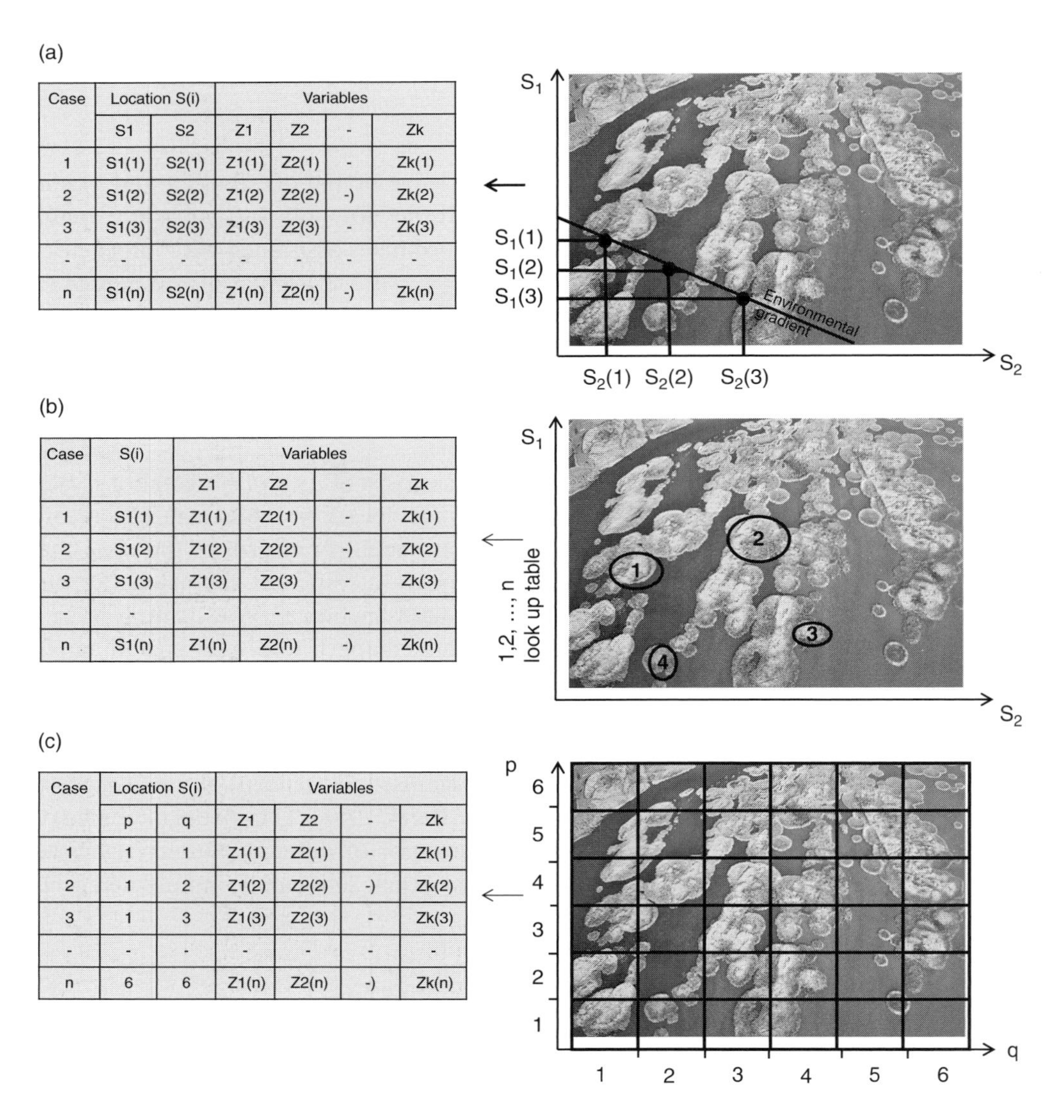

(a)

Case	Location S(i)		Variables			
	S1	S2	Z1	Z2	-	Zk
1	S1(1)	S2(1)	Z1(1)	Z2(1)	-	Zk(1)
2	S1(2)	S2(2)	Z1(2)	Z2(2)	-)	Zk(2)
3	S1(3)	S2(3)	Z1(3)	Z2(3)	-	Zk(3)
-	-	-	-	-	-	-
n	S1(n)	S2(n)	Z1(n)	Z2(n)	-)	Zk(n)

(b)

Case	S(i)	Variables			
		Z1	Z2	-	Zk
1	S1(1)	Z1(1)	Z2(1)	-	Zk(1)
2	S1(2)	Z1(2)	Z2(2)	-)	Zk(2)
3	S1(3)	Z1(3)	Z2(3)	-	Zk(3)
-	-	-	-	-	-
n	S1(n)	Z1(n)	Z2(n)	-)	Zk(n)

(c)

Case	Location S(i)		Variables			
	p	q	Z1	Z2	-	Zk
1	1	1	Z1(1)	Z2(1)	-	Zk(1)
2	1	2	Z1(2)	Z2(2)	-)	Zk(2)
3	1	3	Z1(3)	Z2(3)	-	Zk(3)
-	-	-	-	-	-	-
n	6	6	Z1(n)	Z2(n)	-)	Zk(n)

Figure 2.1 Assigning locations to spatial objects within a riverine marsh landscape: (a) points, (b) polygons and (c) pixels stored in the format of the spatial data matrix. (A black and white version of this figure will appear in some formats. For the colour version, please refer to the plate section.)

Table 2.1 Ten types of geographical analysis that may be undertaken with the spatial data matrix (summarised from Berry, 1964).

Regional analysis	Description
1. Examine the arrangement of cases within a row or part of a row	Spatial distribution studies and maps
2. Examine the arrangement of columns within a single row or part of a row	Study of localised association of variables in a single place
3. Compare pairs of rows or a whole series of rows	Studies of spatial covariation or spatial association
4. Compare pairs of columns or a whole series of columns	Studies of areal differentiation
5. Delineate a subregion for study	Studies using local (as opposed to global) operators
6. Compare a row or part of a row through time	Study of local change
7. Compare a column or part of a column through time	Study of the changing character of an area
8. Study changing spatial associations	Study of the changing character of a spatial association
9. Study changing areal differentiation	Study of the changing character of areal differentiation
10. Compare a submatrix through time	An interplay of approaches 1 to 9

information system (GIS) software. Structured in this spatial matrix, data can be manipulated to develop spatial models that simplify natural processes and patterns while simultaneously accounting for their geographical nature. Table 2.1 summarises 10 approaches to regional analysis that can be undertaken with a spatial data matrix (Berry, 1964).

2.2 Reference Systems, Projections and Datums

Geodesy is the science of measuring and modelling the Earth's shape. This is a necessary precursor to the development of reference systems with respect to which locations on Earth can be defined using coordinates. Scientists' understanding of the Earth's shape has evolved throughout history. The shape of the Earth was considered as far back as 276BC by the Greek philosopher Eratosthenes. Key contributions that have shaped our understanding of the Earth have come from Isaac Newton, Jean Picard and Pierre Bouguer. Werner (1967) provides a detailed review of conceptual developments in the field of geodesy. While it may be tempting to consider the Earth as a sphere, this is an oversimplification. It is actually closer in shape to an ellipsoid because the Earth's sphere is slightly flattened. The distance between the poles is about 1 part in 300 less than the diameter at the equator (Goodchild et al., 2005). One important application of geodesy is that the shapes to which we approximate planet

Earth provide a basis for constructing datums. A datum is a reference surface from which coordinates are measured. One commonly used ellipsoid datum is the World Geodetic System of 1984 (often shortened to WGS84) in relation to which any position on Earth can be located.

Geospatial referencing systems provide a unique scheme according to which every location on the surface of the Earth can be referenced with a pair of coordinates or codes. These codes express distances, areas and volumes and help people to navigate to a specified feature of interest. A Cartesian coordinate system defines a location on the Earth in relation to two distances measured from an origin along two axes. Conventionally these are the x (horizontal) and y (vertical) axes, along which the easting and northing are expressed, respectively. In contrast, the geographic coordinate system defines locations using axes of latitude and longitude. A geographic reference system approximates the shape of the Earth to a sphere and aligns the axes to its axis of rotation, where the Earth's centre of mass lies. Latitudinal lines run parallel to the axis of rotation, and longitudinal lines run perpendicular to it. Lines of constant longitude (running perpendicular to the equator) are referred to as meridians, while lines of constant latitude (running parallel to the equator) are referred to as parallels. The geographical coordinate system is configured in such a way that the spherical Earth is divided up into 360°, with an origin of 0° at the Royal Observatory in Greenwich, London. Each of the 360 degrees of longitude is divided into 60 minutes and each minute into 60 seconds. A degree represents an approximate distance on the ground of 111 km at the equator.

Projections establish mathematical formulae for transforming geographical coordinates onto a flat, two-dimensional surface. A projection can most easily be envisaged as a piece of paper being wrapped around the Earth's surface. The surface detail (e.g. the boundaries of countries) is transferred to the paper before it is unwrapped and laid flat. The points or lines of contact where the paper touches the Earth are called points or lines of tangency (Kraak and Ormeling, 2011). These represent the most accurate part of the projected map, where there is no distortion. Distortion increases with distance away from these points. Any time a round object such as the Earth is represented on a flat surface like a map, features on the round object such as continents will be distorted. Four characteristics of the Earth may become distorted upon application of a projection: area, shape, distance and direction (or angle). Each projection has its own specific patterns of distortion, and, depending on the intended use of the projected map, it is wise to select a projection method that preserves those map characteristics of importance. For example, an equal-area projection maintains the areas of countries, while a conformal projection maintains their shapes at the expense of the areas, and an equidistant projection preserves distances between specified points.

Projections are categorised on the basis of the shape of the projection plane. The three main types of projection are azimuthal, conical and cylindrical (Figure 2.2). An azimuthal projection, sometimes called a planar, polar or orthographic projection, is implemented by placing the flat, projected surface in such a way that a single point of contact, or tangency, is established from this point. This projection maintains true directions and distances from the point of tangency to every other point on the map.

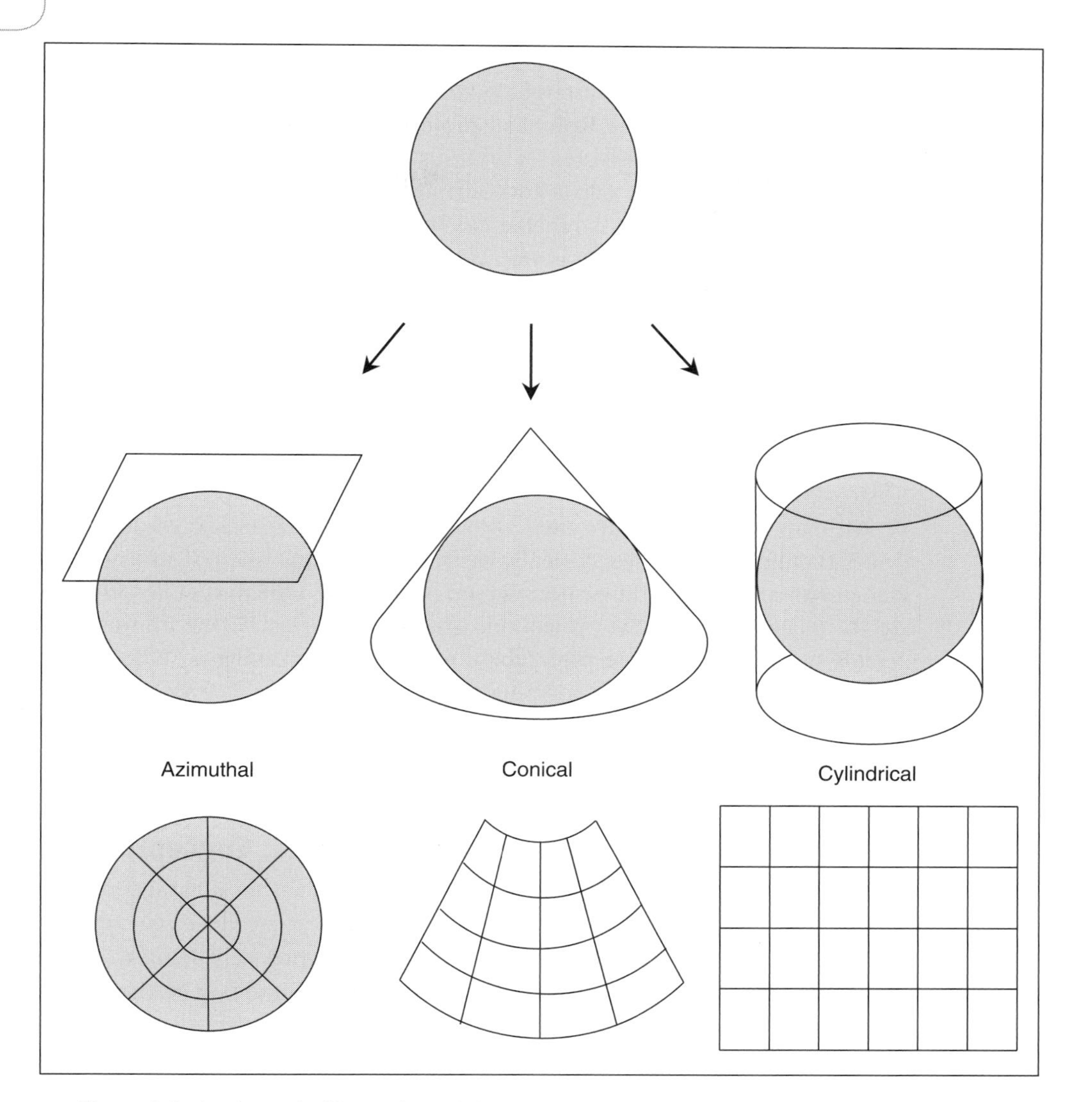

Figure 2.2 A schematic illustration of three common types of map projection: azimuthal, conical and cylindrical.

This projection is most commonly used to map areas that are round in shape or to measure distances and directions from a specific point.

A conical projection fits an imaginary piece of paper around the globe in a cone shape to derive a single standard tangency line of no distortion. Within this line, the projected area will be reliable, whereas areas quickly become distorted outside the conic standard line because the meridians run in circular arcs.

Cylindrical projections fit an imaginary piece of paper around the globe in a cylindrical tube shape. The globe sits inside the cylinder with a line of tangency that stretches all the way around it. Two common types of cylindrical projection are the Universal Transverse Mercator (UTM) projection and the Robinson projection. The

Mercator projection has a single standard line of no distortion that touches the Earth at the equator, with straight parallel and meridian lines running parallel and perpendicular to this. The Robinson projection has two standard lines at 38°N and 38°S. Because distortion increases with distance from the equator, both of these are preferable projections to use at low latitudes, with greater distortion near the poles. The Robinson projection is reliable below latitudes of 45°N and 45°S. To minimise distortion, more frequent standard lines were introduced to the Mercator projection in the form of a global grid dividing the Earth into zones measuring 6° longitude × 8° latitude. This meant that anywhere on Earth could be referenced with respect to a zone in the UTM grid. The Robinson projection does not attempt to preserve a single one of the aforementioned four characteristics that could potentially be distorted. It is neither equal-area, nor conformal. Rather, it abandons both systems for a compromise in which the world appears to look *'right'*, but where no point is free from some distortion.

All map projections have their own specific advantages and disadvantages. They must therefore be applied critically, with an awareness of their utility in relation to the purpose for which the projected map is intended. One particular problem relating to the coastal zone is that projections selected for terrestrial environments rarely coincide with those that are preferable for marine environments. Where two separate datasets are to be analysed together in a common 'coastal' frame of reference, it is therefore necessary to select a single projection and datum that may not be ideal for a given dataset combination (see Section 1.10).

2.3 Precursors to the Collection of Spatial Data: Conceptualisation and Representation

Before spatial data can be collected, it is necessary to precisely define a given phenomenon so that any user of the dataset has a clear understanding of the system or conventions adopted in its production. The steps of conceptualisation and representation largely determine how the geographical world becomes enumerated into a form upon which further analysis can then be carried out. Conceptualisation and representation are important steps of the observation process, as they determine what to include in the spatial data matrix, and in what form (Figure 2.3). Conceptualisation refers to the definition and meaning of the attribute in question. The nature of a particular phenomenon will often provide clues as to whether it should be conceptualised as a continuous raster field or a discrete vector object. A discrete object conceptualisation may be preferable where objects are like points (e.g. a boat mooring), one-dimensional lines (e.g. a river) or two-dimensional polygons (e.g. a lagoon or rock pool). A digital elevation model specifying the height of land pixels above mean sea level is an example of a field representation. These are particularly useful for slowly varying phenomena for which it may be difficult to define boundaries. For example, a wind-blown dune that has slowly built up over time with a steep exposed front, but a low gradient that slopes imperceptibly down in a landward direction.

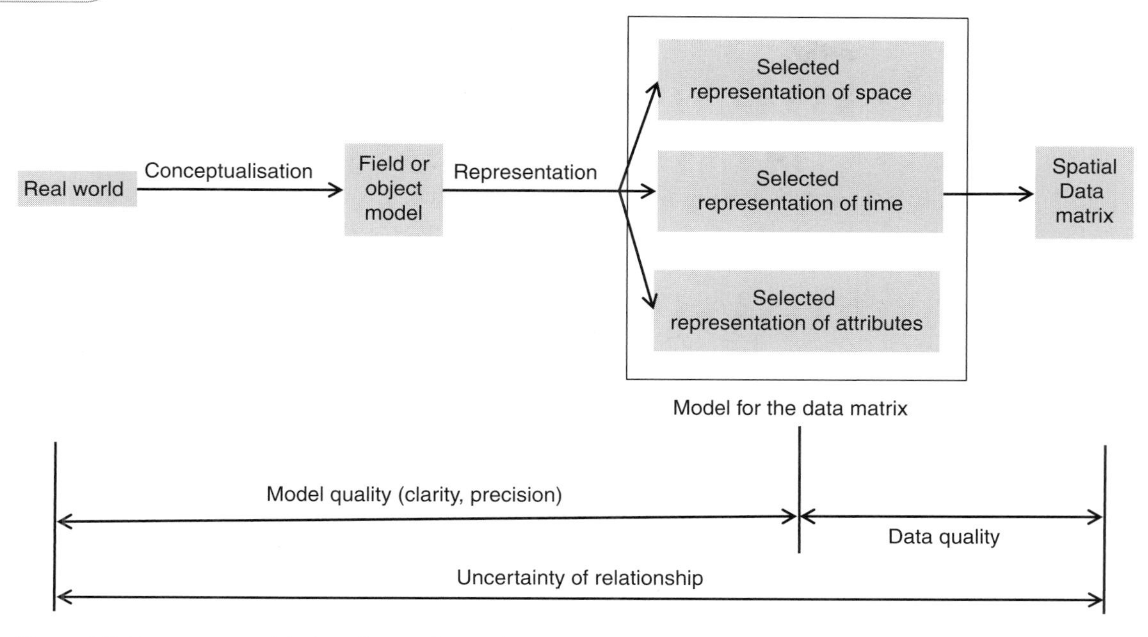

Figure 2.3 From geographical reality to the spatial data matrix, the steps of conceptualisation and representation. Adapted from Haining (2003).

Representation refers to the means by which an attribute is operationalised for the purpose of measurement (i.e. acquiring and storing data on the attribute). Once a data model has been selected (object or field), the appropriate form of representation is sometimes suggested by the attributes in question. This can sometimes be ambiguous. For example, it makes more intuitive sense to represent a stream discharging onto a beach as a line feature than it does to represent the sea itself, which is a larger, more continuous water body, as a line.

2.4 Basic Data Structures: Raster and Vector Datasets

The two most fundamental ways of representing geographic information in digital form are as discrete objects or fields. Fields describe phenomena as a continuous surface that has a value everywhere that falls inside the outer boundary of a dataset. A finite number of values are defined for every possible location within the study area. This would be the case in an elevation map that provides elevation values, rounded off to whole metres above mean sea level, for every location on the map. Field information enables the analyst to answer the question *'what is there?'*, where 'there' can be anywhere inside the study area. In contrast, discrete object representations of the world identify objects with well-defined boundaries that fall inside what would otherwise be empty space. Objects describe individual entities with spatial, temporal and thematic properties and relations. Object information enables the analyst to answer

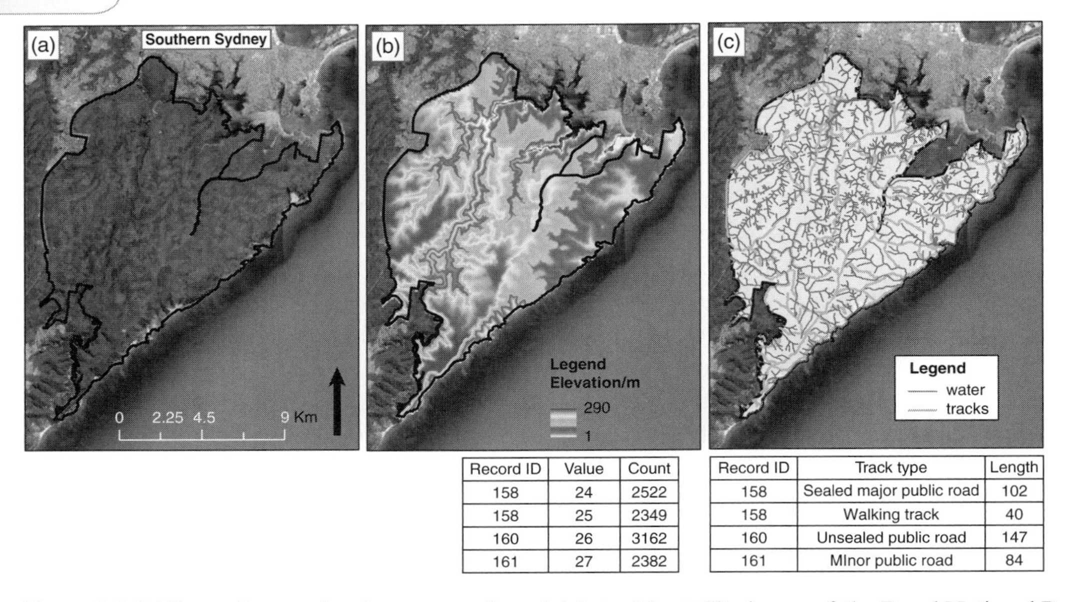

Record ID	Value	Count
158	24	2522
158	25	2349
160	26	3162
161	27	2382

Record ID	Track type	Length
158	Sealed major public road	102
158	Walking track	40
160	Unsealed public road	147
161	MInor public road	84

Figure 2.4 Different formats for the storage of spatial data: (a) satellite image of the Royal National Park, coastal New South Wales; (b) digital elevation model, providing a continuous raster surface of elevation above sea level; (c) polyline vector datasets depicting rivers and roads inside the Royal National Park. Tables beneath (b) and (c) provide illustrative examples of the attribute information associated with raster and vector datasets. (A black and white version of this figure will appear in some formats. For the colour version, please refer to the plate section.)

more direct questions about specific entities, such as: *where is this?, how big is it?, what are its parts?, which are its neighbours?, how many are there?'*

The decision of whether to use a continuous or field dataset corresponds to the step of *conceptualising* a phenomenon. Next it is necessary to identify how this will be represented digitally. Raster and vector data formats are two methods of *representing* geographic data in digital computers (Figure 2.4). A raster representation takes the form of a lattice, tessellation or grid (usually composed of squares) that divides space into an array of square cells. These are sometimes called pixels, which is short for picture elements (Decker, 2001). Raster grid cells are set up in a series of rows and columns, usually aligned with the latitudinal and longitudinal planes to completely cover an area of interest. Raster grids may store one attribute value per pixel, alongside a measure of location that equates to some feature of the grid cell (e.g. the centre or lower left corner coordinates).

Vector data models can be used to represent discrete entities. They do this by delineating points, lines, boundaries and nodes as sets of coordinate values with associated attributes. Vector datasets specify the location (x and y coordinates) of points in zero-dimensional space. These points can collectively form lines in one-dimensional space and polygons in two-dimensional space. A line is defined by a series of vertices (the internal points defining the shape of the line body) and two nodes (end points at either end of the line). It is possible to store additional attribute information with vector features. Vector attribute information may include precise locations for points, lengths for lines, and complete perimeter or section lengths, plus areas, for polygons. However, real-world features need to be generalised in order to be represented as points, lines and polygons because they rarely fit neatly into these categories.

One common technique for the production of vector data is digitisation. This involves an analyst visually interpreting features on a screen, perhaps examining an aerial photograph or map, and tracing over the boundaries of an object to create a digital record (point, line or polygon) to represent that feature. In coastal environments, shorelines are often depicted as simple lines, despite the fact that the location of the actual margin between the land and sea constantly changes with the state of the tide and can be very difficult to define.

Figure 2.4 illustrates the different appearance and information content of raster and vector datasets. The two raster datasets (a) and (b) represent the colour (from a satellite image) and elevation (from a digital elevation model), respectively. The vector dataset (c) represents water and tracks inside the Royal National Park, New South Wales (southern Sydney). The field (raster) elevation dataset stores a value for elevation at every location within the Royal National Park's boundary. This continuous representation of elevation within a grid of pixels is accompanied by a dataset that provides the corresponding elevation of every pixel in the grid, alongside a summary of the count of pixels relating to each elevation value. This is not the case for the object (vector) dataset. This identifies individual entities of water streams or tracks. Each of these entities corresponds to a row in an associated polyline dataset, for which multiple attributes can be listed (e.g. track type and length).

The decision of whether to employ a raster or vector representation depends on:

- the level of precision required for representing features of interest,
- the processing and storage capability of hardware,
- the spatial analysis planned, and
- the nature of the features to be analysed.

It can be difficult to precisely express geographical variation with a raster representation because properties and attributes can only be attached to individual grid cells. Depending on the spatial resolution, this may equate to a large area on the ground relative to the feature or entity of interest. Remote sensing satellites are one of the most common sources of raster data for spatial analysis. These images are composed of raster grids, each of which has an accompanying reflectance value. Because each cell is assigned a single value, all detail on variation in land-cover reflectance within the cell is lost. Individual cells may represent large areas on the ground, for example, a Landsat image is composed of cells (or pixels) of dimension 30×30 m (or 900 m^2). This leads to a lower level of precision for the representation of features than is achievable through a vector representation (Figure 2.5).

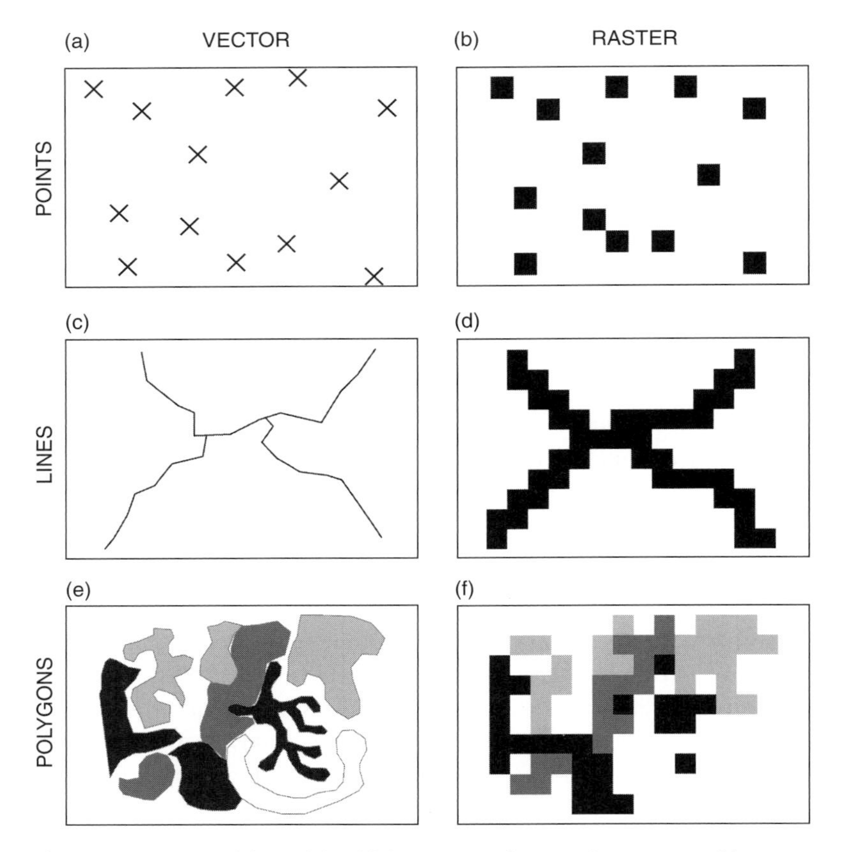

Figure 2.5 The precision with which vector and raster datasets are able to capture point, line and polygon features: (a) vector point features; (b) raster point features; (c) vector line features; (d) raster line features; (e) vector polygon features; (f) raster polygon features.

In computational terms, the storage of vector datasets is simpler and more efficient than a raster representation. Less information is necessary because only features that are of interest are stored, meaning that there is no redundancy in the dataset. Often, vector datasets will be composed of several thousand polygons, each with their own unique geometry. In the absence of a regular grid, datasets can get complicated. This is particularly the case when they are combined with another similar dataset. The manipulation of high numbers of polygons, often with complex individual geometry, requires substantive computing power. This means that simple types of spatial analysis, such as map algebra operations, are generally faster to process using raster datasets.

Finally, where multiple datasets are to be analysed, most techniques for the manipulation of spatial data (e.g. spatial queries or map algebra) work solely with one data format, e.g. raster or vector, but not interactively between the two. If some datasets are already available to the analyst, they may want to acquire further data in a format that aligns with these existing datasets. If this is not possible, then a conversion can be applied to move between these formats. This may result in some generalisation and associated loss of information.

2.5 Sources of Spatial Data

Decker (2001) summarises the basic groups of GIS data as:

- imagery (including aerial photography and satellite imagery),
- digital elevation models,
- digital raster graphics (derived from images), and
- digital point, line or polygon features generated from visual interpretation and digitisation.

Primary data can be collected in a tailored manner to meet the specific objectives of an analytical exercise. Primary data originate from fieldwork, by conducting direct surveys with the global positioning system (GPS), land surveying, the collection of information from environmental monitoring stations and also the acquisition of remote sensing images. Because further analysis is to be undertaken with these data, all data cases must be accurately and carefully georeferenced. Secondary spatial data sources include:

- maps from which relevant information is captured by scanning and digitising,
- archives,
- data generated and stored by public bodies, such as local authorities, and
- commercial datasets generated from field or remote sensing aerial surveys undertaken by geoinformation consultancy companies.

Often an analyst may encounter a situation where useful information is available in the form of old, infrequently used maps archived in a library or filing cabinet. To work with such information using GIS software, it is necessary to take a series of practical steps to bring the data into digital format (Figure 2.6). If the useful information takes the form of a map or aerial photograph, this will need to be scanned or digitised. Archived maps or images are often missing spatial referencing information, so this will need to be added to the scanned or digitised file via the process of georeferencing. Georeferencing involves defining a series of ground control points (GCPs) that are distinctive, observable features for which coordinates are known. A table of GCPs can be constructed that lists each GCP feature against its coordinates. These are obtained either by selecting the corresponding feature on a different, georeferenced image of the same location, or through direct measurement in the field using GPS. For example, it would be possible to georeference an aerial photograph of a beach within two cliffed headlands by visiting both headlands and

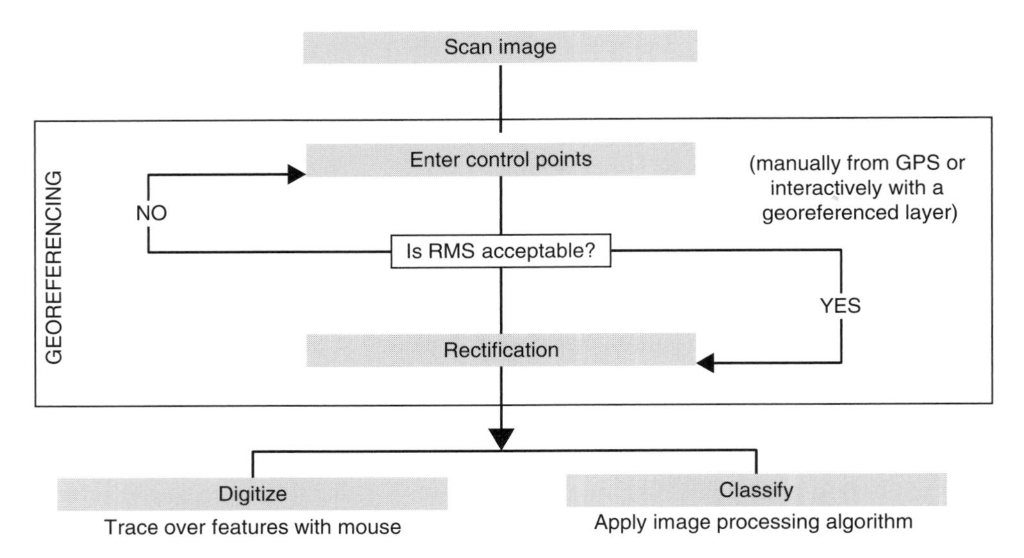

Figure 2.6 A schematic outline of the series of steps that are necessary to get data from paper maps into a computer and georeferenced, such that it can be used for spatial analysis.

storing the waypoint in GPS, then entering these coordinates into specialised software for georeferencing. In practice, more than two GCPs will be necessary for a reliable georectification exercise. An ideal set of GCPs will combine as many points as possible spread evenly across an image to increase the reliability of the georeferencing information. Features used as GCPs need to be highly visible, distinct from their surroundings and immobile through time. The converging branches of a river outlet through variable sand and marsh wetland make an unreliable GCP. This is because the positioning of these branches is likely to change in the time period between acquisition of the source image and the production of the map to be georeferenced. To minimise digitising error, the position of GCPs is most accurately identified at a large scale, i.e. zoomed into a map, such that the features appear large to the analyst.

Orthorectification is a processing step that is often applied to aerial photographs to correct for differential distortion across the image. This differential distortion acts along a gradient moving away from the nadir point or centre of the image, which would have been directly beneath the camera lens when the photograph was taken. Because of the differential distortions, features in the photograph appear to lean outwards from the centre of the image towards the edge to an extent that depends on their height.

Once a map has been scanned and georeferenced, information can be extracted from it. In vector format, this is achieved by digitising observable features. In raster format, this is achieved through the application of an image classification algorithm (see Section 4.10.2).

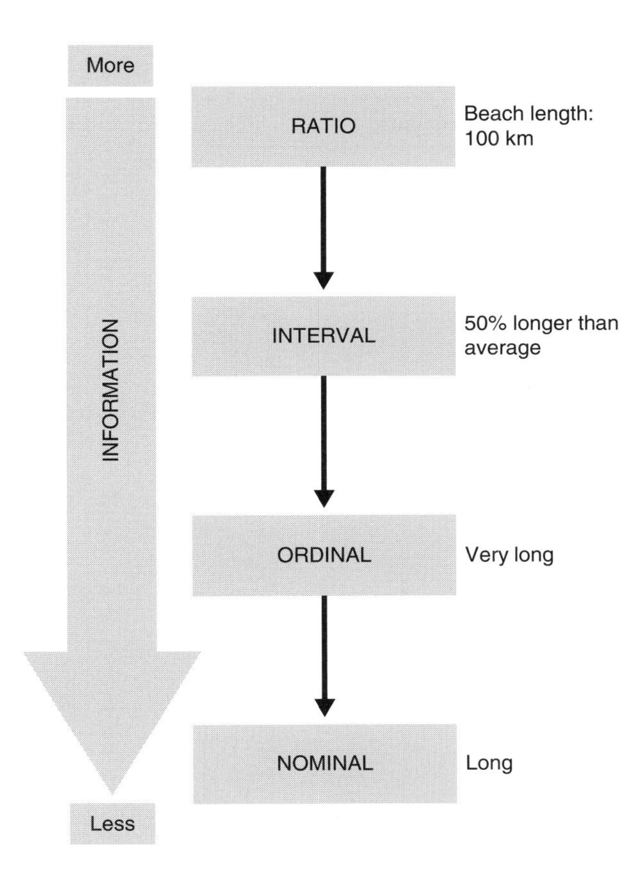

Figure 2.7 The declining amount of information available with a downward transition of measurement levels.

2.6 The Different Levels of Measurement

Measurement involves the assignment of a numerical value to some quality of an object, so as to represent some fact or convention about it, in accordance with definite rules (Chorley, 1967). Figure 2.7 and Table 2.2 summarise the four different levels, or scales of measurement for generating information or data at a nominal, ordinal, interval and ratio level. These are listed in order of increasing information content.

The lowest or least informative level of measurement is the *nominal* scale. This attributes qualitative labels, or uses numbers to designate classes. Where a number is used, the value has no meaning other than to demonstrate equivalence to a nominal class. The same numeral cannot be assigned to different classes and vice versa. Price's classification of coasts provides a good example of a nominal scale, in which the shorelines of the Gulf of Mexico were classified on the basis of their geological characteristics, using terms such as canyon, channel, escarpment and plain (Price, 1954).

Table 2.2 Levels of measurement of spatial data, from high to low down the table.

Description	Discrete object (vector data)			Continuous field (raster data)
	Point	Line	Polygon	
Nominal: A simple classification, designed to establish equivalence between objects in terms of some quality they possess (=)	Coral: live or dead	Shoreline: sandy or rocky	Habitat map (sand, seagrass, coral)	Coastal habitat type (sand, seagrass, coral)
Ordinal: A scale that possesses both equivalence and sequence, designed to allot numbers according to the relative amount of some quality possessed by objects (≤; ≥)	Beach particle grain size scale (log)	Shoreline linearity (high, medium, low complexity)	Index of coral mortality	Beach grain size (coarse, medium, fine)
Interval: A scale that possesses equivalence, order and ratio of intervals, in that there are equivalent intervals or differences between successive orders (≤; ≥; +; −)	Score of sand grain roundness	Shoreline length (m)	Categories of seagrass density	Sea water temperature (°C)
Ratio: A scale that possesses order, equivalence, ratio of intervals and ratio of values with reference to an absolute zero, such that correct ratios are preserved between the magnitudes of numbers (≤; ≥; +; −; ×; ÷)	Beach pebble size	Erosion per annum	Regional per annum incident wave power	Rainfall (cm)

The *ordinal* scale possesses both equivalence and sequence. Numbers or semi-quantitative labels are designated in accordance with the relative amount of some quantity possessed by the object. A relative order or sequence is preserved, although the magnitude of difference between successive numbers is not known. A coastal classification of Norway's coasts provides an example of an ordinal scale, in which segments of coastline were labelled on the basis of wave conditions as large-, moderate- or small-energy wave climates (Price, 1955).

The *interval* scale possesses equivalence and order, and preserves a ratio of equal intervals or differences between successive numbers. Wadell's (1932) index of 'roundness' applied to sand grains is an example of an interval scale. Values of the 'operational sphericity scale' range from zero (non-spherical) to one (a perfect sphere). This sort of scale represents the minimum information content necessary

Table 2.3 Stevens' classification of scales of measurement. After Stevens (1946).

Scale	Basic empirical operations	Permissible statistics
Nominal	Determination of equality	Number of cases Mode Contingency correlation
Ordinal	Determination of greater or less	Median Percentiles
Interval	Determination of equality or intervals or differences	Mean Standard deviation Rank-order correlation Product-moment correlation
Ratio	Determination of equality of ratios	Coefficient of variation

for computing the statistical mean and standard deviation. Nevertheless, it is limited by the arbitrary designation of a zero value. The scale raises the question 'what does a grain that scores a value of zero on Wadell's roundness scale actually look like and how does this shape represent the complete absence of the "roundness" quality?' It is not possible to say that one value is twice or some other proportion greater than another.

Finally, the *ratio* scale is the highest level of measurement that contains the most information about a given feature. In statistical terms, it is the most powerful and versatile of the four measurement levels. It possesses equivalence, order, ratio of intervals and ratio of values. This scale has been designated a zero that has a meaningful position in relation to the quantity represented. For example, a temperature measurement in kelvin units is placed on a scale whereby zero kelvin (0 K) in absolute terms represents the complete absence of any heat energy whatsoever. It is meaningful to multiply such numbers by constants. It is also possible to make statements that draw quantitative comparisons, such as '*the water temperature in a ponded lagoon of 303 K (30°C) in the middle of the day is really twice as hot as the frozen glacier at 151.5 K (−121.5°C)*'. Numbers from these scales of measurement can be inserted into virtually all types of mathematical and statistical procedures.

When developing models using spatial data, the level of measurement of data stored in the matrix is a fundamental determinant of the amount of information available to the analyst. In turn, this determines the statistical and arithmetical procedures that can be applied (see Figure 2.7 and Table 2.3). A higher level of measurement can be thought of as one that incorporates a greater amount of information. This lends more power and versatility to the modelling process (Stevens, 1946). Table 2.3 outlines the different levels at which it is possible to measure a given variable, alongside the corresponding empirical and statistical operations that are possible. It demonstrates the importance of specifying the formal properties of the numbering system that underlies measurement of a variable.

2.7 Spatial Data Quality in the Coastal Zone

The degree to which the quality of spatial data is acceptable depends on the particular purpose for which the data are intended. Aspects of data quality include:

- accuracy,
- completeness,
- consistency, and
- resolution.

Accuracy refers to the difference between the 'true' value of a variable and the value recorded in the database. The terms 'accuracy' and 'precision' are often used interchangeably, although they represent different concepts. It is possible for a dataset to be accurate, but not precise, or to be precise, but not accurate. Precision refers to the spread, or variance, of data values around a central mean. A series of values can be fairly tightly clustered around a mean (and therefore precise), but their absolute magnitude may still be quite different from the 'true' value of a variable (and therefore inaccurate). This situation is represented in the bottom left quartile of Figure 2.8, which illustrates the difference between accuracy and precision. It is also possible that the points may have a high variance that is evenly spread about the nominal 'true' value. Their mean value is therefore accurate in spite of the fact that collectively they are not very precise.

Spatial data accuracy is determined largely by locational and attribute factors. Point location errors are quantified as root mean square errors (See Section 8.2) along the two directions that correspond to the north–south and east–west axes. In the case of locational factors relating to remote sensing data acquisition, sensor geometry is measured by an inertial measurement unit (IMU) coupled with high-precision differential GPS. Fluctuations in the position of the IMU allow the influence of unintended movements of an aircraft during an aerial survey to be measured and corrected for.

Attribute data that are commonly associated with pixels are radiance values. These record the amount of light reflected into the image sensor (see Section 4.6). Radiance values recorded for a pixel are often influenced by light scattering, from an adjacent parcel of land, sea or the atmosphere. Natural conditions such as the Sun angle, geographic location and season, as well as the internal geometry of the sensor hardware, all contribute to these confounding influences that may introduce inaccuracies into the radiance measured for a given pixel (Craig and Labovitz, 1980). In turn, this may introduce uncertainty into the interpretation of pixel content. Processes such as classification might group pixels into thematic classes to which they do not belong as a result of data contamination. The reliability of spatial data requires continual scrutiny. Coincidence between the predictions of a spatial model that is underpinned by sound theory and corresponding measurements collected directly from the field gives us confidence that these closely approximate 'reality'.

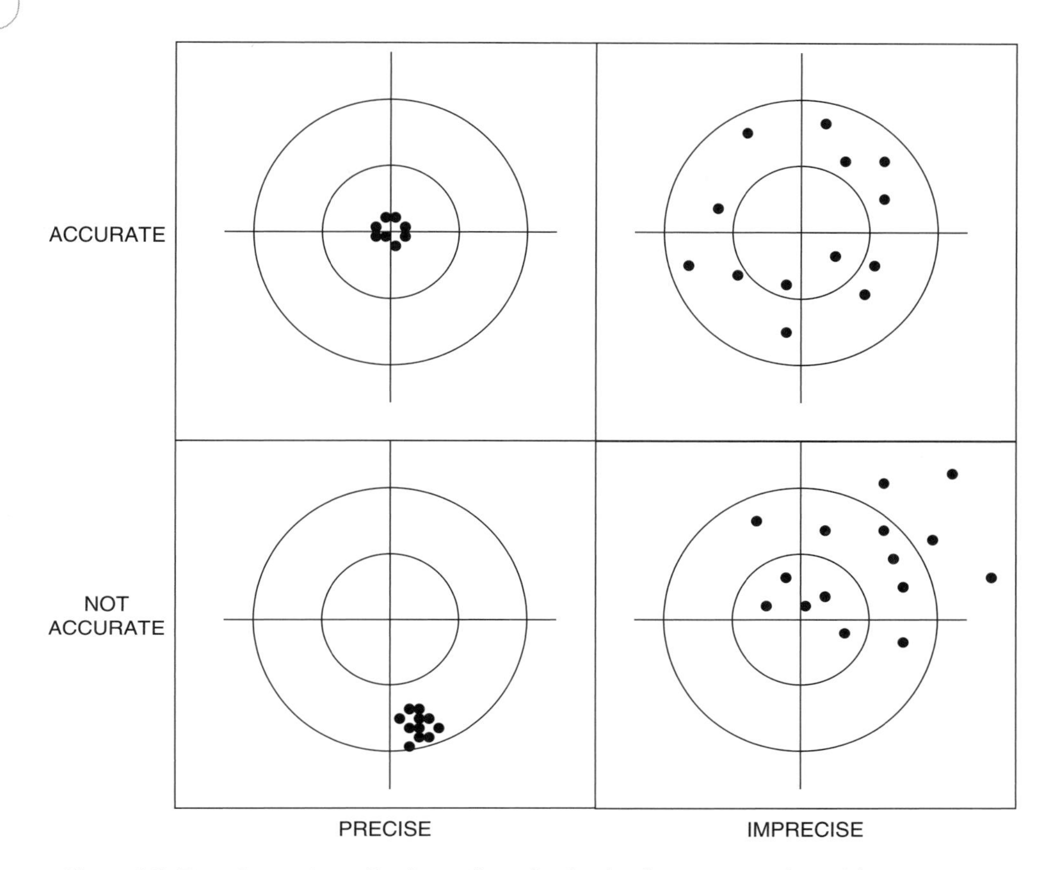

Figure 2.8 Four data point collections of varying levels of accuracy and precision.

Completeness refers to the extent to which the spatial data matrix is populated. Data completeness can be assessed with respect to the presence or absence of relevant attribute variables, or the location in which these have been recorded. Attribute variables can be subdivided into those derived indirectly from a model or remote sensing dataset (i.e. the digital number and its processed derivatives), and those measured directly in the field (e.g. beach width). When conducting spatial analysis, a dataset will encompass information on all variables of relevance to a property of interest. It is impossible to consider all aspects of any physical system. This is because we will never know that all factors influencing the observations are included in the analysis. It seems unlikely that enough factors could ever be considered at once to predict, completely without error, the characteristics of a system (Melton, 1957). The notion of perfect prediction as obtainable through the analysis of a complete, comprehensive dataset is therefore misleading. Nevertheless, it is possible to obtain a dataset for which all values are present, across all attributes and locations measured. Such a dataset can be thought of as 'complete' in the sense that the spatial data matrix is fully populated (Section 2.1).

Consistency is defined as the absence of contradictions in a database, where logical structure and attributes rules describe the internal compatibility of the dataset (Shi et al., 2003). When the objective of spatial analysis is to compare the spatial distribution of phenomena that fall within a study area in a standardised manner, consistency of measurement is important. Given that remotely sensed raster images are composed of a grid, each pixel of which possesses reflectance values within a standardised scale across the same number of wavelengths, such datasets have high internal consistency.

It is necessary to employ measurement tools that decipher information at scales appropriate to processes that are being analysed. *Resolution* refers to the precision of the units employed to capture information about attributes. This is usually a function of the instrument used for measurement. The term 'resolution' can refer to three different characteristics of information: (i) the distance between adjacent measurements on the ground (spatial resolution), (ii) the level of detail to which dataset categories are specified (thematic or descriptive resolution), or (iii) how regularly measurements are taken over time (temporal resolution). In relation to raster and vector datasets, resolution refers to the dimensions associated with the units of measurement. For a raster grid, this corresponds to the length of the side of a pixel. For vector points, this corresponds to the minimum distance that can separate two vector points before they are merged into a single unit.

2.8 Spatial Data Scale

Map scale refers to the ratio between a distance on a map and the distance that represents on the ground (the Earth). A 1 : 250 000 scale map is at a smaller scale than a 1 : 1000 map. This is because 1 divided by 250 000 is the relatively small number of 4×10^{-6}, whereas 1 divided by 1000 is the relatively larger number of 1×10^{-3}. This terminology sometimes causes confusion, because a map made at a large scale usually covers a smaller area of the Earth (i.e. it looks more 'zoomed in') than a map made at a smaller scale (which looks more 'zoomed out').

On a map, the scale is conventionally expressed either as a ratio, a qualitative statement, or on a scale bar. This latter expression is commonly the most helpful for the map reader to interpret distances between points of interest on the map. It is possible to convert a map from a large-scale to a small-scale one by generalising the amount of information available. It is not possible to transition from a small-scale to a large-scale map where additional information or detail is not available. This will produce either a very coarse-looking polygon, or a pixelated image, depending on the data format of the map.

The resolution of vector datasets that have been generated by the process of digitising is a function of the scale at which the digitising was undertaken. Figure 2.9(a) illustrates two vector datasets generated by digitising the coastline of the UK at scales of 1 : 500 000 and 1 : 10 000 000. The vector dataset digitised at the smaller scale (1 : 10 000 000) does not follow the coastline as closely as the one digitised at the larger scale. Errors associated with this dataset will be larger because of human error. Unintentional movements of the digitiser's hands during the digitisation process will

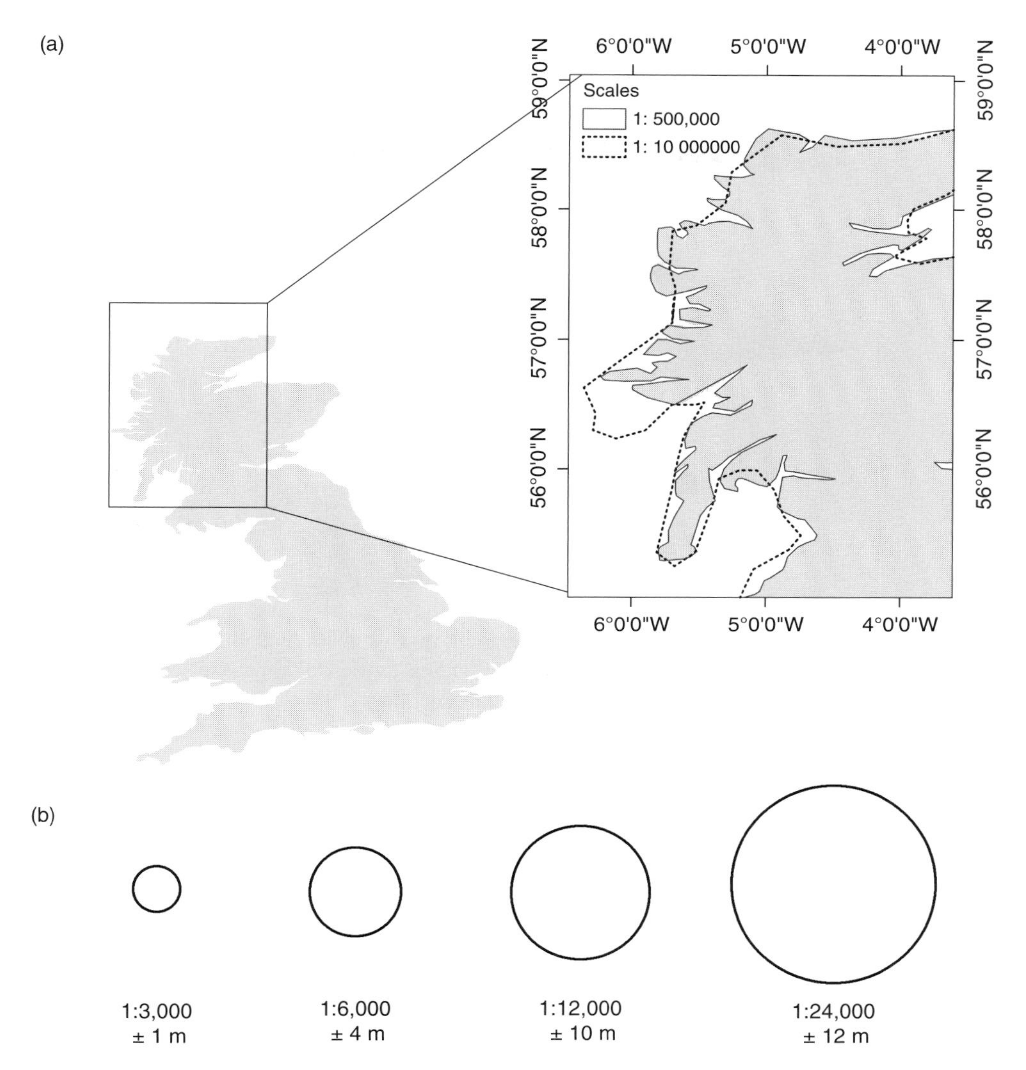

Figure 2.9 (a) The coastline of the mainland UK, showing the area outlining northwestern Scotland digitised at a scale of 1 :500 000 and 1 : 10 000 000. Note the increased tendency for the coastline digitised at the smaller scale (black dashed) to depart from the baseline outline, shown in grey. (b) Root mean square error of vector datasets digitised at several different spatial scales.

represent larger areas on the ground when the digitisation is carried out at a smaller scale than they will at a larger scale. Figure 2.9(b) illustrates the root mean square error of vector datasets that were digitised at several different spatial scales (1 : 3000, 1 : 6000, 1 : 12 000 and 1 : 24 000). It can be seen that the error at the smaller scale of digitisation (1 : 24 000) equates to a greater distance on the ground (i.e. 12 m) in comparison to the larger scale of digitisation (1 : 3000), with a corresponding ground length of 1 m. This highlights the importance of digitising at as large a scale as possible when generating vector data to reduce inaccuracy in the visual interpretation of maps.

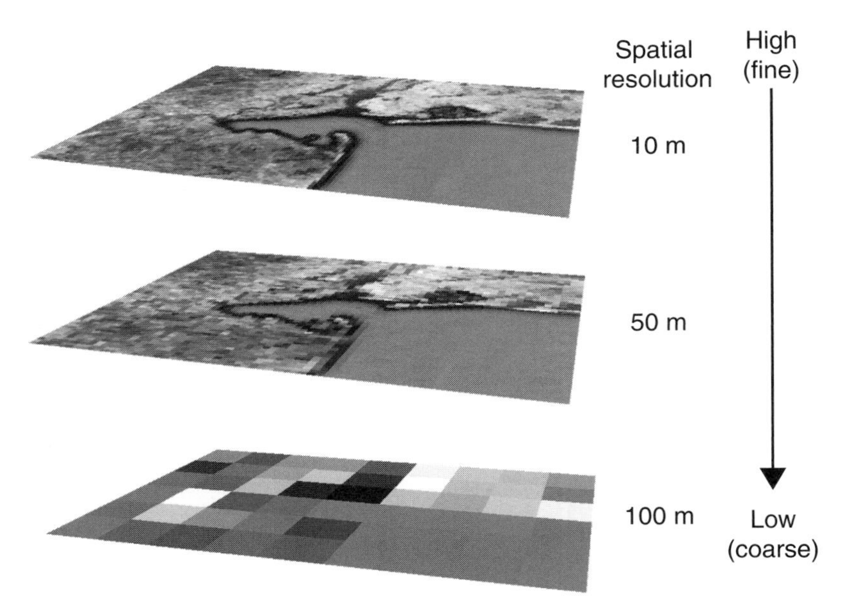

Figure 2.10 A series of raster images of varying spatial resolution showing New York Harbour. (A black and white version of this figure will appear in some formats. For the colour version, please refer to the plate section.)

Raster datasets are primarily acquired from remote sensing technology (see Chapter 4). Remote sensing technology has incrementally improved in terms of greater sensor resolution over the last few decades. This suggests that in the future it will be possible to analyse phenomena at ever-smaller scales, through increased spatial, thematic and temporal resolution of instruments.

As previously mentioned, the spatial resolution of a remote sensing image refers to the length of a side of a component pixel within a raster grid. This determines the geographical scale at which data can be collected. Data are generalised to a single value within a square of these dimensions. Figure 2.10 illustrates a series of raster images of varying spatial resolution of New York Harbour. The general loss of information at a lower or coarser spatial resolution can be seen as the size of the individual pixels increases.

2.9 The Collection of Spatial Referencing Information

Global Positioning Systems (GPS) utilise a global network of Earth-orbiting satellites to locate a position on the Earth. Three subsegments make up GPS as a whole: the space, control and user segments. The space segment comprises a system of 31 satellites in orbit at 20 200 km above the Earth. Each satellite orbits the Earth every 12 h, at a speed of 14 000 km h^{-1}. Collectively, these satellites provide good coverage of the Earth. An individual standing on the Earth's surface is likely to be able to

Table 2.4 Some common sources of error associated with GPS.

Error type (approximate value/m)	Description
Clock (2.1 m)	Timing error in the clock of the satellite or receiver
Ephemeris (2.1 m)	Error in placement of the satellite's position in orbit for a given time
Multipath (1.0 m)	Reflection of satellite signals (radio waves) from objects. The reflected signal takes more time to reach the receiver than the direct signal
Atmospheric time delay: ionosphere (4 m), troposphere (0.7 m)	Time delay of signal due to atmospheric interference
Satellite geometric alignment	Error in the relative position of the satellites

receive a signal from 8–10 satellites at any one time. The control segment is made up of five ground control stations in Hawaii (Pacific Ocean), Colorado Springs (USA), Ascension Island (Atlantic Ocean), Diego Garcia (Indian Ocean) and Kwajalein (Marshall Islands, Pacific Ocean). These control stations monitor the positions of each satellite in orbit. Precise orbits and navigation instructions are continually monitored and coordinated through the master control station at Colorado Air Force Base. Finally, the user segment comprises the device at the location for which the position is to be determined. These devices might include mobile phones, navigation aids, receivers, processors and antennae that enable the community of users to receive satellite broadcasts and convert signals into a precise location on roving devices.

Trilateration is the procedure by which GPS estimates the user's location on Earth. Briefly, the GPS receiver records the travel time of radio signals received from multiple satellites in orbit at a given point in time. Travel times are corrected for delays due to interference from the Earth's atmosphere and combined with precise knowledge of each satellite's location in orbit. This is relayed from the ground control stations to find the distance to each satellite. Once the distance to each satellite is known, a position can be fixed using geometry. Navidi et al. (1998) provide a detailed overview of the procedure of trilateration.

Table 2.4 summarises the sources of error that may occur during the estimation of a position using GPS technology. Differential techniques can be used to minimise errors in estimated locations. For a differential GPS survey, a base station is established for which the position is accurately known. This can be set up above a 'trig survey point' for which local survey data are available. In the absence of local information, a base station can be left to measure its position continuously over time. Under the assumption of stochastic error, the many positions recorded should converge on an accurate, mean estimate of location. The base station simultaneously monitors location while a user segment records the location of a roving unit being mobilised by the user. Given that the position of the base station can be precisely determined, any departures from this position as estimated during the simultaneous survey period can be assumed to be consistent between both the base station and the roving receiver. Information from the base station can therefore be used to correct both location datasets, either in real time (broadcasting corrections) or after the survey has finished (post-processing). Given the relative proximity of the base station and roving GPS receiver in relation to the spatial scales of variability of factors such as atmospheric interference, it is reasonable to assume that signal interference from multiple sources across both receivers will be consistent.

2.10 Applications of GPS Technology in the Coastal Zone

Global Positioning System (GPS) technology is applied in the coastal zone from a range of different platforms. Field-based surveys can be undertaken on foot with a roving GPS receiver. Where a high degree of accuracy is required, this can be coupled to a base station for differential survey quality of data positions. Figure 2.11 illustrates a range of applications of GPS technology in coastal environments for the purpose

Figure 2.11 Some different applications of GPS technology for the collection of spatially referenced data. (a) A snorkeler towing a GPS unit along in a floating waterproof bag while taking ground referencing photographs above a coral reef. (b) Research volunteers carrying GPS drifters into the surf zone as part of a study on rip currents. (Photo credit: Patrick Rynne.) (c) A drone collecting imagery from a GoPro above mangroves in Minnamurra River, New South Wales. (Photo credit: Mark Newsham.) (d) A coupled GPS monitor and side-scan sonar instrument display attached to a boat. (Photo credit: Frank Sargeant.) (A black and white version of this figure will appear in some formats. For the colour version, please refer to the plate section.)

of ground referencing, or 'ground truthing', a satellite image to make a map. Hand-held GPS units can be towed along by snorkelers who are simultaneously collecting photographic records of the seafloor composition. Tracking receivers can be released and monitored remotely, as in the use of GPS drifters to map surface currents. GPS units can also be attached to a platform, e.g. a drone or a boat, from which greater spatial coverage can be achieved for collecting information. GPS records collected from a boat-based platform require additional correction in the vertical domain for the influence of the tidal cycle.

Mobile phone applications have made GPS technology more widely available to the general public. Often the GPS technology in mobile phones incorporates additional information, such as the locations of Internet Protocol (IP) addresses and wireless networks, to further increase the positional accuracy of the user. Mobile phone applications have been developed to make use of these improvements for field survey. For example, Open Data Kit provides a customisable and user-friendly interface for the collection of spatially referenced metadata and photographs (Brunette et al., 2012).

2.11 Sampling Spatial Patterns of Coastal Phenomena

The purpose of sampling for spatial analysis is to acquire a series of data cases that will allow an analyst to make robust inferences about a population on the basis of a subset of individuals drawn from that population (Haining, 2003). Because it is spatial analysis, each population member (or sampled data point) has a geographical reference.

Geographic data are only as good as the sampling scheme used to create them. To estimate a geographical distribution, Ripley (1991) identifies three main classes of sampling plan as random, stratified random and systematic (Figure 2.12). In accordance with these schemes, sites are chosen either in a manner such that every location has an equal chance of being sampled (random), or with respect to a partitioned scheme in which sites are either randomly located (stratified random) or systematically placed (systematic sampling). Increasing degrees of stratification or systematisation can be introduced to accommodate the unequal abundance of different phenomena on the Earth's surface. Resultant observations of spatial patterns will depend upon a number of factors (Dale and Fortin, 2014), including:

- the sample size, i.e. the number of observations,
- the spatial extent of the study area,
- the grain size and shape of the sampling units,
- the sampling strategy or spatial layout of the sampling units used to collect the data (e.g. transect, lattice, random), and
- the distance between sampling units.

Where processes are stationary (i.e. their statistical properties do not change as a function of space), a random sampling design and use of global spatial statistics are appropriate. In contrast, non-stationary processes (those that exhibit direction-specific variation across a study area) call for stratified sampling, use of local statistics and restricted randomisation tests (Wagner and Fortin, 2005).

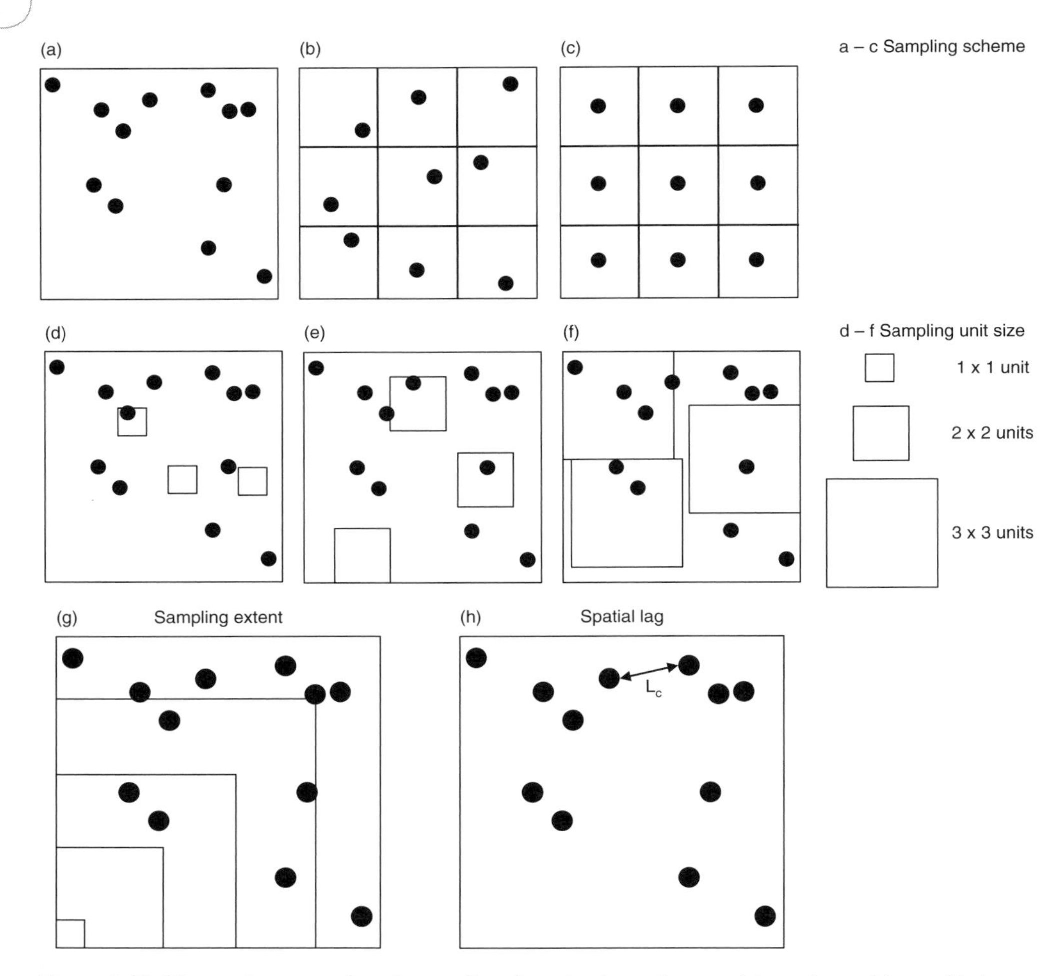

Figure 2.12 Illustrative examples of sampling site selection schemes: (a) random, (b) stratified random, (c) systematic, (d)–(f) varying the sampling unit size, (g) varying the sampling extent, and (h) introducing a spatial lag (L_c = minimum permissible distance between points).

Spatial dependence in a dataset is the tendency for point samples to be more alike the closer they are together in space (see Section 6.2). The presence of spatial dependence in a dataset implies that points close together will carry duplicate information. An effective sampling scheme, particularly one that provides the basis for statistical testing, must account for this by adopting an optimal distance between samples. This must avoid them being located so close together that spatial dependence results in them being too alike. To this end, a spatial lag can be introduced to a sampling scheme to provide a minimum distance for spacing apart sampling points. Chapter 6 outlines how advanced statistical tools such as the semivariogram can be used to empirically define these dimensions. Dunn and Harrison (1993) tested the statistical efficiency of systematic, random and stratified random sampling schemes on two-dimensional datasets and found that large gains in statistical power were apparent when spatial dependence was accounted for using lag functions that reflected the varied and complex spatial dependence of the data. It is also important that sampling distances should not be so far apart that they are unable to adequately capture the natural scales of variation intrinsic to a given phenomenon. It is therefore critical to have some understanding of these scales of variation before embarking upon a sampling exercise.

2.12 Topology and Database Structures for Storing Spatial Data

The topology of a GIS database determines how spatial features relate to each other. Qualitative relationships can be stored within spatial databases. These rules are often used to ensure data quality control, or to edit and optimise data queries (Goodchild et al., 2005). For example, when geographic data are collected via the procedure of digitising (see Section 2.4), it can be useful to test the topological integrity of data. This is particularly the case when datasets run to several thousand points, lines or polygons. To ascertain the logical validity of a dataset, tests might pose questions focused on the properties of a dataset. For example, with respect to the connectivity of a network of lines, the query *'do all sections of rivers in a catchment connect?'*

In addition to validating the geometry of vector data, topological relationships can also be interrogated to assess polygon adjacency, or trace networks through polyline features. The topology of raster datasets is calculated based on the relative position and value of each cell within a grid. For example, the value of a cell can be checked against the value of the surrounding cells to identify groups of cells that are next to each other and share the same value. By interrogating relative positions, it is possible to build topological relationships based on contiguous areas of similarly valued cells. These are sometimes referred to as zones. Depending on the scale of variation of the feature of interest and the spatial resolution of the raster grid, it can be difficult to capture fine-scale topological relationships from raster grids.

The topology of vector data keeps track of how the points, lines and polygons link by storing additional information on how individual features are defined. As mentioned previously, a line is defined by a series of vertices (the points defining the shape of the line body) and two nodes (end points at either end of the line). Where point features are part of a line, the topological information will track which line they define. Where lines are made up of many smaller lines (i.e. running between vertices), topological information will include the identity of the node with which the line starts (FNode) and ends (TNode). These are designated as 'from' and 'to' nodes. Where lines form polygons, topological information includes the identity of the polygon to the left (Lpoly) and the polygon to the right (Rpoly) of each line. In many GIS software packages, topological information is hidden from the analyst's view (e.g. in Esri's ArcGIS software, it is stored in a geodatabase repository).

The overall structure of a database determines the organisational principles that guide the manner in which data are stored, organised, queried and manipulated. There are different types of database structure for storing spatial information. These may need to accommodate several different types of relationship depending on the relative

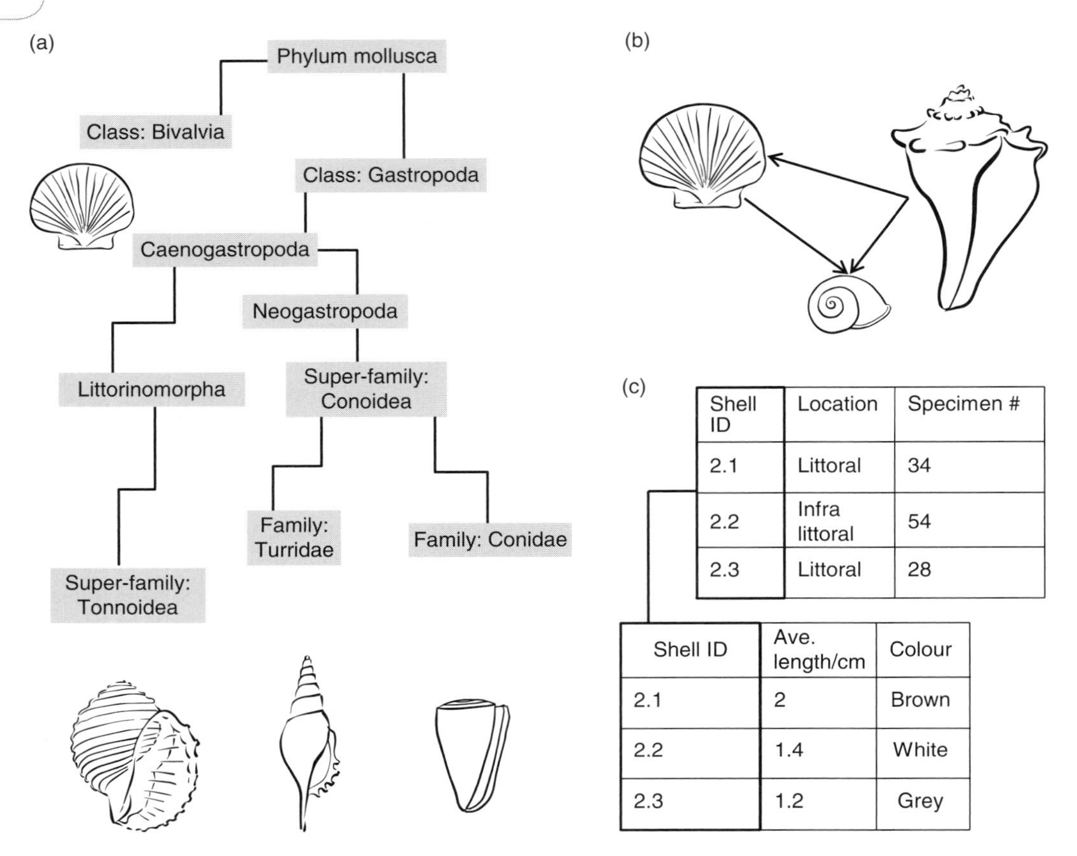

Figure 2.13 A schematic illustration of the three common database structures: (a) hierarchical database (a taxonomy of seashells); (b) a network database (crustacean food web); and (c) a relational database (examples of linked seashell records with attribute data).

number of features and attributes in the dataset. These include one-to-one (one feature to one attribute), many-to-one (many features to one attribute), one-to-many (one attribute associated with many features) and many-to-many (numerous features associated with numerous attributes). The three most common database structures for storing these relationships are hierarchical, network and relational types, each of which differs in the kind of relationships that can be stored between spatial features and their attributes (Figure 2.13).

A hierarchical database stores features in such a way that they are classed into unique combinations of attribute values. These unique combinations can be used to sequentially distinguish between features through a layered (hierarchical) series of levels providing information about the different features. At the bottom of a hierarchy, features are placed into classes that uniquely identify their combination of attributes across levels that relate to different criteria. A taxonomic key such as that shown for shells in Figure 2.13(a) is a common example of a hierarchical database. Network databases can store many-to-many relationships. This is because, although each feature is only stored once, they can link to many attributes and vice versa (see Figure 2.13(b)). A food web is an example of a network database. Relational databases are tables that operationalise the concept of the spatial data matrix, in the sense that tables contain rows of information relating to a single object (point, line polygon or raster grid cell) for which the location is stored. Columns relate to characteristics or attributes of that object. Different classes of objects can be stored as separate tables and joined together to create new views of the database. These joins can be used to transfer information on particular attributes between different object classes. Relational databases are the structure that is most commonly employed by GIS and spatial analysis packages. This is because they represent an optimal compromise between the three structures discussed here in relation to practical criteria including ease of use, search flexibility, data storage, redundancy and whether the four types of data relationship can be easily handled.

2.13 Summary of Key Points: Chapter 2

- Spatial data on different characteristics at different locations are most efficiently stored and analysed in the form of a spatial data matrix.
- To cross-reference different spatial datasets, they need to be in the same frame of reference. To achieve this, it is necessary to use a standardised geographic coordinate system, or a projection and datum. Three common types of projection are azimuthal (planar), conical and cylindrical.
- Before spatial data can be collected, the phenomenon under consideration must be conceptualised (*what does it mean?*) and represented (*how will the information be stored?*).
- Raster (grid cells) and vector (point, line polygon) data structures are the two basic formats for storing digital spatial data. Raster data are easy to manipulate, but can generate large files and be imprecise. Vector data are simple, efficient and precise, but complex to manipulate.
- There are four levels of measurement for collecting spatial data (nominal, ordinal, interval and ratio). These hold different amounts of information and determine what analysis can be performed.
- The four aspects of spatial data quality are accuracy, completeness, consistency and resolution.
- Map scale refers to the ratio between a distance represented on a map and the same distance on the ground (the Earth). A large-scale map holds more information than a small-scale map.
- The global positioning system (GPS) is a global network of Earth-orbiting satellites for locating a position on the Earth. Positional information collected from GPS is subject to errors; these can be reduced using differential methods.
- The topology of spatial data governs how the information, along with the relationships between different data components, is stored. In GIS databases, spatial data are most commonly stored in a hierarchical structure.

3 Basic Geographical Analysis With Spatial Information in Coastal Environments

3.1 Basic Geometric Operations With Spatial Data

The geometric information associated with spatial data makes it possible for basic analytical operations to be carried out that make explicit use of the geographical distribution of a dataset. These operations include measurement of distance, area, proximity, buffer, distance decay, overlay and the application of contextual operators. If Cartesian coordinates identify the location of two points, simple metric rules can be used to determine the distance between the points. A hypothetical right-angled triangle can be generated around these points, which are situated at either end of the hypotenuse. Pythagoras' theorem can then be used to calculate the square of this length as the sum of the squares of the lengths of the other two sides. This approach uses the topological information associated with the points, and will only work for points that have been projected to a flat surface. This simple rule can be scaled up from a single pair of points to multiple points making up a polygon. Once the distances between several points have been established, the collective properties of those points belonging to a single polygon can be further used to calculate areas from distances. The area of a polygon can be calculated from the coordinates of its component points by subdividing it into a series of triangles, then calculating and summing their areas (Lloyd, 2010). Because of the simpler geometry associated with raster grids, it is generally easier to calculate areas for data that are stored in this format. The area of a raster surface can simply be calculated by multiplying the total number of raster cells by the area of each individual cell (which is usually the square of the length of a pixel side).

A buffer operation builds a new object in relation to an existing set of objects (e.g. points, lines, polygons) by identifying all areas that are within a certain specified distance of the original objects (Lloyd, 2010). A buffer may be created around a vector object such as a point by establishing a circle with the required radius (set to the buffer distance) around the feature point. Similarly, if the buffer is to be established around a line or polygon, the distances are sequentially calculated along the outer periphery of that feature until a continuous border around the feature has been established, then any internal boundaries are dissolved.

Distance decay refers to the manner in which the effect of a phenomenon at a given location attenuates with distance from that phenomenon's location. A raster proximity map can be generated from vector feature data, or from raster grids that are coded with '1's and '0's to depict objects of interest (which coincide with the value 1). In the latter case, the output grid will record, for every cell in the grid, the distance to the cells of interest in the input grid. The polluting effect of an oil spill on the sea surface decreases in a predictable fashion with distance from the point source of oil. More complex mathematical functions can be used to characterise the manner in which a phenomenon decays with distance from the source. This may take the form of a continuously smooth linear decrease, or a less continuous negative power or exponential distance decay. Distance decay functions are commonly expressed as regular functions that are isotropic (uniform in every direction), although geographic variation is often not regular in form. Returning to the oil spill example, this would be the case if an island presented a barrier to the spread of oil in a given direction, or wind-driven waves encouraged the spread of oil in one direction over another.

Overlaying two or more maps or image layers registered to a common coordinate system involves superimposing one over the other. An overlay operation using multiple vector datasets compares the location of points, lines or polygons in each data layer to those in one or more additional data layers to produce a new data layer that records the result. For a comparison to be possible, all data layers need to be represented at the same scale and in the same frame of reference (i.e. projection and coordinate system). Sites of intersection between all the input lines, vertices and nodes are recorded as a new topological feature. Overlay analyses might identify areas where two or more layers *intersect*. In this case, the layer generated only identifies areas of the study region where both input layers are present. Alternatively, a *union* overlay identifies areas where at least one of the layers is present. In this case the layer generated will include a broader region of the study area indicating the presence of one or other of the input layers. For example, a coastal conservationist may wish to compare the location of proposed beach site developments with vulnerable or endangered dune vegetation along a beach hinterland. They could overlay a shapefile identifying areas where proposed developments are planned with a shapefile of mapped, endangered dune vegetation. An intersect function could be employed to identify areas of potential concern where planned development coincides with vulnerable vegetation. Deciding which overlay function to use (intersect or union) will depend on what relationship is to be captured between multiple vector datasets. These relate to the Boolean conditions of AND and OR, respectively (see Section 3.7).

3.2 Linking Data Through Spatial and Attribute Joins

Much of the spatial analysis in this book describes multivariate datasets where all the information is present in the same attribute table. Often the analyst is faced with a scenario in which they wish to analyse the relationship between two datasets from disparate sources. Different variables may need to be brought together as a basis for comparison by joining them. A join can be established on the basis of either the spatial location of the data (a *spatial* join) or some attribute of the data (an *attribute* or *table* join).

A spatial join utilises one of the most powerful features of a geographic information system (GIS): the ability to link datasets on the basis of their common geographic location. This fundamental basis for relating data from different sources is very widely used in spatial analysis. The following illustrative examples all have in common the fact that multiple datasets have been linked together solely on the basis of their geographical location:

- linking of heavy-metal concentration in mussel tissue to ambient pollution from adjacent port cities and river discharge (Chapter 1);
- overlay of multiple maps (e.g. showing distance to footpaths, geology and topographic slope) to identify areas meeting criteria relating to risk of visitor injury from rock falls (Section 3.9); and
- the linking of live coral cover to water depth and terrain characteristics of rugosity and bathymetric position index (Chapter 7).

An alternative way to link spatial datasets is through an attribute, or relational, join. These joins use common columns of attribute information as a basis for cross-referencing and linking two datasets. The subject of the attribute columns must align in a way that includes all rows in the dataset. An example of an attribute join would be linking point locations of sediment samples taken from beach compartments along a length of coastline on the basis of a unique compartment identifier, such that other attributes of the sediment samples stored separately can be analysed as a single dataset. The choice of whether to employ a *spatial* or an *attribute* join to link datasets will be determined by what information is available to provide the basis for establishing a link.

3.3 Visualising and Exploring Spatial Point Patterns

Points are one of the simplest forms of spatial data. They represent a set of locations that correspond to objects (e.g. the locations of barnacles, mangrove shrubs or outlet pipes along a coastline). Because many phenomena are represented as points within a GIS database, it is important to be able to explore the geographical patterns that emerge when these points are mapped across a two-dimensional space.

Point-based approaches to confirmatory modelling seek to explore the degree to which point events are clustered, dispersed or spatially structured in some way, such that they are more likely to occur in some locations than in others. They allow the analyst to pose, and answer, the question: *'are the points clustered or dispersed?'* Point pattern analysis can compare the spatial arrangement of a set of points to a random arrangement to determine whether a statistically significant difference occurs between these arrangements. This comparison provides insight into the processes underpinning the point distributions. Where individual points represent particular events such as shark attacks (see Section 3.3.4), or incidences of exposure to environmental pollution, such comparisons may help to explain reasons for attacks or possible sources of environmental pollution.

If point datasets have values associated with them, they are referred to as *marked* points. Associated patterns can be explored and evaluated using measures such as

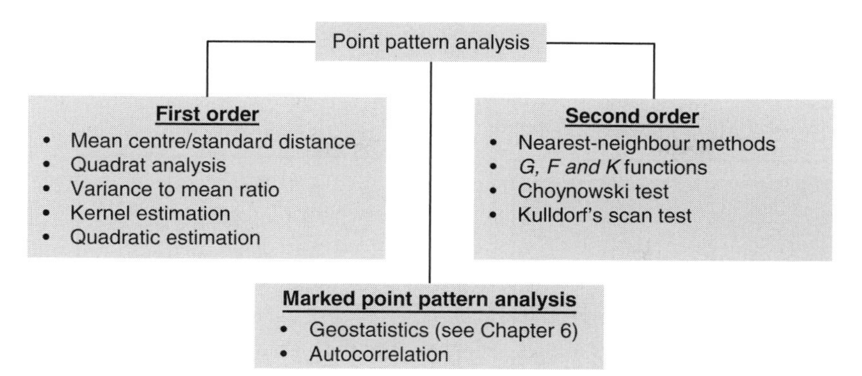

Figure 3.1 Techniques available for the analysis of first-order and second-order effects and marked point patterns, as described in Section 3.3.

autocorrelation (Chapter 6). Marked point pattern analysis focuses on the spatial behaviour of the *values* of the attribute itself. It is the basis of many geostatistical techniques covered in Chapter 6, and is distinguished from other types of point pattern analysis that focus more broadly on *whether* and *where* events occur. Different types of point pattern analysis can also be distinguished on the basis of whether they focus on first- or second-order effects (Figure 3.1).

First-order effects of a point pattern relate to the simplest, or most fundamental, level of organisation, i.e. relating to patterns caused by the distribution of *individual* points. The first-order effects of a point pattern can be explored by calculating the intensity or mean number of events per unit area at a given location. First-order effects are generally interrogated through the visualisation and exploration of datasets.

Second-order effects refer to a secondary level of organisation, that is, the relationship between *paired events* within a study region. Second-order effects and marked point patterns can be interrogated using inferential or confirmatory statistical methods. These approaches are not exclusive. They can be applied in a sequential manner, and confirmatory statistical methods are often applied after the initial exploration of first-order effects. Bailey and Gatrell (1995) classify the analysis of spatial point patterns into visualisation, exploration and modelling. This provides a useful framework for discussing the different techniques available for point pattern analysis.

3.3.1 Visualising Spatial Point Patterns

Visualisation allows the analyst to get a feel for data events by plotting them on a map to interpret any obvious patterns related to their distribution and present these to a broader audience. Basic descriptive measures of spatial point patterns include the mean centre (mean of the x and y coordinates) and standard distance (standard deviation of the x and y coordinates). These are calculated on event coordinates to indicate the central tendency of the points and the group dispersion around that centre, respectively. Figure 3.2 illustrates the mean centre and standard distance of clustered and dispersed point patterns of barnacles photographed on a rock platform.

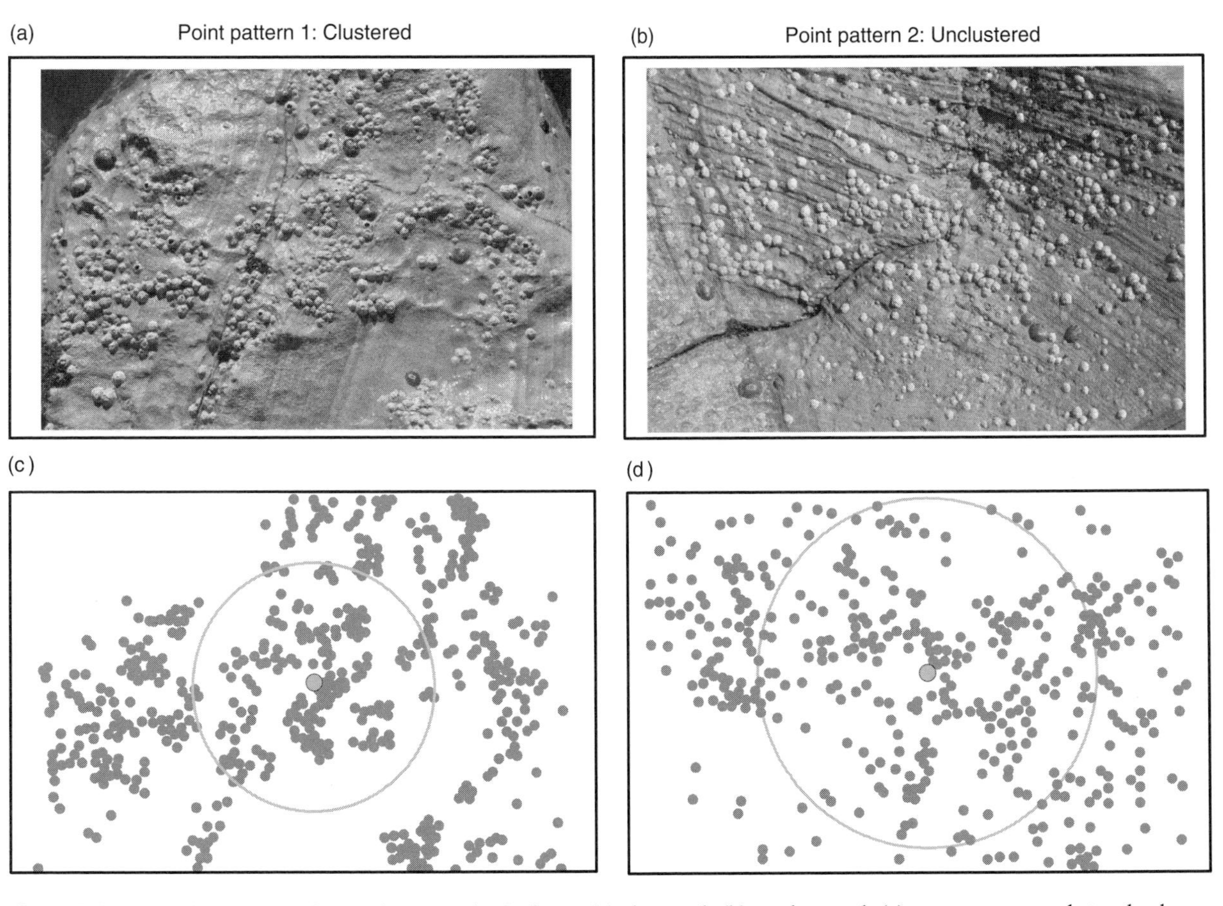

Figure 3.2 Barnacle patterns observed on a rock platform: (a) clustered; (b) unclustered; (c) mean centre and standard distance of the clustered pattern in (a); and (d) mean centre and standard distance of the unclustered pattern in (b) shown by the large point and circle, respectively. (A black and white version of this figure will appear in some formats. For the colour version, please refer to the plate section.)

3.3.2 Exploring Spatial Point Patterns: First- and Second-Order Methods

First- and second-order event distributions can be explored through methods that quantify their intensity or density. Quadrat analysis and kernel density estimation are examples of first-order methods. These methods superimpose areal units over the observed distribution to compare regional sample statistics to an overall baseline calculated from the whole area sampled. Quadrat methods derive their name from the historical use of square sampling frames in field sampling. They proceed by placing a regular grid over the point pattern and counting the number of events falling within each square. Density estimates will depend on the dimensions of the sampling square and the origin of the quadrat grid. Because estimates of density are expressed as a number, it is possible to compute spatial autocorrelation metrics to assess the degree of clustering or regularity in a dataset. Quadrat counts generalise much of the spatial detail present in the observed pattern, particularly where large quadrats are defined. Smaller quadrats tend to have a higher variability relative to the size of the overall, global population of points. The appropriate dimensions of the quadrat or kernel will depend on the scales of variation inherent in the point patterns themselves.

Kernel estimation methods provide a more sophisticated approach for investigating point density. These are often used in a confirmatory manner to identify statistically significant clusters or 'hotspots' of events (see Sections 3.3.3 and 3.3.4). They do so by calculating a probability density of events from a sample of observations (i.e. an expected density). This is then compared to an observed density for a given region of the total study area. Rather than using a grid of quadrats to sample sub-areas, a set of points arranged into a regular grid is overlaid onto the observed point pattern. The number of events within a radius r (or otherwise defined neighbourhood, also referred to as a 'moving window') around each point within the overlaid regular grid is counted. Density is then calculated by dividing this number by the area of the circle (πr^2). By centring the moving window on a grid of points, a smoother estimate of point pattern variability can be derived.

Kernel estimations can be expanded to employ a geographical weighting scheme (a kernel function) that varies the influence of the events in relation to how far they are from the centre of the grid point defining the analysis window. A variety of functions exist that will derive a smoother kernel estimate of point density for a given location. One commonly used kernel density estimate function is the quartic function, defined as

$$\hat{\lambda}_\tau(x) = \sum_{d_i \le \tau} \frac{3}{\pi \tau^2} \left(1 - \frac{d_i^2}{\tau^2}\right)^2 \tag{3.1}$$

where the kernel estimate of density, $\hat{\lambda}_\tau(x)$, is based on the distance (d) between each event and the centre of the kernel (x) for a given bandwidth or kernel radius (τ) that sums all points falling within this bandwidth ($d_i \le \tau$) to provide the overall density estimate. Note that the distance between each event and kernel centre is subtracted from 1. This adds more weight to points closer to the kernel centre and less to those

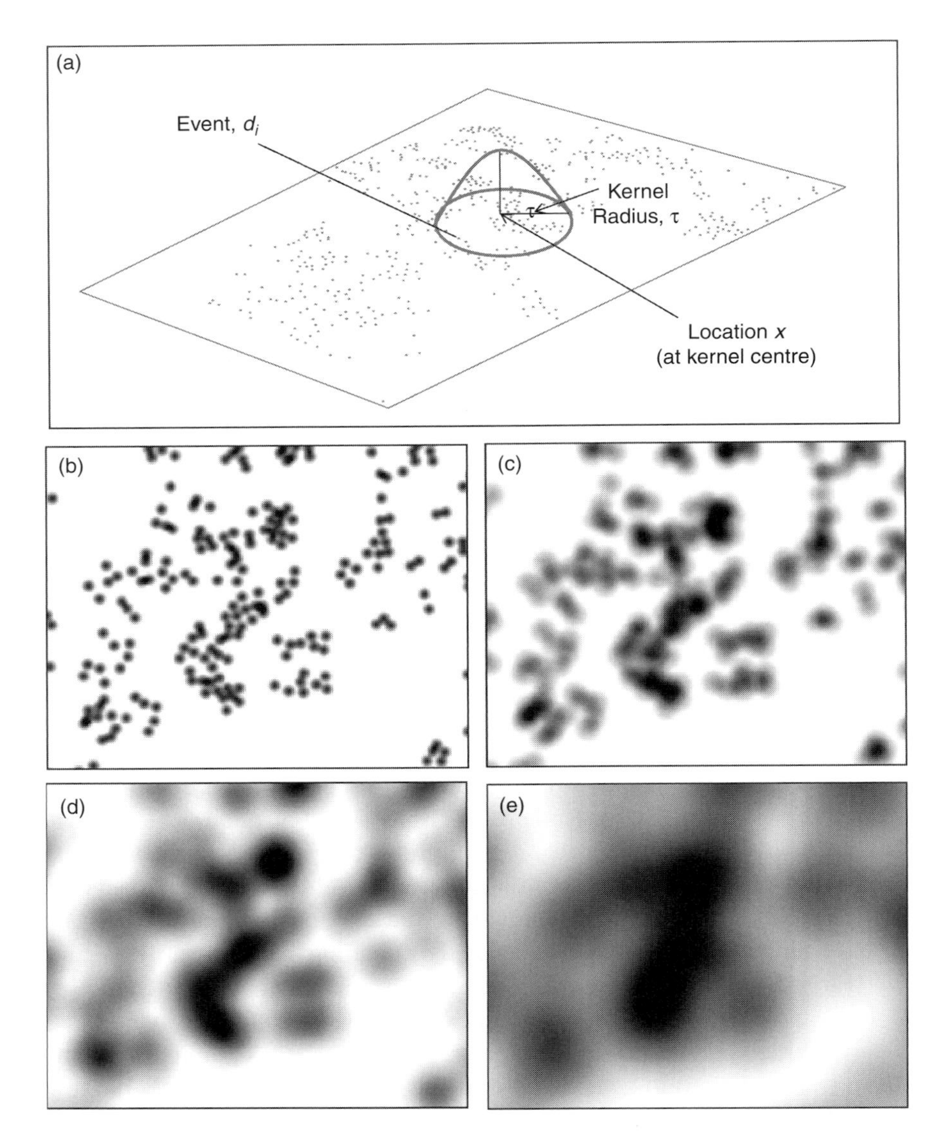

Figure 3.3 (a) An oblique view of barnacle point pattern 1 (Figure 3.2), showing the key features of a kernel estimation of the clustered barnacle point pattern 1, with estimates of intensity for the entire point surface derived from a range of bandwidths (kernel radius lengths) $\tau = 40$ cm (b), $\tau = 80$ cm (c), $\tau = 160$ cm (d) and $\tau = 320$ cm (e). All panels (a) to (e) depict the entire point pattern, although the points in panel (a) have been tilted so that they can be viewed at an oblique angle for illustrative purposes.

further away. Figure 3.3 illustrates how this function is derived from the clustered barnacle point pattern. The radially symmetric function is centred on the location at the centre of the kernel (x), where the maximum weight is applied, $3/\pi\tau^2$, and this drops off smoothly to a value of zero at the outer perimeter of the bandwidth at distance τ. The effect of increasing the bandwidth, τ, is to expand the region around the kernel centre, x, within which observed events influence the density estimate at x. A very large distance, $\hat{\lambda}_\tau(x)$, will appear flat and devoid of local detail. As τ is reduced, this will tend towards a spiked pattern approximating the points themselves. This is illustrated by the range of bandwidths used for the calculation of density estimates in Figure 3.3(b)–(e).

Second-order intensity estimates of a point pattern are based on the distance between points (hereinafter referred to as an 'event'). This is measured as either the distance between a randomly chosen event and the nearest event (*event–event* approach) or the distance between a randomly selected location in the study region and the nearest-neighbouring observed event (*point–event* approach). Lloyd (2010) provides a detailed description of nearest-neighbour methods for exploring second-order point patterns. The *mean nearest-neighbour distance* calculates the average distance between a series of paired events inside a given area. Event neighbours are not necessarily restricted to pairing a single event and the nearest single event. Distance bands can be defined, within which any events that fall are incorporated into the neighbourhood for calculation of regional density values.

The approaches summarised so far are informal techniques for the exploration of spatial point patterns. They explore whether or not observed distributions of points are clustered. Where the analyst wishes to further construct a model to test a hypothesis about the nature of the point pattern, a more formal framework for statistical induction may be needed (see Section 3.3.3).

The mean nearest-neighbour distance forms the basis of the G and F functions, calculated using *event–event* or *point–event* approaches, respectively. These functions are the cumulative frequency distributions of the nearest-neighbour distances across a given distance band. They are calculated as the proportion of nearest-neighbour distances $d_{\min}$ that is less than a specified distance d away (Lloyd, 2010). As d is increased, it follows that G will also increase until the value of 1 is reached, which corresponds to a distance that incorporates the maximum nearest-neighbour distance for a collection of events. A plot of G against distance can be used to identify a clustered point pattern, because the rate of increase in G will be large in comparison to the increase in d. Likewise, for a regular pattern, the increase will be more gradual (see Lloyd, 2010, for further detail).

Ripley's K function is the most commonly used second-order estimate of point density. This function operates on a similar principle to the G and F functions, although it simultaneously assesses the distances between all events in the study region. It is conditional on an event at one point and is defined as πd^2. Ripley's K function stipulates an expected value of K under a null hypothesis of complete spatial randomness. In practice, a transformed version of this function is calculated as $\sqrt{K / \pi}$. Figure 3.4 illustrates a plot of the transformed Ripley's K function for the barnacle distribution shown in Figure 3.2(a). The number of events falling inside a specified distance, or radius, from each event is counted. Counts are then summed across all events within the study region and scaled. This value can be calculated for a range of distances (radii) to derive a plot of distance against observed K function, which in turn can be compared to a plot of the expected K (Figure 3.4). Where the observed number of events exceeds the expected number of events, this indicates event clustering. A geographical

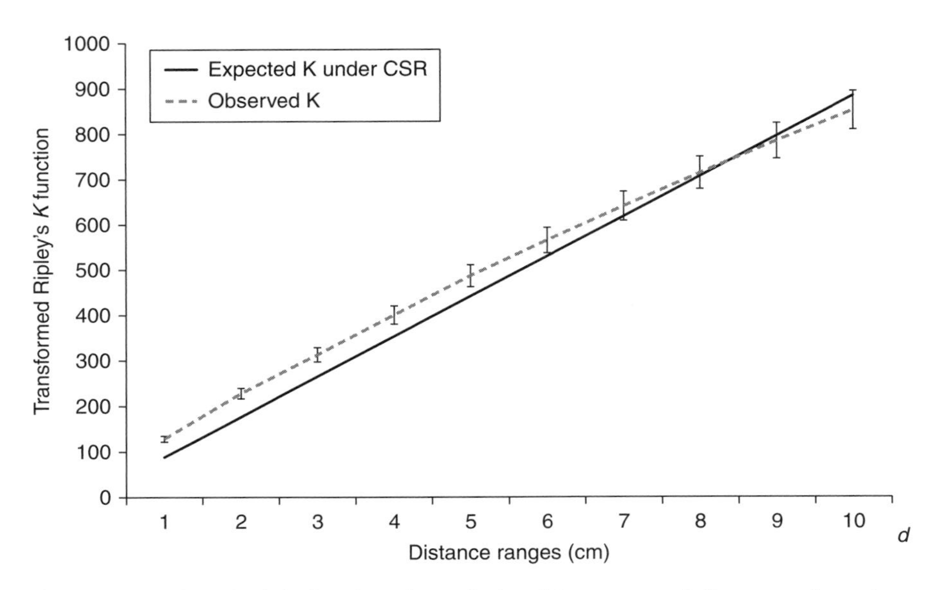

Figure 3.4 A plot of Ripley's *K* function calculated for a range of distance values, *d*, across the clustered barnacle point pattern illustrated in Figure 3.2(a). Error bars indicate higher and lower confidence intervals for the observed value of *K*, after transformation. The higher-than-expected calculated values of *K* at smaller distance bands (i.e. < 8 cm) indicate statistically significant clustering within this smaller distance range.

weighting term can be incorporated. Additional measures can be introduced to address 'edge effects' for those points that fall close to the periphery of the study area.

3.3.3 Point-Based Approaches to Confirmatory Modelling

In most cases, the analyst wishes to establish whether the observed point pattern distribution is clustered or random. Probabilistic statistical testing invites the analyst to consider the probability of observing a number of events given an expected distribution. Observed and expected values of the various density measures can then be compared under different hypothetical models. The null hypothesis is that of complete spatial randomness (CSR). This is usually represented as a homogeneous Poisson distribution over the study region to formally express the probability of a given number of events occurring within a specified distance from each event. It represents a hypothesis against which to test whether observed patterns are regular, clustered or random (Bailey and Gatrell, 1995). For example, an expected value of the *K* function can be derived from the average intensity of the point patterns and the distance, *d*, at which the *K* function is calculated (πd^2). This can then be compared to an observed value as an indication of whether clustering exists. Where the observed value of $K(d)$ exceeds the expected value, a clustered point pattern is suggested. If this is not the case, the point pattern is likely to be regular. With any of these tests, a simple classification of regions can be achieved by mapping the outcome of this comparison between observed and expected functions.

A number of different inference tests exist for calculating the probability that the difference between the observed and expected distribution of events is statistically significant. These include the calculation of relatively simple indices of dispersion and variance-to-mean measures, or more formal frameworks for testing significance against theoretical or simulated distributions. A variance-to-mean ratio can be used to indicate the degree to which a point pattern departs from that predicted on the basis of chance alone. For a given average point pattern intensity, a distribution, such as the Poisson distribution, can be calculated to express the probability of a given number of events occurring in a unit area of measurement (e.g. a quadrat). Associated empirical tests (e.g. a chi-squared test) then provide a basis for accepting or rejecting the null hypothesis (that the point pattern adheres to complete spatial randomness) (Lloyd, 2010). Monte Carlo randomisation procedures assess point patterns relative to complete spatial randomness by generating multiple point patterns as a basis for comparison against which significant departures from real point patterns can be detected.

Two examples of more sophisticated inference frameworks for detecting and locating statistically significant clusters are the Choynowski test and the Kulldorff's scan test. The Choynowski test allows the analyst to calculate the probability of observing more than a specified number of cases, given an expected distribution. It is assumed that the observed number of events approximates a Poisson distribution of the expected number of events, such that the probability of getting more than or equal to a specified number of events, given a Poisson distribution, can be calculated as:

$$\begin{aligned} \text{Prob}\{O_i \geq x \mid E_i\} &= \sum_{s=x}^{\infty} e^{-E_i} E_i^{\,s} / s! \\ &= 1 - \sum_{s=0}^{x-1} e^{-E_i} E_i^{\,s} / s! \end{aligned} \tag{3.2}$$

where $\text{Prob}\{O_i \geq x \mid E_i\}$ means the probability of observing a count of events given an expected count, s denotes the observed number of events and E_i denotes the expected number of events for location i. The fact that probabilities sum to 1 is used to convert an infinite number of outcomes to a finite sum. Thus, for an expected number of events $\text{Prob}\{O_i \geq x \mid E_i\}$, it is possible to calculate the probability of observing a specified number of events. This is then compared to a frequency distribution derived from many observations of event counts (e.g. 999 permutations) to ascertain the level of significance of the result (see Figure 3.5). Table 3.1 provides a worked example of the use of Choynowski's test for detecting statistically significant clusters of points. For an observed and expected number of events of 6 and 3, respectively, the test finds the probability of obtaining the observed number of events, given the expected number (see (3.2)).

Conventionally, the probability of rejecting the null hypothesis (complete spatial randomness) when that hypothesis is true needs to be less than 0.05 to justify rejecting the null hypothesis. In this example, $\text{Prob}\{O_i \geq x \mid E_i\} = 0.0839$. We therefore conclude that the null hypothesis of complete spatial randomness is retained.

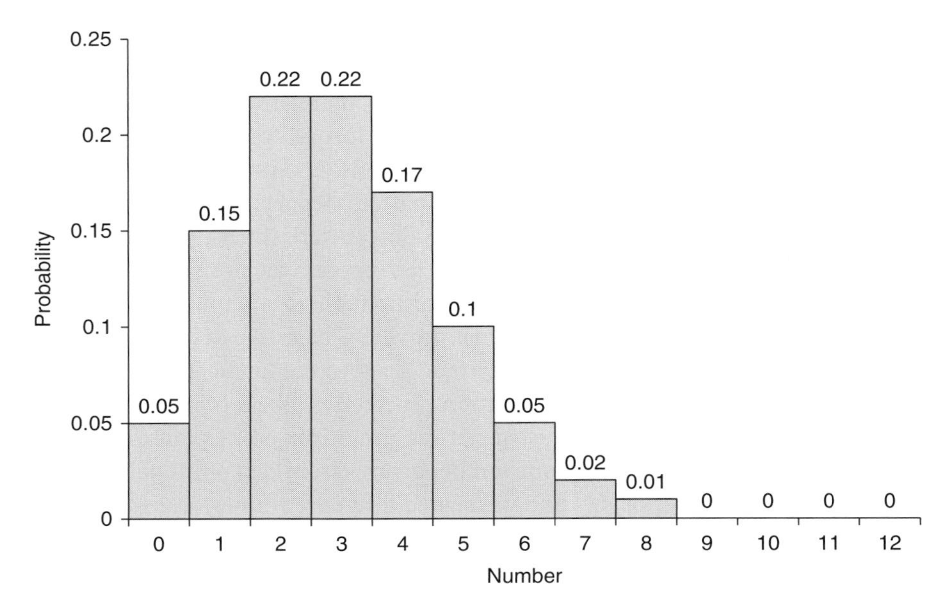

Figure 3.5 Overview of Choynowski's test for detecting statistically significant clusters of points.

Table 3.1 A worked example of Choynowski's test for detecting statistically significant clusters (or 'hotspots') of point events. Values are calculated for an observed and expected event density of 6 and 3, respectively. See text for a definition of terms.

s	e^{-E_i}	E_i^s	$s!$	$E_i^s / s!$	$e^{-E_i} E_i^s / s!$
5	0.049787	243	120	2.025	0.1008
4	0.049787	81	24	3.375	0.1680
3	0.049787	27	6	4.5	0.2240
2	0.049787	9	2	4.5	0.2240
1	0.049787	3	1	3	0.1493
0	0.049787	1	1	1	0.0498
$\sum_{s=0}^{x-1} e^{-E_i} E_i^s / s!$					0.9160
$1-\sum_{s=0}^{x-1} e^{-E_i} E_i^s / s!$					0.0839

It is important to be aware of the problems or limitations associated with statistical testing. In the case of the Choynowski test, there is a non-zero probability that the observed count could fall within the significance level because of chance. This problem is known as 'multiple testing'. If a region is divided into enough subregions (n), where n is large, there are bound to be some regions that are statistically clusters.

This probability will converge to 1 as the number of regions approaches infinity. The detection of clusters is also dependent on the relationship of the spatial framework of data analysis to the entity itself. These problems can, to a varying extent, be addressed through the application of alternative testing methods, such as Kulldorff's scan test.

3.3.4 Detecting Clusters of Shark Attacks Along the California and Florida Coastlines

This case study uses the two heavily populated coastlines of Florida and California in the USA to demonstrate the Kulldorf's scan test. This is a scan-based test that detects the presence of statistically significant clusters in point-based datasets. In this analysis, the points represent locations where shark attacks have occurred. The objective of the analysis is to find out whether a cluster or 'hotspot' exists. This represents a location where it is statistically more likely that attacks will occur (Figure 3.6). For such an analysis, it is necessary to have records of positive 'cases', i.e. the spatial coordinates associated with known shark attacks. These should be accompanied by 'control' records that represent the coordinates associated with shark sightings, but where attacks have not occurred. Shark attack records were taken from the *Global Shark Attack* file. Control records were extracted from local shark tagging initiatives, including the *Tagging of Pacific Pelagics* (*TOPP*) programme of the *Census of Marine Life* for the California case study (Jorgensen et al., 2010) and the *R.J. Dunlap Marine Conservation Program* tagging programme of the *University of Miami* for the Florida case study (Hammerschlag et al., 2012).

The scan test works by moving a circular analysis window across the geographic area of interest. The observed and expected numbers of cases are compared within each circular window in turn. Complete coverage of the area of interest is achieved by establishing a series of grid points covering the entire study region. The circle is centred on each point in turn to ensure that all data points will fall inside an analysis window. Within each window, the probability of being a 'case' (i.e. that a shark attack has occurred) is computed for all points (both case and control) falling inside the test window. The null hypothesis is that there is an equal probability of a shark attack at every location within the study area along the coastline (and therefore no significant spatial cluster of attacks will exist). A global likelihood function is calculated, based on the total number of case and control points and the area they covered to express the probability of a shark attack occurring across the entire study area (i.e. the expected number of shark attacks). This is compared to a localised likelihood function, which is expressed as the product of the individual Bernoulli probabilities within that window (Kulldorff, 1997). A likelihood ratio statistic is then calculated as the function inside the test window (i.e. the observed number of shark attacks) divided by its maximum value under the null hypothesis. For each analysis window, the alternative hypothesis is that there would be an elevated chance of a shark attack occurring within the window as compared to outside.

Conventionally, the most likely cluster is the zone that maximises the likelihood ratio statistic (Kulldorff and Nagarwalla, 1995). The distribution of the likelihood

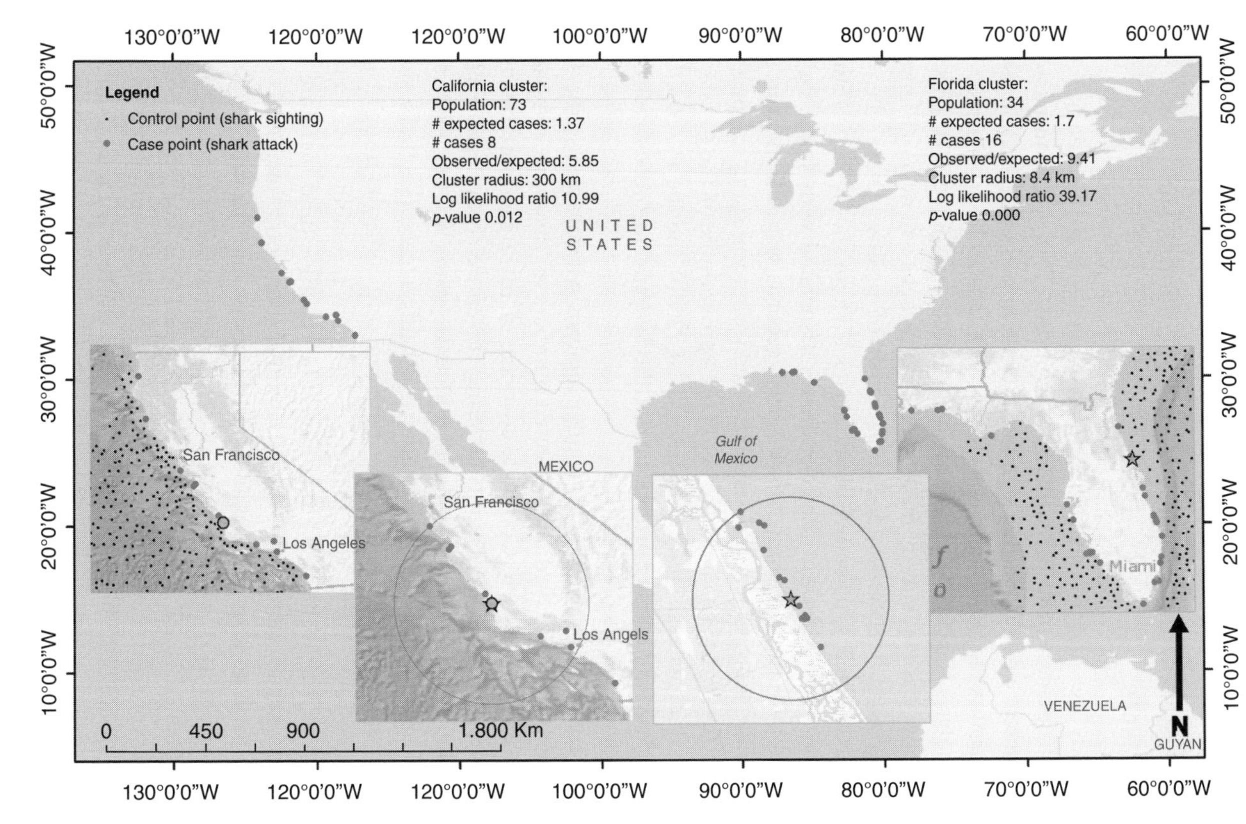

Figure 3.6 A map of shark attack records along the coasts of California and Florida, USA. Insets show the location of attacks in relation to known locations of sharks where attacks have not occurred, derived from tagging programmes. Red circles indicate statistically significant clusters or 'hotspots' of shark attacks, where it is statistically more likely that an attack will occur. Details of clusters are provided in the inset text. (A black and white version of this figure will appear in some formats. For the colour version, please refer to the plate section.)

function depends on the underlying population distribution. This cannot be determined directly, so a Monte Carlo simulation is therefore used to test the significance of the likelihood ratio statistic against a null distribution. This simulation consists of 999 replicates, each of which involves identifying 31 cases at random from the population of 512 cases. For each replicate, the value of the Bernoulli likelihood function is calculated. Simulated likelihood functions are then ordered and ranked, with the highest value being assigned 1. A significant result is therefore obtained at the 5% probability level if the observed value for the data falls among the highest 50 of these values.

Statistically significant clusters of attacks were identified in this case study for both the California and Florida coastlines, providing grounds for rejecting the null hypothesis of complete spatial randomness at both locations. The California cluster was centred on an area just north of Santa Cruz, although its dimensions (300 km radius) were large enough to encompass attacks that occurred in the southern San Francisco Bay area and the northern suburbs of Los Angeles. A much smaller cluster was identified for the Florida site (radius 8.4 km). This was centred near Edgewater, approximately 60 km northeast of the city of Orlando. Here, 17 attacks had occurred out of a total of 34 points that fell inside the test window.

The practical value of this cluster detection technique is that it identifies geographical areas where resources can best be focused to prevent future attacks. Cluster locations are likely to reflect the places where the most people enter the water, the movements of sharks in those areas and the level of protection afforded to those areas (e.g. through airborne shark patrols and netted beaches). The absence of clusters associated with other highly populated coastal cities in Florida (Miami, Tampa and St Petersburg) may indicate the successful use of measures to safeguard against attacks. The cluster at Edgewater might therefore represent a stretch of coastline that falls within reach of a populated city and is therefore subject to high visitor numbers, but perhaps not an immediate focus of prevention measures, as this city is not situated directly at the coast.

3.4 Line Analysis: Networks

A network is a group of interconnected elements in a landscape, usually represented on a map or in a GIS dataset as an arrangement of intersecting lines. Connectivity is central to space and spatial information and the structure of a network captures connections among arbitrary features of a landscape (e.g. nodes, vertices and arcs; see Chapter 2). Nodes represent the beginning or end point of a line and can be connected by any form of relationship. For example, they may all belong to a single river or stream. Network information is primarily composed of arcs and vertices that define the shape of the line and enable an analyst to answer questions about connectivity, such as *'are nodes m and n connected?, what is the shortest path from m to n?, where are the sources and sinks in the network?'*

Network connectivity can be modelled using a matrix that specifies for all arcs whether they are connected or not. Summary metrics such as the α and γ connectivity indices can then be derived from those matrices (Chou, 1997). Network analysis has been applied in coastal environments to compare nitrogen cycling across multiple eutrophic coastal ecosystems, including lagoons, deltas and estuaries (Christian et al., 1996). Neural networks have been employed to model algal bloom dynamics along Hong Kong's coastal waters (Lee et al., 2003) and seasonal fluxes in the trophic structures of benthic and pelagic populations of Maspalomas coastal lagoon (Canary Islands) (Almunia et al., 1999). Connectivity networks have been used to model the risk of outbreaks of crown-of-thorns starfish among the many reef platforms of the Great Barrier Reef (Hock et al., 2014).

3.5 Contextual Operators

Contextual operations in GIS assess the relationship of a given spatial feature to its surrounding region. For example, the steepness of a sand dune face can only be assessed in relation to surrounding areas of the dune face. For the contextual analysis of raster grids using the Esri software ArcGIS, three types of operator can be distinguished: local, focal and zonal. These are collectively distinguished from global operators, which simultaneously analyse entire raster grids or data layers and do not therefore take into account the localised context of pixel subgroups or raster cells within the grid. A local operator will generate a new output grid in which the value of a given cell is dependent on that same cell represented in multiple input grids. In contrast, a focal operator will generate a new output grid in which the value of a given cell is dependent on the value of that same cell in other input grids, but also the cells surrounding it, i.e. within a given focal area. Similarly, a zonal operator will generate a new output grid in which the value of a given cell is dependent on the combined value of a group of cells falling within a zone defined by an independent data layer. A global operator scales up to the broadest geographical context possible by including all the cells within the input grids.

Contextual operators use moving windows to specify the context (i.e. area of cells) to be taken into account at different locations of a study area. Contextual operators that are commonly applied to coastal environments include those that generate new derivative characteristics from digital elevation models, such as slope, aspect and flow properties. These are sometimes referred to as 'terrain derivatives'. A slope contextual operator uses a moving three by three pixel window to calculate the rate of change in the slope (defined as the ratio of cell height to ground distance) in all directions, before assigning slope to the maximum direction. Aspect defines the direction towards which the elevation changes the most. This is calculated from the slope dataset, again using a moving three by three pixel window to identify the downslope direction of the maximum rate of slope change. Flow tools can be used to model how water will flow through a landscape after a rain event, given the topographic character of the landscape. Such tools are useful for defining drainage catchments and identifying the paths that rivers may take through the landscape to flow out into estuaries. A contextual operator might define flow accumulation by iteratively comparing each grid square to its immediate neighbours to determine the maximum drop in elevation. Across a grid, this will determine the direction in which water will drain or flow between cells. Individual directions can then be uniquely coded to determine, for any given location, the direction in which that water will flow.

3.6 Trend Surface Analysis

Trend surface analysis can loosely be defined as any method for characterising spatial patterns that partitions out the effects of space as a variable. This allows properties of interest to be explicitly expressed in relation to spatial considerations. The objective of trend surface analysis is to characterise general tendencies within a sampled data surface by fitting a mathematical function to a surface by least squares regression through the sample data points. In this way, a surface can be derived that minimises the variance of the surface in relation to the input values. Relationships between samples and geographical location (e.g. sediment grain size–environment or species–environment) are captured by explicitly including space as a variable in any function specification.

Figure 3.7 illustrates the results of a trend surface analysis undertaken on benthic invertebrate communities sampled across a border between seagrass and sand in the mudflats of Moreton Bay (Barnes and Hamylton, 2013). Communities were sampled along a spatial axis running between bare sand and seagrass at intervals of 25 cm. This explored whether transitions in four summary metrics of benthic invertebrate community assemblages (total number of individuals sampled, mean estimated and observed species density and inverse Simpson index) could accurately and consistently be captured in a simple spatial model. In all cases, community assemblage metrics were accurately expressed by second- and third-order polynomial models that operated along a north–south aligned axis of transition from seagrass to sand.

3.7 Analysis of Grids and Raster Surfaces

3.7.1 Map Algebra

Map algebra refers to the processing and analysis of gridded raster data (i.e. in the form of a map) through the application of algebraic operations. This forms the basis of some common routines for processing remote sensing images, such as 'masking' an image to generate a new image that only has data in a user-specified geographical area of interest. Such a manipulation might involve multiplying a map in the form of a raster grid that expresses a variable of interest as a continuous field by another 'mask' grid. The mask grid has values of 1 for areas where data are required and values of 0 where they are not required. Map algebra also provides the means to implement Boolean logic or conditional statements with raster datasets (see remainder of Section 3.7).

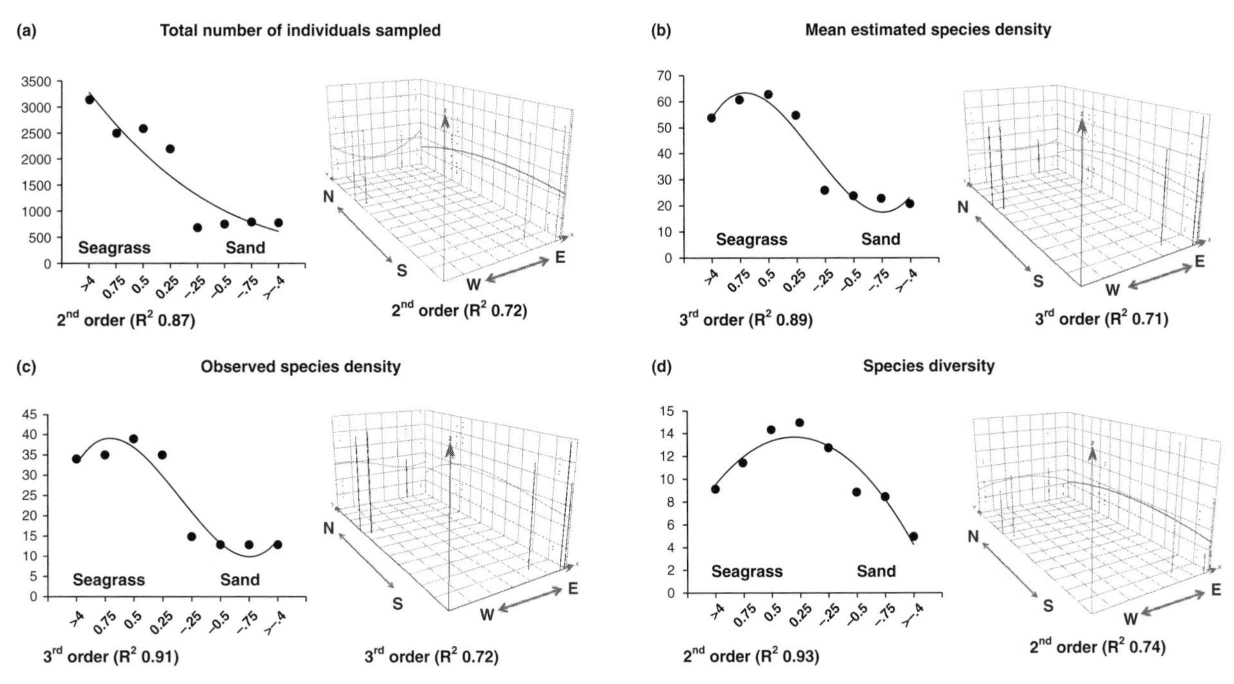

Figure 3.7. Trend surface analysis plot of the assemblage metrics for benthic invertebrate communities sampled across a sand/seagrass interface in Moreton Bay, Queensland. Mean values for each sample lattice are shown on the left-hand side as distances (m) from the seagrass/sand interface and values are plotted in their geographical (x, y) locations on the right-hand side with the z axis depicting the value of the metric. The x axis on the plots is a proxy for distance from the seagrass/bare sand interface. Green and blue lines indicate lines of best fit along the spatial (north to south, east to west) axes. (A black and white version of this figure will appear in some formats. For the colour version, please refer to the plate section.)

3.7.2 Specifying Models Using Boolean Logic and Conditional Statements

A Boolean function provides a mathematical method for determining a given output based on some logical calculation from Boolean operators. In 1984 George Boole, a Professor of Mathematics at Queens College, Cork, Ireland, published *The Laws of Thought* in which he set out a discourse on algebraic logic that defined a series of digital operators which became the basis of Boolean logic (Boole, 1854). Boolean operators are simple words (AND, OR, NOT or XOR) used as conjunctions to combine or exclude spatial data records from a data query (see Figure 3.8). These operators represent the four possible combinations of relationships between a pair of records stored in a dataset. For a given geographic area or attribute characteristic, it is possible to use Boolean logic to establish which subsections of the dataset exhibit these combinations of relationships. Given that vector or raster datasets are commonly composed of hundreds of thousands of records, implementing dataset queries using these operators can greatly reduce the amount of records returned. This helps to successively identify records that meet specific analytical criteria.

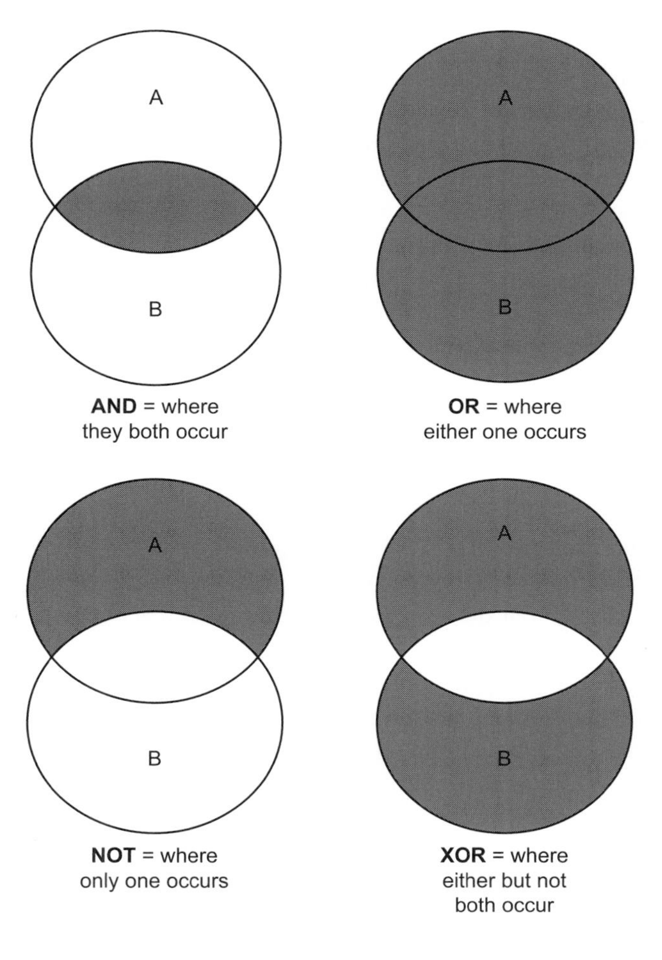

Figure 3.8 Venn diagrams illustrating the four possible Boolean relationships between two datasets.

Boolean logic employs simple mathematics to express logical concepts by reducing all queries to a simple binary form. This takes the form of a conditional statement that can be answered as yes/no (or true/false). For example, rather than ask *'How long is the beach?'*, a Boolean query would ask *'Is the beach greater than 1 km long?'* Structured Query Language (SQL) is the standard database query language that is used to select entries in the database that meet a particular criterion. Conditions are subdivided into attributes, operands and values. In the previous example, the attribute is 'beach', the operand is 'greater than (>)' and the value is '1 km'. It is possible to apply more than one question at a time by linking them together with Boolean operators:

'beach' > '1 km'

AND

'sediment' = 'siliciclastic'

Conventionally, Boolean data consist of two binary values corresponding to records where the conditions are met (i.e. yes or true, coded 1) or records where the conditions are not met (i.e. no or false, coded 0).

The two types of query that can be implemented with Boolean conditional statements are attribute and location queries. Attribute queries define a given attribute and ask *'where is that?'* Location queries take a different approach to define a particular location and ask *'what is there?'* Before implementing a query, it is necessary to define the criteria that set the conditional parameters of the query itself. For example, if we seek to identify beaches that are 'long', it is necessary to decide what constitutes a long beach in the first place.

Conditional statements made with Boolean logic are implemented differently with raster and vector datasets because of the different topology associated with these data formats. Key factors that define how the query will proceed include whether the query is an attribute or location query, how many questions are being asked and how many data layers are being queried.

3.7.3 Implementing Conditional Statements With Raster Datasets

The attributes of many raster datasets are queried simply by reclassifying their values in line with the parameters of the query. Raster cells that meet the criteria will be reclassified to 1 and those that do not meet the criteria will become 0. For example, if the cells have an associated number that represents beach length measured from one end of a beach and we define long beaches as those that exceed 1 km, then all cell values above 1000 (assuming the units are metres) will become 1 and those below 1000 will become 0. Several criteria can be queried by establishing multiple value ranges (for example, we could perform an additional query to further subdivide the raster cells to include 'very long beaches', defined as those over 2 km in length).

Where multiple queries are performed on more than one raster dataset (or map), raster overlay functions can be used to apply Boolean conditional statements. As with any

Table 3.2 Simple algebraic expressions used to implement conditional statements with raster datasets using Boolean logic.

Boolean operator	Algebraic expression
AND	Multiplication
OR	Addition (followed by reclassification of 2 to 1)
NOT	Subtraction (followed by reclassification of −1 to 0)
XOR	Addition (followed by reclassification of 2 to 0)

comparison of multiple datasets, first it is necessary to ensure that the raster grids overlap in spatial extent and are in a comparable geometric frame of reference. Data may then be reclassified in line with the necessary criteria and then overlaid using map algebra to apply a given Boolean operator to the combination of map grids (see Table 3.2).

Locational queries use tools to identify individual (or groups of) pixels on the basis of their location and their characteristics (e.g. mean, median, etc.), or by using raster overlay or masking techniques (see Section 3.7.1).

3.7.4 Implementing Conditional Statements With Vector Datasets

In the case of vector datasets, information does not conform to regularly spaced grid cells. Queries are therefore implemented by combining complex vector geometry from several datasets and rebuilding topological relationships. An overlay function will identify all intersections between lines that comprise each of the input layers by rebuilding topology and associated tables prior to the query. Two types of overlay function exist for both attribute- and location-based queries that correspond to different Boolean operators. These are the *intersect* function (corresponding to the AND operator) and the *union* function (corresponding to the OR operator) (see Section 3.1).

3.8 Multi-Criteria Decision Analysis With Multiple Spatial Datasets

One of the defining strengths of GIS is that they allow the exploration of relationships between data that have been integrated from a diverse range of sources. Multi-criteria decision analysis is a process that combines different geographic datasets and transforms them into tailored maps of geographic areas that meet a predefined set of criteria. For example, a sand dune may be defined as an area within a specified distance of the coastline, with a steep slope and primarily composed of sand. A multi-criteria model might employ three individual maps that contain information relating to each of these individual criteria (distance to the coastline, terrain slope and surface sediment characteristics) to identify which map areas meet all these collective criteria. The result would be a map of sand dunes defined from these three criteria. As with the Boolean queries, areas meeting specified criteria will be coded 1 and areas not meeting them will be coded 0. Results are then overlaid to identify geographic areas meeting all three criteria. Where priorities or preferences are apparent, these can be built into the analysis in the form of a weighting scheme that affords variable levels of importance to different criteria.

Multi-criteria decision analysis (MCDA) is one of the most intuitively appealing and direct ways in which spatial analysis can generate information of value for coastal management decision-making. Examples of MCDA analysis in coastal environments include the following:

- Use of maps indicating negative impacts on the coast, such as erosion, dredging, water quality, habitat change and recreational activities to prioritise areas for integrated coastal zone management (Fabbri, 1998).
- Selection of suitable carp aquaculture sites on the basis of water, soil and infrastructure in Bangladesh (Hossain et al., 2009), or the identification of suitable aquaculture sites for Japanese scallops in Funka Bay (Japan), based on biophysical variables (sea temperature, chlorophyll, suspended sediment and bathymetry) and infrastructure characteristics (distance to town, pier and land-based facilities) (Radiarta et al., 2008).
- Mapping the priorities of different stakeholders to define an optimal spatial arrangement of different conservation protection levels in Asinara Island National Marine Reserve (Italy) (Villa et al., 2002).
- Mapping sections of shoreline that are vulnerable to the impacts of climate change, such as flooding from sea-level rise because they meet a set of predefined indices (Raaijmakers et al., 2008).
- Coastal vulnerability analysis and the assessment of multiple dimensions of risk from coastal erosion (Roca et al., 2008).

3.9 A Multi-Criteria Decision Analysis of the Risk of Injury From Rock Falls at Worbarrow Bay, Dorset, UK

This case study outlines the key steps taken in a hypothetical multi-criteria analysis to identify areas where members of the public may be at risk from rock fall at Worbarrow Bay, near Lulworth Cove, UK. Worbarrow Bay lies on the southern coast of the UK at the centre of the Dorset and East Devon Coast World Heritage Site, otherwise known as the Jurassic Coast. The bay receives over 17 million 'visitor nights' each year (May 1993). The area is popular because of its historical, spiritual and cultural significance, as well as the unique geological features observable here that are often visited by school and university field-trips (Figure 3.9). In recent years, numerous rock falls and landslides along this section of coastline have led to hazard assessment becoming a management priority for this coastal landscape.

Figure 3.10 illustrates a multi-criteria decision analysis to identify geographical areas that pose a high risk of injury to members of the public from rock fall events. High-risk areas are defined as those that meet three equally weighted

Figure 3.9 Worbarrow Bay, east of Lulworth Cove, Dorset, UK. Headland and bay coastline exposing strata from Portland stone to chalk. Features include (a) Iron Age hill-fort on chalk ridge, (b) recent landslides in chalk cliffs with talus extending below sea level, (c) coastal path at risk from cliff-top retreat, (d) stacks resulting from erosion of limestone steeply dipping strata which continue as a submerged ridge to the headland in the left foreground, (e) incised dry valley in chalk, (f) landslides in chalk, sands and clays, and (g) shingle beach with occasional landslides. (A black and white version of this figure will appear in some formats. For the colour version, please refer to the plate section.)

criteria relating to proximity to footpaths, topographic slope and the nature of the underlying geology:

- criterion 1 – areas that were located within 50 m of footpaths,
- criterion 2 – areas that had a slope gradient that exceeded 10%, and
- criterion 3 – areas that were underlain by softer, less consolidated chalk deposits.

Geographic areas that met all three of these criteria were identified as being of high risk. The practical value of such an analysis is that it identifies geographical areas where management activities may be required. These might include the introduction of signage and handrails to reduce risk of injury. As the Jurassic Coast is 155 km long, and a lot of this length is subject to intense pressure from tourism, identifying areas in which to focus management resources can produce significant practical efficiencies.

3.10 Exploratory Spatial Data Analysis (ESDA)

Exploratory data analysis (EDA) is defined as a collection of techniques for detecting patterns, errors and outliers, and identifying distinguishing or unusual features in data by summarising descriptive statistics (e.g. mean, mode, median) (Tukey, 1977). Haining (2003) distinguishes exploratory data analysis from exploratory spatial data analysis

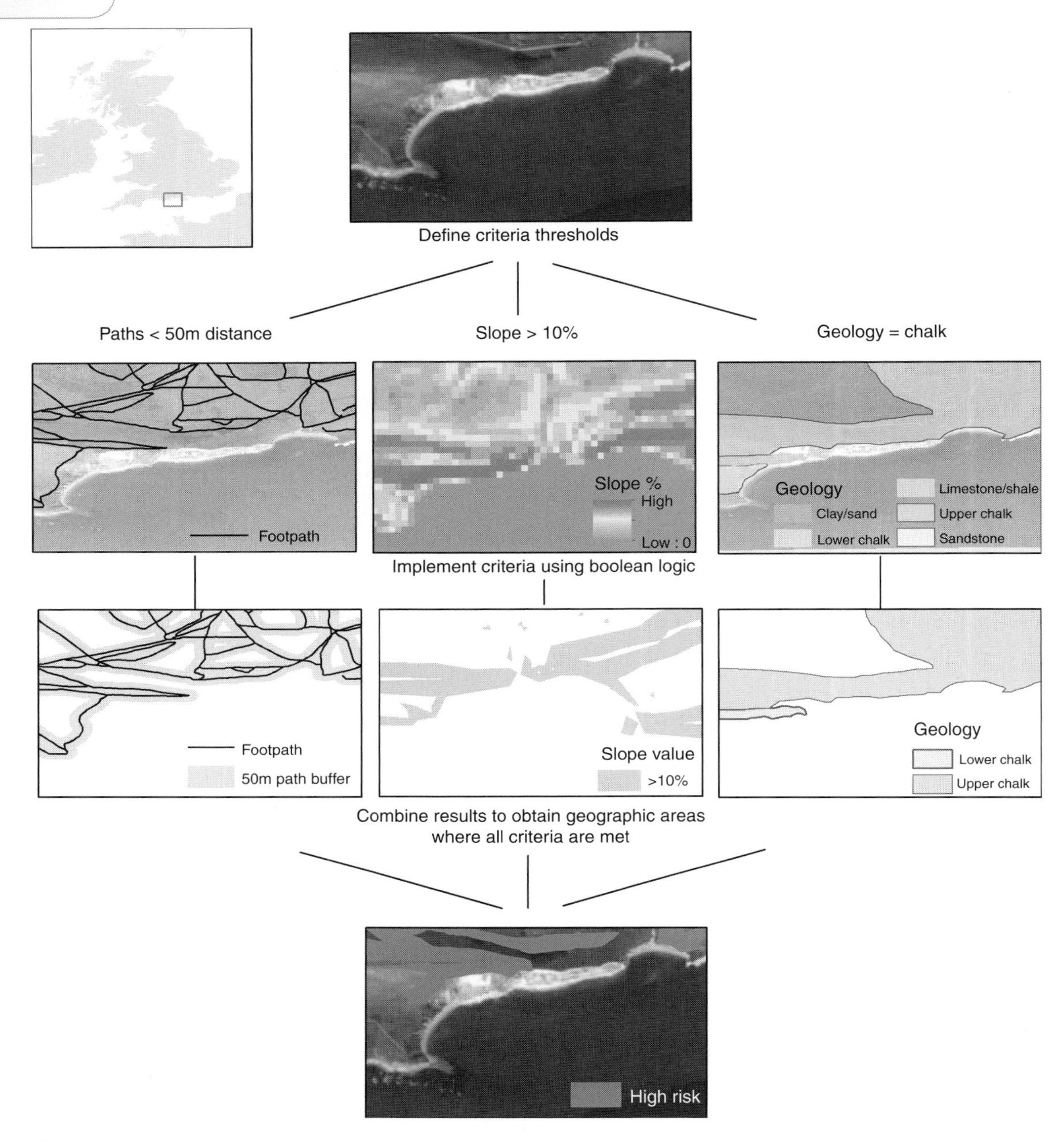

Figure 3.10 A hypothetical illustrative example of multi-criteria spatial analysis at Worbarrow Bay near Lulworth Cove, Dorset, along the coastline of southern England. The risk of injury from rock falls can be modelled based on distance to footpath, topographic slope and underlying geological character. Text boxes indicate key parts of the process of multi-criteria analysis. (A black and white version of this figure will appear in some formats. For the colour version, please refer to the plate section.)

Table 3.3 Some key distinctions between the aims and methods of exploratory data analysis (EDA) and exploratory spatial data analysis (ESDA).

	EDA (non-spatial data)	ESDA (spatial data)
Aims	Pattern identification Hypothesis formulation Model assessment	Spatial pattern detection Hypothesis formulation based on the geography of the data Spatial model assessment (e.g. a spatial regression model)
Methods	Visual Numerical Graphical	Visual Numerical Graphical Cartographical

(ESDA) by virtue of the explicit spatial methods employed in ESDA; particularly through the cartographic representation of data values (see Table 3.3). ESDA provides a formal set of statistical exploratory tools to make use of digital maps to detect and interrogate patterns across landscapes. Exploration is undertaken with the objective of formulating hypotheses from data. Exploratory analysis (whether spatial or non-spatial) is often carried out as a precursor to more prescriptive confirmatory data analysis. It can also be applied to examine model results post-analysis and provide evidence on whether model assumptions are met and whether diagnostics (e.g. the amount of variance explained) are subject to spatial dependence.

Statistical methods employed in ESDA are primarily descriptive and can be thought of as occupying the analytical territory between a simple visual assessment of map patterning and more formal inferential methods that extend statements about the sample to statements about the wider population from which the sample was drawn. Exploratory spatial data analysis allows the analyst to ask questions such as *'does there appear to be a relationship? And do these relationships appear to vary across the study area?'* Mapping data points is a crucial step in the exploratory process because it allows trends and outliers to be examined in map form as well as more traditional statistical plots (Cressie, 1993). Spatial data may be explored in the form of a lattice or a set of points (Diggle, 2013). Datasets can be presented in the form of a box plot, for example, and then individual cases can be highlighted from this representation and simultaneously highlighted on the map, enabling the analyst to ask *'where do we have extreme incidences?'*

In ESDA, the link between the map and the information that underlies it becomes crucial. Valuable insight is generated from the ability to view both extreme incidences and general trends in their geographical context. Methods are primarily visual (i.e. charts, graphs and figures) and numerical (i.e. employing descriptive statistics and box plots). An emphasis is placed on the use of *resistant* statistics. These are statistical descriptors that are not sensitive to the presence of outliers (i.e. the median as opposed to the mean, and the interquartile range as opposed to the variance). In practice, ESDA can be achieved through a coupling of GIS software with spatial statistics and implemented through programming interfaces such as those that are available through R software. Alternatively, spatially explicit shapefiles can be input into spatial statistics software such as GeoDa and used to define a spatial weights matrix and visualise the results of spatial analysis.

The analyst's viewpoint throughout ESDA is dependent on the areal partition, the scale of the partition and its configuration with respect to the underlying reality that it represents. It is therefore important to distinguish between the use of global and local tools for reporting statistical properties. The term 'global' relates to the entirety of a dataset. Global tools calculate values that simultaneously take all the data into account. Conversely, local tools run a window across the map to compute statistics across localised subsets in a constant, repetitive manner. Stationarity refers to the characteristic of a spatial process in which statistical properties such as the mean, variance and autocorrelation do not change over space or time. The objective of local tools is to reveal non-stationarities within the data. There is no 'correct' size or configuration to the window employed for data subsetting.

Computer visualisation techniques commonly used in ESDA include brushing, dynamic brushing, interactive modification, dynamic interactivity and dynamic projection mechanisms (Buja et al., 1996; Cook et al., 1996, 1997). Behind each of these techniques is the objective of rendering spatial patterns visible by presenting them in a manner that is amenable to detection. Buja et al. (1996) provide a classification of ESDA techniques in relation to the analytical objectives of focusing individual views of a dataset, linking multiple views and arranging many views. Focusing describes any operation that determines how a dataset is presented and viewed, including choice of variables, mode of presentation, map projection, zoom, aspect ratio and pan. Linking can occur any time that multiple visualisations of a dataset (box plot, scatterplot, map) are viewed simultaneously. When data cases are selected in one view, establishing a live link will automatically trigger the same selection in the other windows. Brushing is a dynamic form of linking whereby data cases selected in a graph or map are continuously updated by moving a 'brush' over the observations in a given view of the data view. The arrangement of many views for a dataset may be useful where comparison of data across a given semantic theme could potentially add insight. For example, this could be the case in the arrangement of a pairwise scatterplot matrix that compares all possible variable combinations for a multivariate model (Fotheringham and Rogerson, 2013). Choropleth mapping is a common way to express values attached to area or point data in which ranges or values are assigned a particular shading or colour (Robinson, 1958). Localised indicators of spatial association (LISAs) are a related class of exploratory data analysis techniques that break down global measures of association to identify local pockets of non-stationarity (e.g. hotspots) and assess the influence of particular locations on global statistics (Anselin, 1995).

Three-dimensional, oblique views are a popular method for representing mapped spatial patterns of raster datasets. Figure 3.11 illustrates the application of some ESDA techniques to city population data in an exploratory analysis of the how population settlements are related to the coastline. In Figure 3.11(a), the height of pixels representing cities is plotted as a function of the population of those cities. A dynamic link is illustrated in Figure 3.11(b) between a scatterplot and the location of cities plotted on

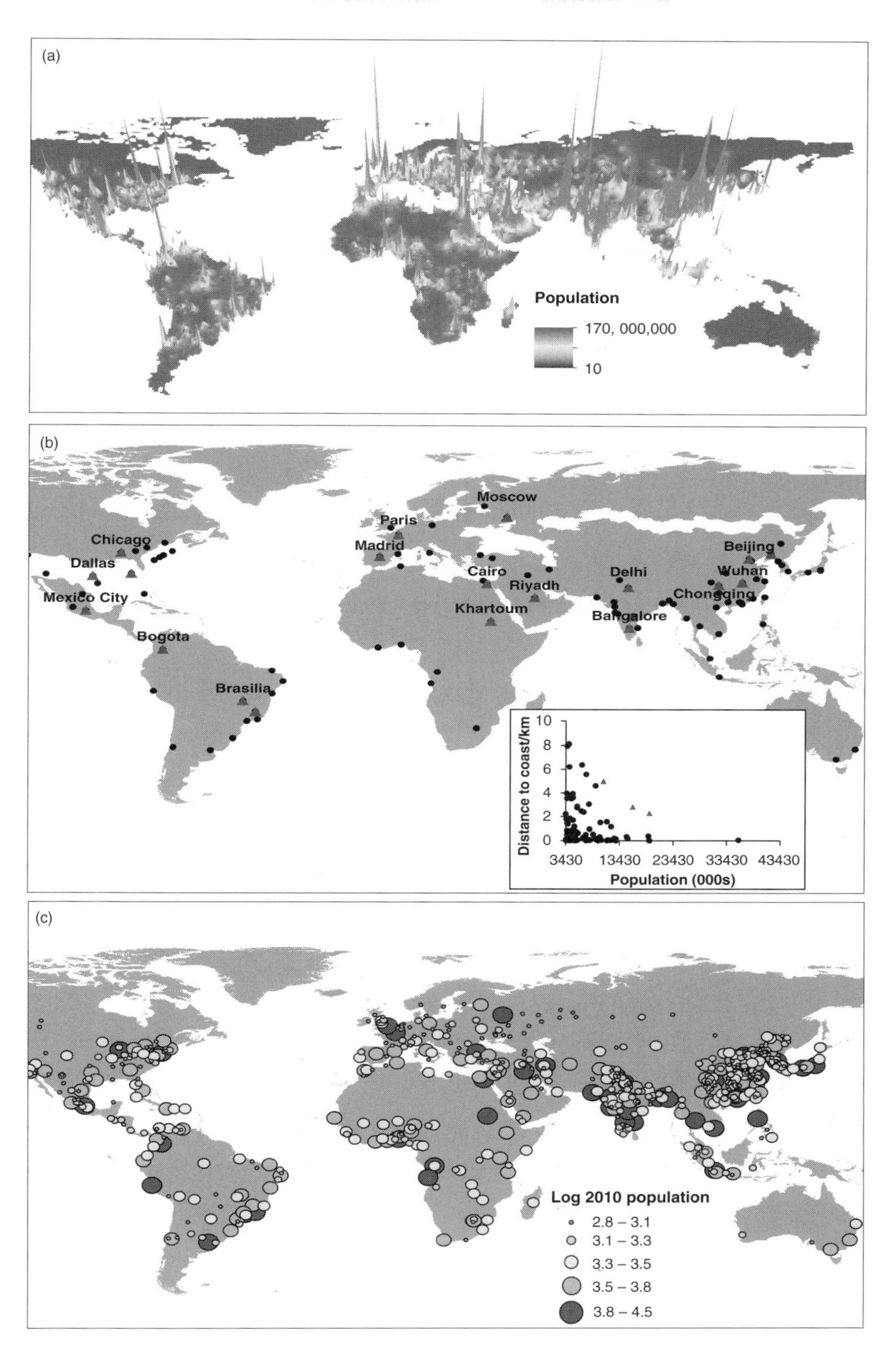

Figure 3.11 Some ESDA techniques. (a) Focusing: an oblique, three-dimensional view of world population centres, showing height above ground as a function of population (cell size, 10 km). (b) Linking: a scatterplot of distance to coast vs population (lower left), with outliers highlighted and plotted as triangles on the map; the identity of outlying cities is shown on the linked attribute table (right-hand side). (c) Arranging Views: a choropleth map of coastal cities with the symbology (size and colour of circles) linked to population as an example of an arranged view of a point dataset. (Data source: Natural Earth public domain dataset.) (A black and white version of this figure will appear in some formats. For the colour version, please refer to the plate section.)

a world map. Each data point represents a city, and their corresponding populations are plotted against distance to coastline. By using this link to simultaneously highlight points on the map and in the plot, it is possible to identify those outlier cities that go against the general trend of highly populated cities being located at the coast (e.g. Delhi, Moscow, Chicago, etc.). The choropleth map shown in Figure 3.11(c) adopts a symbology, specifically the size and colour of the circles making up the map of cities, which is a function of city population. In doing so, any outlier cities with large populations are rendered highly visible on the map.

While ESDA is not used in formal hypothesis testing, it nevertheless plays two important roles in the development of spatial models. Firstly, it renders statistical patterns visible. In doing so, it invites the analyst to consider postulates that might not otherwise have been included at the model specification stage. Secondly, in a staged approach to model building that sequentially tests different combinations of variables, it provides clues as to the performance of existing variables and the identity of missing covariates. This is particularly the case where model residuals are mapped to reveal geographical areas where the model is under and over predicting an outcome. The case study in Section 8.8 provides an illustrative example of how mapping residuals enables a model to be re-specified, using the spatial properties of the data themselves as a guide to the appropriate specification.

3.11 Summary of Key Points: Chapter 3

- Basic geometric operations with spatial data include measurement of distance, area, proximity, buffer, distance decay, overlay and the application of contextual operators.
- Different spatial datasets can be linked by virtue of their common *locations* or *attributes*.
- Spatial point patterns can be explored to see whether they are clustered, dispersed or spatially structured in some way.
- Inferential statistical methods test the significance of patterns detected (e.g. the significance of a cluster or 'hotspot' of points).
- Contextual operators, trend surface analysis and map algebra are all basic forms of analysis undertaken with raster datasets.
- Conditional statements can be specified that use Boolean logic to answer questions from spatial data. This forms the basis of multi-criteria decision analysis for coastal management.
- Exploratory spatial data analysis offers a formalised collection of statistical tools and techniques for detecting patterns, errors and outliers and identifying trends or unusual features in geographical data.

4 Mapping Coastal Environments

4.1 What Is Mapping?

At the most basic level, mapping is the process of making a 'graphic representation on a plane surface of the Earth's surface or part of it, showing its geographical features positioned according to pre-established geodetic control, grids, projections and scales' (Stiegeler, 1977). Making and studying maps is the central focus of the discipline of cartography. Many of the conventional cartographic rules that shape how maps are constructed were developed following European campaigns of discovery in the eighteenth century and are summarised by Arthur Robinson (1958) in his seminal book *Elements of Cartography*. This seven-part book established widely adopted cartographic standards relating to sources of data, Earth–map relations, data processing, perception and design, cartographic abstraction and map execution. The process of making a map can be expanded beyond the map itself to incorporate the wider activities of data collection, decisions of cartographic data generalisation and design, and also the reception of the map by the user (Board, 1967). A wider definition of mapping therefore takes into account the whole process of information gathering, understanding and shaping by the cartographer, encoding in the map, decoding and interpretation by the user. This wider definition therefore incorporates associated intellectual–cognitive and perceptual issues, as much as questions of data representation (Crampton, 2011).

4.2 Why Map Coastal Environments?

Most of the analytical exercises discussed in this book begin with the construction of a digital map to represent the structural form of a coastal environment. The act of mapping an area forces the person making the map to express their understanding of landscape features and the processes that shape them. They are required to develop classification schemes within which different land surface cover types can be organised and identify their spatial distribution. Mapping is therefore an activity that helps us to express our understanding of different coastal environments, and to communicate this understanding to others. Maps are a fundamental part of coastal spatial analysis. They often provide the starting point of, inspiration for and means of analysis. Digital maps and their composite vector or raster structures (see Chapter 2) represent valuable datasets in their own right. They can be explored for patterns and interrogated for clues as to how the features depicted by the map are arranged in space, how they relate to external influences and potentially how they might change through time.

The thematic content of a coastal map can be biophysical or socioeconomic in character. It may draw on information across timescales ranging from seconds to years. Phenomena featured in a map and described in the legend can include the following:

- biological habitats, such as kelp beds, coral reefs and different sandy bottom types (Luczkovich et al., 1993);
- geomorphological landforms, such as sand dunes, cliffs and different types of atoll (Andréfouët et al., 2001);
- vegetation communities, such as intertidal saltmarshes (Belluco et al., 2006), seagrasses (Kirkman, 1990) or mangroves (Heumann, 2011);
- benthic and coastal fauna, such as oyster beds (Allen et al., 2005);
- human uses across land areas, or zonation of activity types, including fishing, boating, diving and aquaculture (Day, 2002); or
- coastal processes, such as sediment plumes (Clark, 1993) and water currents (Richardson and Walsh, 1986).

While maps represent the building blocks of spatial analysis, they also provide a visually appealing and intuitively accessible means of presenting the results of any analysis undertaken. Indeed, some analytical techniques are specifically designed to create an output (e.g. the identification of land areas) that is best expressed in map form. Examples of such techniques include habitat suitability mapping of cold water corals or dugong populations (Davies and Guinotte, 2011; Hamylton et al., 2012b) and species distribution modelling of marine benthos in the North Sea (Elith and Leathwick, 2009; Reiss et al., 2011). It is useful to distinguish between mapping observable coastal phenomena with the objective of creating a representation of features (albeit features with different characteristics, such as landforms or processes) and mapping the results of analysis, sometimes referred to as a model.

4.3 The Historical Importance of Coastal Mapping: Marie Tharp

Mapping is an activity that helps us to express our understanding of coastal environments. To this end, maps have provided important evidence that has shaped scientific debates about the processes by which coastal landscapes evolve. One classic example of the role of maps in making visual arguments is the impact of a map produced by oceanic cartographer Marie Tharp. This provided the first bathymetric representation of the Mid-Ocean Ridge in the Atlantic Ocean.

Marie Tharp was originally from Bellefontaine, Ohio. She developed an early interest in mapping through accompanying her father on field survey outings for the US Department of Agriculture. After studying geology and mathematics, she joined the Lamont Hall Geological Observatory at Columbia University. Here she began to map the deep ocean floor from soundings collected by ocean-going research vessels. In partnership with Bruce Heezen, a specialist in the gathering of seafloor seismic and topographic information, Marie produced the first physiographic map of the North Atlantic Basin in 1957. Until this point, very little was known about the character of the ocean floor. This map revolutionised both the scientific and public understanding of this environment by revealing mountains, ridges, canyons and rifts on what was previously thought to be a flat seabed.

For many years, Marie plotted soundings of the ocean floor from her drafting table (Figure 4.1). When Tharp started piecing together the North Atlantic profiles, she noticed a cleft running down the centre with peaks on each side. Other data showed large numbers of earthquakes occurring along the feature and a deep notch in the centre of the ridge. This confirmed Tharp's hunch that it might be a rift valley like the one found in East Africa. One of the principles of uniformitarianism, central to the discipline of geology, states that the present is the key to the past and that contemporary features observable on the Earth's surface are part of a uniform and continually repeating set of processes. The idea of a rift valley in the Atlantic basin would support a controversial budding hypothesis that the continents move across the surface of the Earth. At this time, it was known by geologists that the Earth was expanding but the mechanism through which this was occurring was unknown. Within this debate, the idea that continents were drifting was considered by many (including Heezen) to be 'scientific heresy'. Plate tectonics emerged as the most revolutionary and important geological theory from an era characterised by scientific controversy. Marie Tharp once remarked about this controversy (in 1949): *'I was so busy making maps I let them*

Figure 4.1 Marie Tharp sketching the ocean floor from sounding profiles at her drafting table. (Image produced with permission from the Lamont-Doherty Earth Observatory.)

argue. I figured I'd show them a picture of where the rift valley was and where it pulled apart. There's truth to the old cliché that a picture is worth a thousand words and that seeing is believing' (see Tharp, 1999). The scientific significance of the Heezen–Tharp North Atlantic map is that it provided a key piece of evidence that was in agreement with the theories of plate tectonics and continental drift. This allowed the concept of plate tectonics to move into the realm of legitimate debate and later into the mainstream of Earth science.

Heezen and Tharp (1977) later collaborated with artist Heinrich Berann to produce the first World Ocean Floor Panorama for *National Geographic* magazine in 1977 (Figure 4.2), showing the true extent of the mid-ocean ridge feature. *'Establishing the rift valley and the mid-ocean ridge that went all the way around the world for 40,000 miles – that was something important. You could only do that once. You can't find anything bigger than that, at least on this planet'* (see Tharp, 1999).

Making maps as a woman scientist during the 1950s, Marie's work lacked the recognition it deserved in publications and presentations. Nevertheless, the significance of her maps has now been recognised through numerous awards.

4.4 Defining Areas to Be Mapped

At the beginning of any coastal mapping campaign, it is important to establish an area of interest that will define the geographical scope of the map itself. Some coastal landforms, such as islands, beach compartments and estuaries, are naturally organised into easily distinguishable units that can be delineated and mapped accordingly. Other features, such as seagrass beds or sloping continental shelves, are less clearly defined because they may gradually disappear over a continuum of reductions in shoot density, or an imperceptibly deepening sandy shelf. In such cases, it can be useful to ascertain the spatial scales that are relevant to *processes* underpinning the observable features to be mapped. These are likely to be linked to the objectives of the mapping campaign. For example, if the objective was to measure the spread of an oil spill, then the processes responsible for spreading oil across the sea surface would provide a useful indication of the appropriate geographical area to be mapped. These might include local weather conditions, wind and wave energy, and their associated spatial scales of variation.

The importance of processes occurring within environments is evident from the various definitions of 'landscape' or the allied concept of the 'seascape' that exist. These include:

- a heterogeneous land area composed of clusters of interacting systems repeated in similar form throughout (Forman and Godron, 1986),
- ecological and geographical entities, integrating all natural and human patterns and processes (Naveh, 1987), and
- the major structural features of coastal areas collectively, alongside the physical and ecological processes that transcend individual habitats (Ogden, 1997).

An emphasis on processes occurring within the landscape calls for a redefinition of the landscape as the physical space within which patterns, organisms or processes

Figure 4.2 World Ocean Floor Panorama (Bruce C. Heezen and Marie Tharp, 1977). Copyright by Marie Tharp 1977/2003. Reproduced by permission of Marie Tharp Maps LLC, 8 Edward Street, Sparkill, New York, NY 10976. (A black and white version of this figure will appear in some formats. For the colour version, please refer to the plate section.)

interact. Such a definition requires processes to be identified that are relevant to the subject of the map and the associated temporal and spatial scales over which they act. Once these have been identified, a corresponding geographical area of observation can be delineated that is appropriate to the spatial scales of variation inherent to the processes under scrutiny. For example, to observe spatially explicit ecological phenomena such as feeding patterns or the spread of disease, O'Neill et al. (1999) suggest that the study area should be at least two to five times larger than the spatial extent of the largest process under study. This represents a trade-off between too small an area, which might not include enough pattern, and too large an area, which may include several processes and thereby incorporate variable processes that act differently across subregions of the study area (Dale and Fortin, 2014).

It is important to remember that a landscape also encompasses structural characteristics that are the time-integrated result of a range of temporally and spatially sequenced processes. Figure 4.3 illustrates a range of physical and biological spatial scale ranges that may be pertinent for the assessment of climate change impacts on ecology and coastal morphodynamics (Cowell and Thom, 1994; Forman, 1995; Woodroffe, 2002; Slaymaker and Embleton-Hamann, 2009). Identifying dimensions of the process ranges associated with a feature of interest can be a difficult and subjective process. For example, if we consider the importance of water movements that govern the development of sandy islands and spits, the temporal and spatial scales associated with 'water movement' encompass a wide range of variability. These include the occurrence of interglacial high sea-level episodes (10^5 a, 100 km), which are significant in terms of island evolution, to the passage of individual waves (1–10 s, 1–10 m), which are significant determinants of sediment

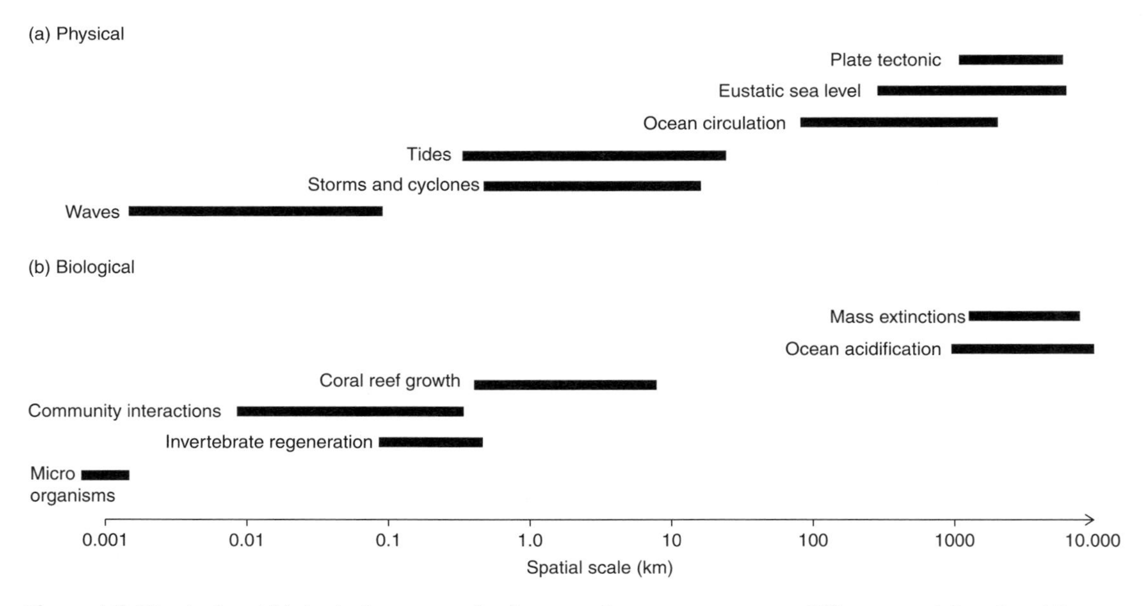

Figure 4.3 Physical and biological processes in the coastal zone occur across different spatial scales. After Woodroffe (2002).

transport and deposition (Gourlay, 1988). Thus, when defining a geographical area to be mapped, it is important to be mindful of the objectives of the mapping exercise and how these relate to any ecological, geomorphic or anthropogenic processes that may underpin the features to be represented by the map.

4.5 Traditional Field-Based Approaches to Mapping Coastal Environments

Most mapping exercises require some *in situ* information to be collected directly from the area that is to be mapped. Table 4.1 summarises a range of methodologies for field mapping that have been widely adopted since the seventeenth century. Their application to a low wooded island is illustrated in the Low Isles case study (Section 4.14).

Table 4.1 A summary of traditional field-based approaches to mapping in coastal environments.

Technique	Summary description	Reference
Theodolite triangulation	A network of high-accuracy topographic control points is established with a theodolite to precisely measure angles in the horizontal and vertical planes from either end of a fixed baseline. Triangulation fixes the location of the point via trigonometry	Ganci and Shortis (1996)
Plane-table survey	A smooth, flat table mounted on a sturdy base is used as a foundation for a drawing sheet, on which objects of interest are sighted and plotted. An alidade, a ruler with a telescopic sight, is used to draw the direction and angle of objects. A staff is held at each point plotted to determine distances	Milne (1972)
Compass traverse	Horizontal angles are measured using a survey compass and distances are measured using either a fibron tape or calibrated paces. Traverses are undertaken as closed circuits. Large environments are subdivided into sectors and related to a primary network	Stoddart (1978)
Tacheometric methods	Tacheometry measures distance to a vertical staff established in the centre of the area mapped. Horizontal distance is calculated from the vertical angle sighted inside two well-defined points on the staff and the known distance between them	Mussetter (1956)
Optical rangefinder methods	An optical rangefinder is used to measure the distance between an operator and an object. The rangefinder can be coupled with a compass and vertical circle to establish relative horizontal and vertical angles	Nason (1975)
Horizontal sextant methods	The horizontal sextant allows precise measurement of angles, from which positions can be related to the location of fixed reference points. This method is often used to establish planimetric control in conjunction with other methods for plotting further detail	Lušić (2013)

Coastal environments impose some specific practical constraints on *in situ* mapping activities. Some of the more traditional techniques summarised here have now been replaced by more recent technologies that overcome the need for intensive field sampling. Nevertheless, it is important to understand how maps were historically produced. This is because maps that are several centuries old can be used as a baseline against which landscape change is determined. Such determinations should be made with a sound understanding of the methodology by which the baseline itself was produced.

4.6 Remote Sensing

Remote sensing is the acquisition of physical data relating to an object by a sensor device separated from it by some distance (i.e. 'remotely') (Lintz and Simonett, 1976). Optical remote sensing instruments operate in the visible portion of the electromagnetic spectrum. They provide a synoptic portrait of the Earth's surface by recording numerical information on the radiance measured from a series of picture elements (pixels) across a number of spectral bands (Green et al., 2005). Often these instruments are mounted on aircraft or satellites. From these vantage points, their distance from the Earth enables them to survey large geographical areas (for example, images acquired from a Landsat satellite are typically 150 km × 150 km). This is advantageous in coastal environments because changing weather conditions, inaccessible locations and the logistical challenges of carrying out underwater fieldwork all compromise the ability to directly survey large areas.

Figure 4.4 provides a classification of remote sensing systems. Two pertinent distinctions can be drawn from this classification: the difference between *active* and *passive* systems and, with respect to passive systems, the difference between *imaging* and *non-imaging* systems. Active systems illuminate the object of study with their own supplied radiation (e.g. a laser pulse of light) and record a return signal. Passive systems detect naturally occurring radiation that is already passing through the atmosphere, usually in the form of sunlight. Radar and LiDAR (light detection and ranging) instruments are examples of active sensors because they emit and measure energy in the microwave and in the visible to near-infrared wavelengths, respectively. Passive remote sensing can be more precisely defined as the practice of deriving information about the Earth's land and water surfaces using the Sun's electromagnetic radiation reflected from the Earth's surface in one or more regions of the electromagnetic spectrum (Campbell, 1996). An imaging system measures the intensity of radiation as a function of position on the Earth's surface in a two-dimensional pictorial representation of intensity (Rees, 1999). This usually takes the form of an image acquired from an overhead perspective, composed of a raster grid of pixels (see Chapter 2), for which spatial coordinates are available. The remainder of this chapter is primarily concerned with such imaging systems.

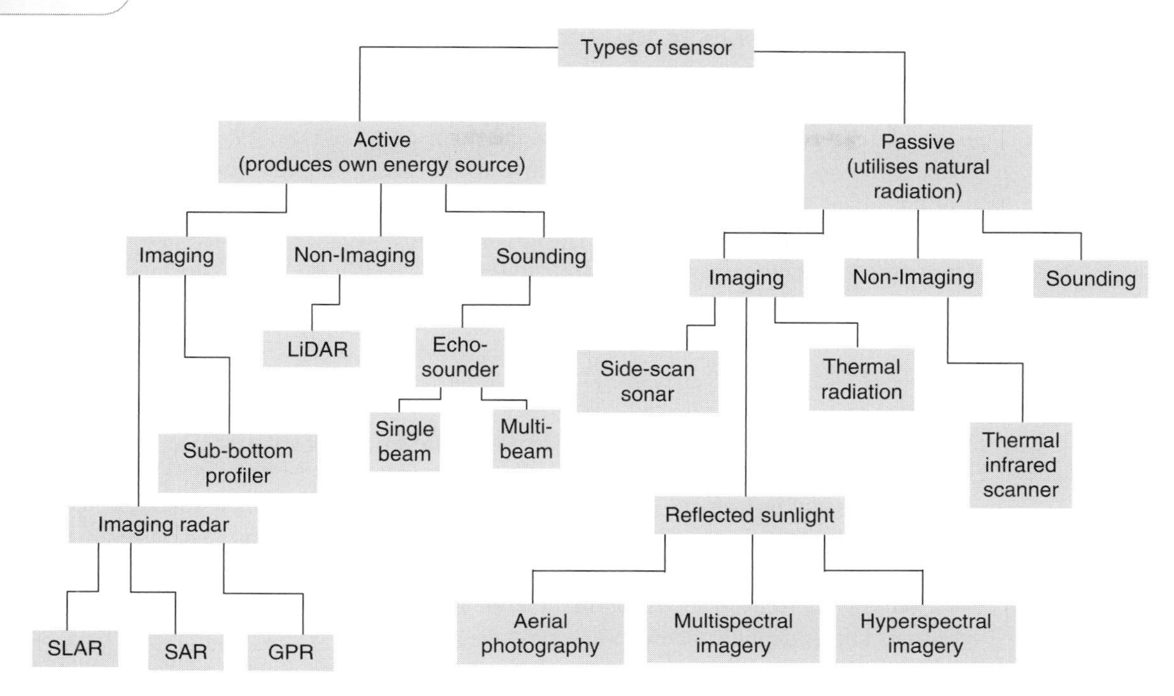

Figure 4.4 Taxonomy of remote sensing systems for coastal mapping. Adapted from Rees (1999).

4.7 Optical Remote Sensing of Shallow Marine Environments

Aerial techniques for coastal surveying became widely adopted at the end of the Second World War. This was a time when photography of many coastlines, including European continental shorelines and remote oceanic islands in the Pacific and Indian Oceans, was acquired for military purposes (Yonge, 1929; El-Ashry and Wanless, 1967; McLean et al., 1978). It was not until the 1970s that satellites first hosted sensors intended for Earth observation, including those with the capability to resolve sub-aquatic environments and their associated features. Satellite remote sensing for coastal mapping traces its roots back to the Landsat programme that was established in 1972. Over the following decades, this was extended across several satellite sensors, including SPOT, Ikonos, QuickBird and Hyperion (see Table 4.3 for further details). Subsequent developments in sensors have continued to increase the amount of information that can be acquired in both the spatial and spectral domains.

The uses to which coastal remote sensing instruments have been put can be broadly divided into mapping biological features (e.g. coral reefs, kelp, seagrass and oyster beds) or geomorphic features (e.g. subsurface sand or channels, basins and lagoons). Bathymetric mapping from satellites has also been an area of substantive research focus. Information on water depth can be extracted from optical images by drawing on the variable attenuation of light at different wavelengths through the water column (see Section 4.13.3). The condition of seafloor features has also been an area of growth in mapping applications. Characteristics such as coral reef health, extent of coral bleaching and the presence of dredge plumes have increasingly become a focus of features resolvable from space (Holden and LeDrew, 1998; Elvidge et al., 2004; Yamano and Tamura, 2004; Evans et al., 2012). Finally, the multi-temporal production of maps across a given period enables the detection and quantification of changes to marine environments at the landscape scale. This is particularly useful when maps are made over time periods that coincide with known disturbance events, for example monitoring changes to eelgrass beds as a result of coastal development (Macleod and Congalton, 1998).

4.8 Sensor Specification for Mapping Coastal Environments

A remote sensing instrument (or spectrophotometer) consists of a light source, a system to disperse the light into wavelength bands and a detection system. The detection system should be linked to hardware and software for digitisation and storage of information. Sensing instruments focus incoming radiation onto a detector to create an electronic response. This is digitally recorded in a pixel array (Figure 4.5a). For example, in the case of the CASI sensor, as white light passes through a diffraction grating, it is subdivided into a coloured spectrum. This is then focused by a lens onto a charge-coupled device (CCD) array. In this way, multiple wavelengths of light (approximately 19 bands ranging from infrared to blue wavelengths) are simultaneously detected by a CCD array of 512 × 288 pixels (Figure 4.5b). Output digital numbers recorded in the image are dependent on the incident photons received by each detector within the array. These are converted to radiance values, which specify the amount of light entering the sensor field of view that corresponds to a given area of ground, having been reflected from a given solid angle ($W m^{-2} sr^{-1}$).

Most remote sensing images are not a single image; they can be subdivided into a series of layers or wavebands. Each of these wavebands relates to a different sampled section of the electromagnetic spectrum. The spectral resolution of an image, sometimes also referred to as radiometric resolution, refers to the number of wavelengths sampled by the sensor that has acquired the image. Most sensors are designed to detect light specifically within one region of the spectrum (e.g. optical sensors detect visible light in the wavelength region 400–700 nm). Within this region, they detect multiple wavelengths of visible light, such as green, blue and red, in smaller windows of this wavelength range. A sensor such as MODIS, which has 36 bands, is therefore said to have a higher spectral resolution than Landsat, which has eight bands.

Sensors differ in their ability to differentiate signals from different wavelengths of light by virtue of their spectral resolution. They can broadly be divided into hyperspectral and multispectral sensors on the basis of the band windows in which they sample the electromagnetic spectrum. Multispectral sensors measure radiance data for several discontinuous spectral bands for each pixel of the image (Rees, 1999). Hyperspectral sensors typically sample many narrow sections to provide continuous coverage across the spectrum (Mather and Koch, 2011).

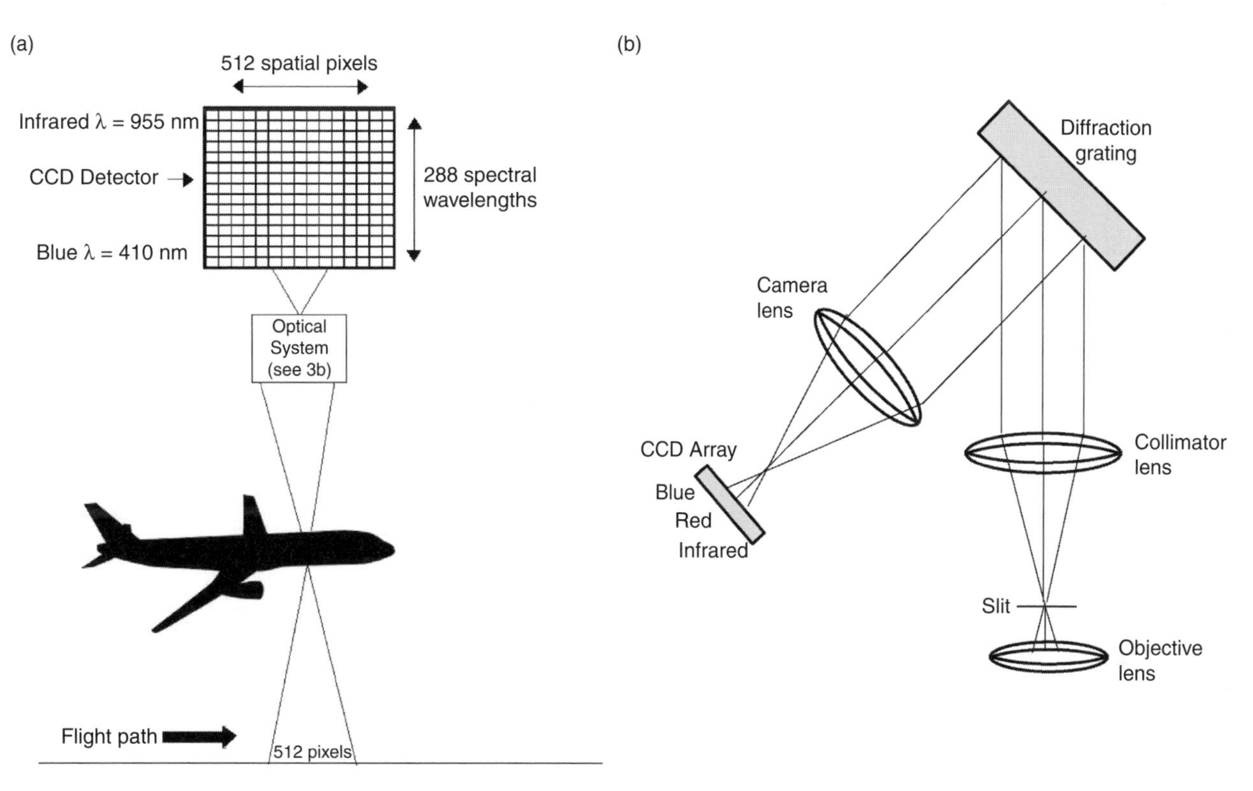

Figure 4.5 The pushbroom scanner (a) and sensor head optical system (b) associated with the Compact Airborne Spectrographic Imager (CASI) airborne remote sensing instrument.

Several studies have isolated single wavelengths or portions of the electromagnetic spectrum that are useful for distinguishing between features of interest in coastal environments (see Figure 4.8). A common finding among these studies is that there is a considerable amount of spectral reflectance variability within all measured features; for example, a spectral signature of the same species of coral can vary markedly (Hochberg and Atkinson, 2000). Several well-placed, narrow (10 nm) spectral bands are therefore necessary to detect subtle differences in reflectance between coastal communities (e.g. seagrass vs algae, brown algae vs green algae) (Hochberg et al., 2003). Separation of the major seafloor features in coastal environments therefore requires a band combination that emphasises the distinct spectral characteristics that are apparent. These are located in areas of the spectrum where the reflectance characteristics of the different seafloor-cover types diverge. One major challenge to the detection of subsurface features is the absorption of light through the water column at wavelengths towards the red end of the visible spectrum (> 650 nm). Successful band combinations must therefore also fall within atmospheric windows. These wavelengths of light must be able to travel through the atmosphere, cross the water surface and penetrate down the water column before being reflected back to enter the sensor field of view and be detected.

The spatial resolution of an image refers to the length of the side of one of the component pixels that make up the raster grid. This is the digital form in which most images are stored (Section 2.4). A higher-resolution image will have smaller pixels with shorter dimensions that represent smaller areas on the ground. For example, the DigitalGlobe WorldView-2 satellite sensor has a higher spatial resolution (2.4 m) than the Landsat satellite sensor (30 m). Because the pixels are smaller, the image will have a greater number of pixels for a given area on the ground, which means that it has a higher information content. Spatial resolution is an important property of a remote sensing image because it determines what can be seen and identified. This is the case for both a person visually interpreting the image and an automated computer algorithm designed to identify features from the image. As a general rule, the smallest feature that can be distinguished in an image will be double the size of the spatial resolution of the image. Spatial resolution and map scale both depend on the level of detail that was recorded when the data were originally collected. Table 4.2 indicates the resolution and corresponding level of detail collected at different map scales. A jetty of length 10 m will only be visible using imagery of spatial resolution around 5 m. Smaller features such as rock pools therefore need to be mapped using high-resolution imagery. Geomorphic landscape features that occur at broader scales can be mapped using lower-resolution Landsat (30 m spatial resolution) or SeaWiFS (1 km spatial resolution) data.

The spatial resolution of a remote sensing image is a function of the distance between the sensor and the Earth. Sensors that are closer to the Earth are able to produce images at a much higher spatial resolution (smaller pixels) than those that are further away. Image data acquired from airborne platforms typically have spatial resolutions of 1–3 m, with aerial photographs achieving resolutions of less than 1 m. This high resolution usually comes at the cost of overall geographical coverage, as airborne

Table 4.2 A summary of the relationship between map scale, resolution and the size of feature that can be mapped.

Map scale	Resolution	Size of smallest feature detectable
1 : 10 000	5 m	10 m
1 : 24 000	12 m	24 m
1 : 50 000	25 m	50 m
1 : 100 000	50 m	100 m
1 : 250 000	125 m	250 m
1 : 1000 000	500 m	1000 m

sensors are not able to simultaneously survey such a large area. They have a smaller field of view because the plane is closer to the ground. This smaller field of view means that the sensor must be transported along along multiple pre-planned flight survey lines to generate imagery of comparable geographical extent. Some satellite sensors have spatial resolutions commensurate with airborne imagery, but these have comparably limited spectral capability (typically one to three broad water-penetrating bands). With respect to commercially available remote sensing image datasets, there is generally a trade-off between spectral and spatial resolution for coastal mapping. Sensor choice will ultimately depend on the objectives of a given mapping campaign and how the specifications of available sensors can meet these.

Coastal mapping ranges markedly in terms of descriptive resolution, that is, the number of classes that can be mapped. The accuracy of a coastal map quantifies the difference between a measured value and its 'true' value (Mather and Koch, 2011). Estimating the accuracy of classifications of remotely sensed data generally involves a comparison between remotely derived information (e.g. the output digital map) and datasets collected independently as a validation exercise (e.g. a field dataset, see Section 8.11). The performance of remote sensing instruments for producing accurate coastal maps is inherently tied to the spectral and spatial resolutions of the sensor. These differ widely between instrument configurations and dictate the types of land-cover or benthic classes that can reliably be discriminated. In general, increasing spatial and spectral resolution of the sensor results in accuracy improvements (Green et al., 2005). Figure 4.6 illustrates that greater spectral and spatial resolution allow for increased accuracy in discriminating between different land-cover types. It shows that, as the number of classes mapped increases, there is a corresponding reduction in map reliability. At a coarse descriptive resolution (i.e. four classes), coastal maps tend to be more accurate, with a corresponding drop in accuracy as more classes are added. This effect is observed across a range of sensors, although this disadvantage is offset by an increase in map detail. Pre-processing imagery often improves the accuracy of a map (Section 4.10.1).

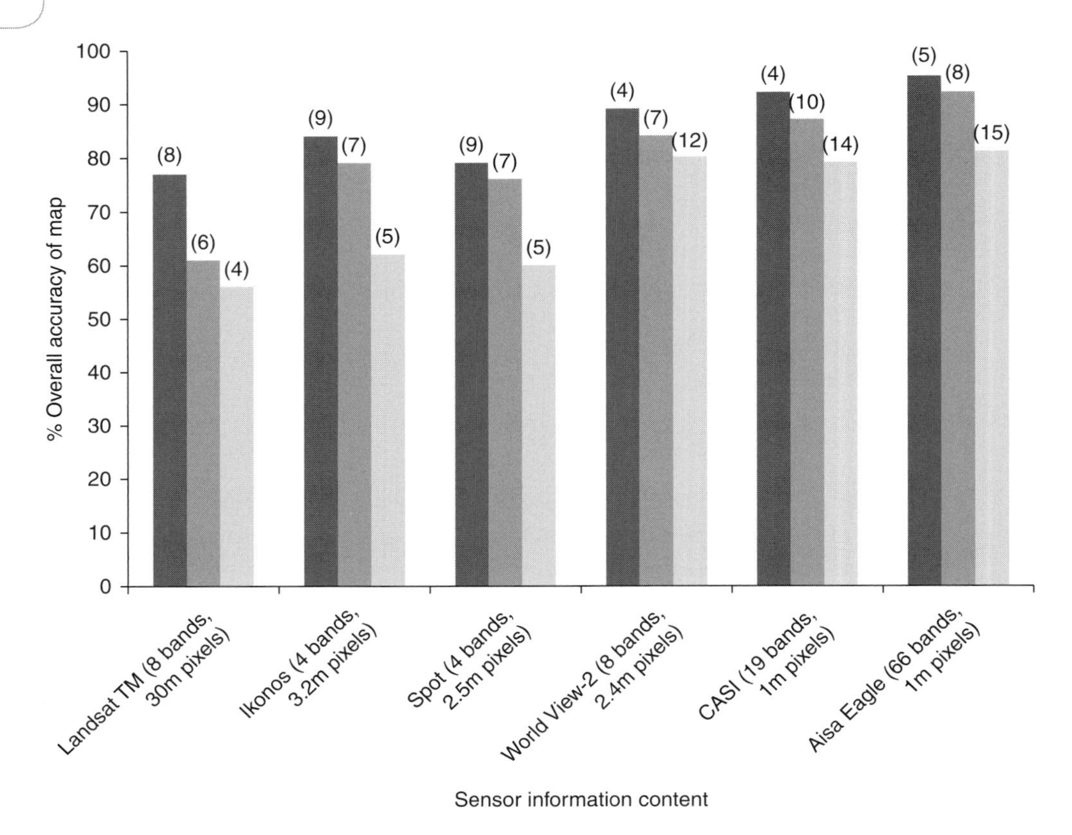

Figure 4.6 The relationship between descriptive resolution and accuracy of maps derived from optical remote sensing instruments of varying spectral and spatial specification. Bar colours indicate the map descriptive resolution and transition from dark (low resolution or detail) to light (high resolution or detail). Numbers in parentheses above bars indicate the number of classes mapped.

4.9 Optical Satellite and Airborne Images for Coastal Mapping

Landgrebe (1999) describes three fundamental 'domains of space' within which information is held by remote sensing datasets. These are image space, spectral space and feature space (Figure 4.7). Image space allows pixels to be viewed within their geographical context or raster grid. Spectral space refers to the response of an individual pixel to the ground cover as a function of the different wavelengths. Feature space views the response of all pixels across an n-dimensional space that is defined by the spectral wavebands of an image (Figure 4.7). For example, the feature space of an image composed of three wavebands is defined by plotting these three bands against each other in three-dimensional space. These different domains of remote sensing images represent ways in which they can be interrogated. They provide a useful framework for reviewing the range of Earth observation satellite images and their suitability for application to coastal mapping, as outlined in Table 4.3.

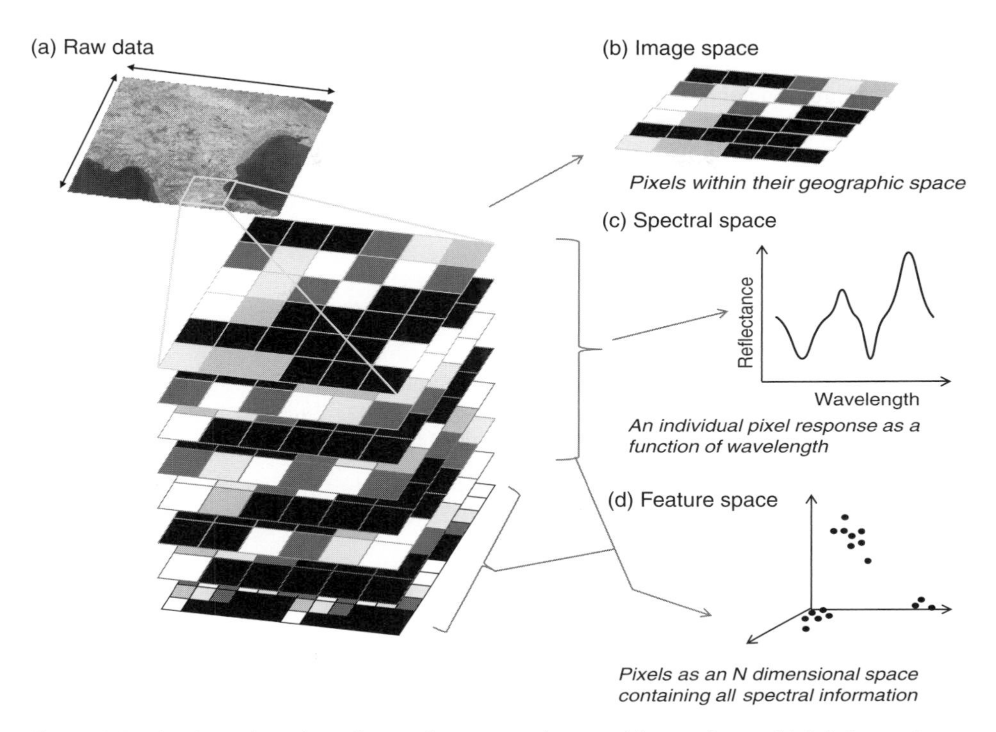

Figure 4.7 The three domains of space in a remotely sensed image from which information can be drawn.

The ability to differentiate different coastal ground-cover types relies on the positioning of the wavebands not only relative to each other in the electromagnetic spectrum, but also relative to the spectral signatures associated with the features being mapped. An image classification algorithm will be more likely to successfully differentiate between pixel groups relating to different ground covers if the input wavebands are placed in regions of the spectrum that maximise the differences between the spectral signatures of coastal features. Figure 4.8 illustrates the unique spectral reflectance signatures for sand, algae and coral. These plot reflectance of each seafloor-cover type as a function of wavelength for sand, algae and coral. It can be seen that coral and algae have spectral signatures that are more similar than sand. Across different wavelength regions of the electromagnetic spectrum, both of these signatures are consistently different to sand, which is much brighter and therefore has higher reflectance values. Conversely, deep water consistently has a low reflectance across all wavelengths (and is seen as dark pixels in the image). Spectral wavebands that discriminate between these types of seafloor cover will therefore be placed in regions of the spectrum where the difference between reflectance properties is maximised (i.e. above 650 nm). While the 730nm wavelength appears to maximise the spectral differences between coral and algae, the reflectance values of sand and coral are similar at this wavelength, which would likely result in spectral confusion and a low mapping accuracy for these cover types. When calibrating a remote sensing instrument, the ideal placement of spectral bands is therefore a function of instrument specification, atmospheric properties and the land-cover types to be mapped.

Table 4.3 A summary of the image, spectral and feature space characteristics of some remote sensors that are commonly applied in the coastal zone.

Image	Spatial resolution; footprint (image space)	Number of bands; position (spectral space)	Band width (feature space)
SeaWiFS	1.1–4.5 km; 2801 km^2	8 bands; 402–422, 433–453, 480–500, 500–520, 545–565, 660–680, 745–785, 845–885 nm	20–40 nm
MODIS	1 km; 23 300 km^2	36 bands (9 ocean bands); 405–420, 438–448, 483–493, 526–536, 546–556, 662–672, 673–683, 743–753, 862–877 nm	10–15 nm
Landsat TM	15–30 m; 185 km^2	7 bands; 450–520, 520–600, 630–690, 760–900, 1550–1750, 2090–2350, 10 400–12 500 nm	60–1800 nm
Ikonos	3.2 m; 11.3 km × 360 km	4 bands; 445–515, 505–595, 632–698, 757–853 nm	56–403 nm
SPOT	5–10 m; 3600 km^2	5 bands; 480–710, 500–590, 610–680, 780–890, 1580–1750 nm	70–230 nm
WorldView-2	2.4 m: 138 km × 112 km	8 bands; 400–450, 450–510, 510–580, 585–625, 630–690, 705–745, 770–895, 860–1040 nm	40–180 nm
QuickBird	2.44 m; 360 km × 18 km	5 bands; 430–545, 466–620, 590–710, 715–918, 405–1053 nm	115–1048 nm
GeoEye-1	1.54 m; 360 km × 15.2 km	4 bands; 450–510, 510–580, 655–690, 780–920 nm	35–70 nm
Hyperion	30 m; 7.7 km × 6.2 km	220 bands; 357–2576 nm	10 nm
CASI	1 m; user-defined	18 bands; 430–900 nm, user-defined	5–10 nm
AISA-Eagle	1 m; user-defined	128 bands; 400–994 nm, user-defined	5 nm

4.10 Mapping Methods

4.10.1 Image Pre-Processing Challenges in the Coastal Zone

The interpretation of remotely sensed images in terms of coastal features is complicated by a range of environmental factors. These include the presence of clouds obscuring features of interest, absorption and scattering of light in the atmosphere, limited and variable depth of light penetration into the water column, variation in

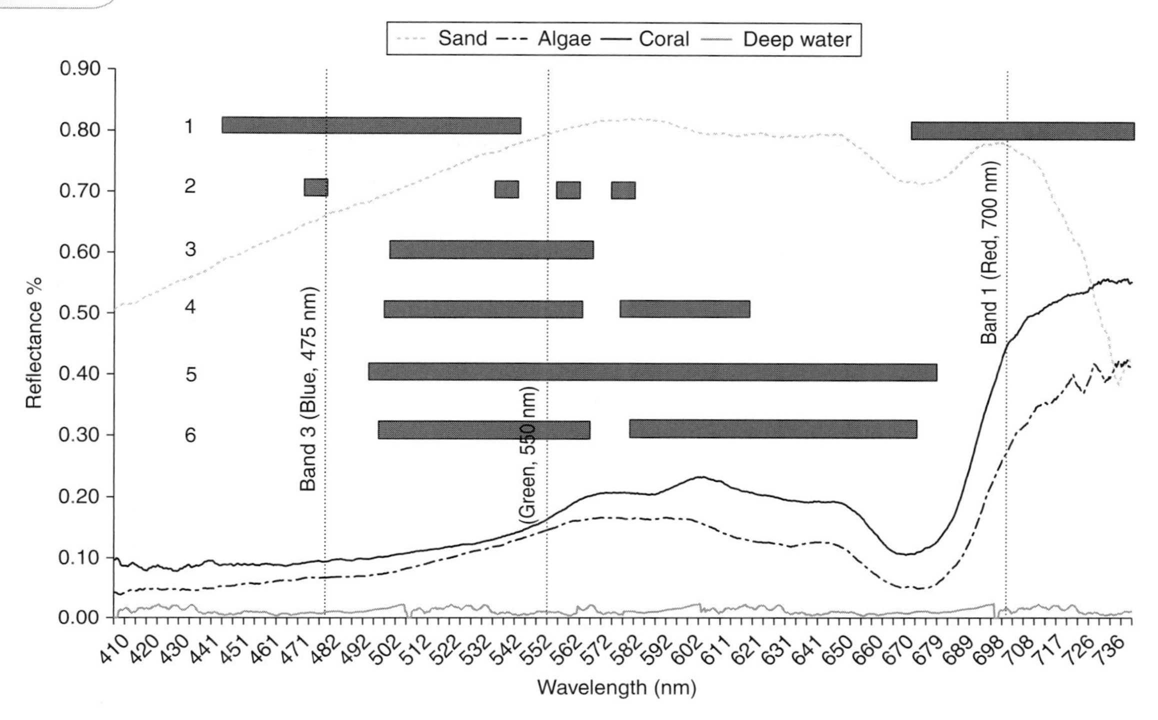

Figure 4.8 Spectral signatures of common coastal seafloor types (sand, algae, coral and deep water) collected with a GER1500 spectrometer from Aldabra Atoll, Seychelles, and associated published recommendations for the placement of wavebands within the electromagnetic spectrum for discriminating between seafloor types. Red numbers indicate the following publications: 1, Myers et al. (1999); 2, Hochberg and Atkinson (2000); 3, Wettle et al. (2003); 4, Holden and LeDrew (2000); 5, Kutser et al. (2000a, 2000b); 6, Holden and LeDrew (1999). The positions of the three different bandwidths of the satellite images employed for mapping in Section 4.10.3 are shown as vertical lines (after Joyce, 2005). (A black and white version of this figure will appear in some formats. For the colour version, please refer to the plate section.)

water quality across a single image scene, and backscatter of light causing sun glint from the sea surface. To overcome these challenges, it is necessary to pre-process imagery so that the nature of the electromagnetic radiation emitted from the seafloor can be accurately interpreted in terms of the different types of ground cover present.

Several methods have been developed to address the variable absorption and scattering of light in the atmosphere that influences remotely sensed images. One of the main problems is the addition of scattered light from the atmosphere into the signal received by the sensor. An atmospheric correction estimates the magnitude of this component of the signal and removes it. A simple but effective way to achieve this is through undertaking a dark pixel subtraction. This assumes that the darkest pixels in an image contain no ground-cover reflectance signal and can therefore be used to quantify atmospheric contributions across wavebands (Mather and Koch, 2011). It involves extracting the mean radiance value for each waveband from a collection of dark pixels and subtracting this from each waveband independently across the entire image. More sophisticated approaches attempt to derive atmospheric properties such as surface pressure, water vapour column, aerosol and cloud characteristics. Their effects on the image are subsequently estimated and removed. This can be achieved by incorporating them into a correction matrix for inverting 'radiance at detector' measurements into 'radiance at surface' values. The Fast Line-of-sight Atmospheric Analysis of Spectral Hypercubes (FLAASH) module within the image processing software ENVI (Environment for Visualising Images) is an example of such an approach (Cooley et al., 2002).

Mapping features that are underwater introduces additional challenges because of the presence of an overlying water body. The differential attenuation of light radiation at different wavelengths as it passes through the water column within a shallow depth range (<25 m) results in both a decreased ability to discriminate between different underwater features with increasing depth and different spectra being recorded for the same feature at different depths (Green et al., 2005). Several techniques can be used to characterise and correct for the effects of the water column on benthic reflection. These include the use of image waveband ratios to develop depth-invariant images (Lyzenga, 1981), incorporation of light attenuation into linear equations for spectral unmixing of seafloor bottom types (Hedley and Mumby, 2003), multi-step optimisation (Goodman and Ustin, 2007) and radiative transfer modelling (Kutser et al., 2000a, 2000b). Green et al. (2005) review the different methods available for water column correction.

Light undergoes specular reflection as it hits the water surface and a single incoming ray of light is reflected as a single outgoing ray. This mainly occurs during windy conditions that result in an uneven water surface, when the Sun is lower in the sky. Longer wavelengths of light, such as those that lie towards the infrared end of the optical section of the spectrum, do not penetrate the water surface. In some instances they return to the sensor and add speckles of light commonly known as 'glint' to an image. Several algorithms have been developed that use infrared reflectance above deep water to estimate the magnitude of glint and remove it from the signal prior to further processing (Hedley et al., 2005). The application of pre-processing techniques that address the effects of the atmosphere, water column and water surface in light transfer have increased the accuracy to which shallow water marine environments can be mapped using airborne and satellite remote sensing (Mumby et al., 2004).

4.10.2 Classification of Remotely Sensed Images

Classification in remote sensing involves the categorisation of response functions recorded in imagery (i.e. detected light that has reflected from the Earth's surface) as representations of real-world objects. Pixels or collections of pixels known as objects are subsequently placed into user-defined groups (Mather and Koch, 2011). The groups into which pixels are classified depend on a combination of the resolving capability of the instrument employed and the objectives of the mapping exercise. Algorithms for classifying remote sensing images can be divided into supervised or unsupervised. Supervised classifications are guided by externally derived information such as field data (i.e. they are 'supervised'). Unsupervised classifications simply categorise pixels based on the internal statistical properties of an image. These operate in 'feature space', in which the response of an individual pixel is plotted as a function of reflectance corresponding to the different wavelengths of light detected. Pixels are plotted within an *n*-dimensional space that is governed by the spectral resolution of the sensor. Within this space, algorithms group clusters of pixels together based on the principle that similarities in their reflectance properties indicate similar ground-cover types (see Figure 4.10b).

Another important distinction that relates to image classification is whether the unit of classification is a pixel or an 'object'. Object-based classifiers subdivide or segment pixel grids into groups of statistically similar pixels or objects prior to classification. This segmentation process is based on the spectral properties, shape, texture and neighbourhood context of pixels within the grid (Darwish et al., 2003). Segmented objects are subsequently classified according to semi-automated rule sets (Leon and Woodroffe, 2011). Object-based approaches have been shown to improve performance for mapping coral reef environments across different spatial scales (Benfield et al., 2007; Phinn et al., 2012).

4.10.3 Mapping Aldabra Atoll, West Indian Ocean

This case study demonstrates how three bands from a QuickBird remote sensing satellite image can be used to make a digital map of Aldabra Atoll. The map distinguishes between four ground-cover classes: sand (class A), coral (class B), terrestrial vegetation (class C) and deep water (class D). The unique spectral reflectance properties (or spectral signatures) of three of these classes across the optical section of the electromagnetic spectrum (400–700 nm) are shown in Figure 4.8. The positions of the wavelengths of the satellite imagery are shown as vertical lines labelled bands 1 to 3 within the red, green and blue subsections of the spectrum, respectively.

Figure 4.9(a) shows the three raw data greyscale images that make up bands 1 to 3 of the western side of Aldabra Atoll. The three-axis plot of reflectance in each band for all pixels in the image (Figure 4.9b) corresponds to the 'feature space' illustrated in Figure 4.7(d). Clusters of pixels with similar ground-cover characteristics can be seen here, and it is within this feature space that a classification algorithm assigns class types on the basis of each pixel's location. In the case of a supervised classification, the algorithm can be guided by field records collected *in situ*. Ground referencing photographs taken of the seafloor character at specific locations (Figure 4.9b) can be plotted onto the image. Corresponding reflectance values in each of the three bands can then be extracted for each pixel. In this way, a collection of 'training data' can be built up to guide the classification algorithm by delineating approximate areas within feature space that belong to different seafloor-cover categories. The plot in Figure 4.9(b) shows multiple pixels plotted in this way for different seafloor-cover types. The breadth of each pixel cluster in feature space reflects the inherent natural variability in reflectance across the

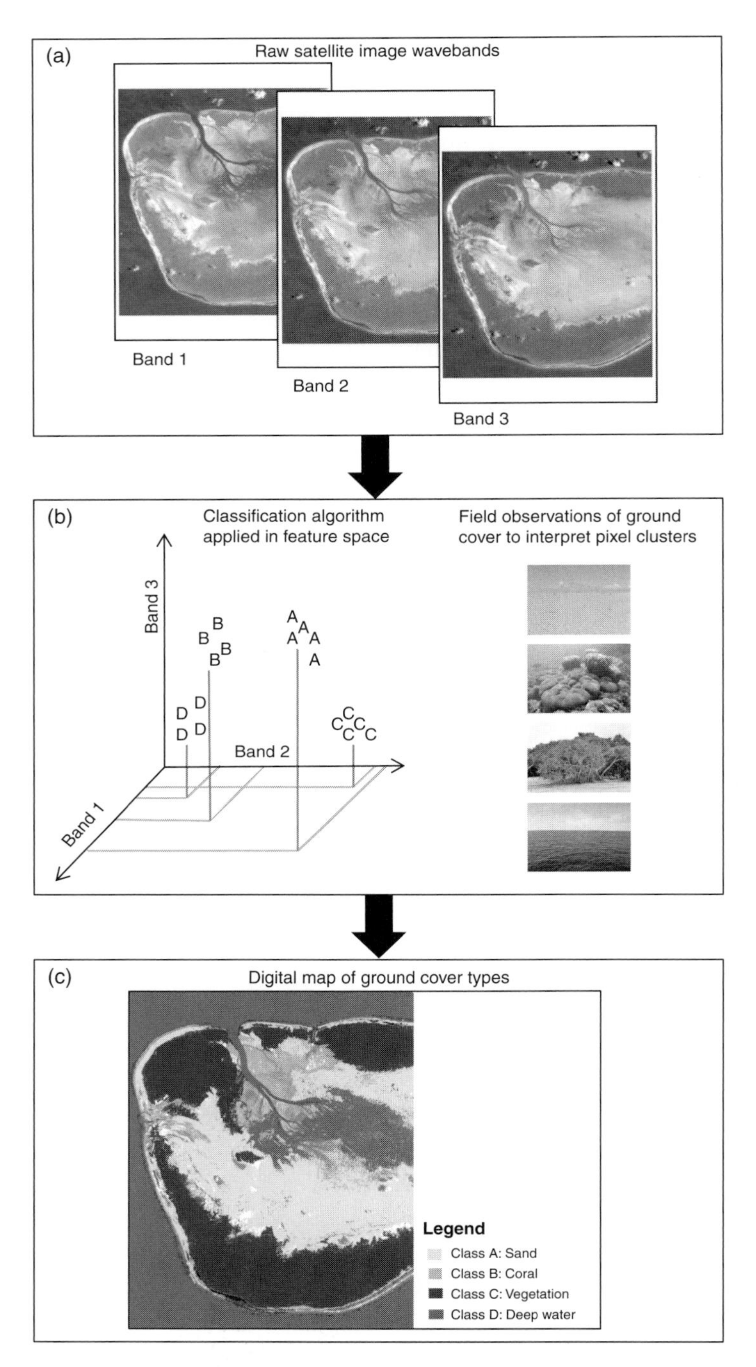

Figure 4.9 A flow diagram illustrating the application of a supervised classification algorithm to the raw bands of a satellite image of the western section of Aldabra Atoll to distinguish four ground-cover classes in a digital map: sand (class A), coral (class B), terrestrial vegetation (class C) and deep water (class D). (A black and white version of this figure will appear in some formats. For the colour version, please refer to the plate section.)

different seafloor-cover surfaces. Groupings may also be differentiated based on environmental factors. For example, deep coral pixels falling in deep water may form a separate cluster to coral pixels in shallow water.

The classified output is a raster image in which each and every pixel has been assigned a nominal class. It is no longer a multilayered dataset of radiance values, but a single layer of nominal thematic classes (see Figure 4.9c). Before this digital map is used, it must be validated against independent *in situ* field records of the different seafloor-cover classes. Such a validation exercise might include some form of accuracy assessment against which the reliability of the map can be gauged (see Chapter 8).

4.10.4 Different Types of Digital Coastal Maps

The level of detail into which coastal environments are mapped ranges considerably, depending on the resolving capability of the instrument employed. Communities are often hierarchically broken down to accommodate user requirements, field data availability and sensor ability (Williams and Lance, 1977). Hierarchical schemes allow some areas to be mapped in greater detail than others due to the availability of more data. For example, top-level descriptions might identify broad geomorphological context, while lower tiers describe relative coverages of benthic communities discriminated from ground verification data. The structure of a combined hierarchical classification scheme that includes both geomorphological and biotic substratum cover is systematic and tier-level descriptors are not used interchangeably. This allows for geomorphological classes (e.g. beach dune) to be coupled with ecological ones (e.g. sedges and grasses). In practice, many coastal mapping strategies employ coarse mapping techniques at the regional scale, augmented by finer descriptive resolution maps at specific locations of interest (Spalding et al., 2001). This aligns well with common scenarios of data availability, whereby some areas associated with features of interest have been more intensively surveyed than others. Similarly, there may be particular zones of management jurisdiction for which more information is available and maps with greater detail are required.

Classification techniques can also influence the level of detail incorporated into mapped schemes. Image-based or 'sensor-down' approaches allow the classification scheme to be driven by the statistics associated with the image itself. By contrast, ground-up classifications use detailed spectra collected *in situ* with field spectrometers to establish an alternative, field-led link between substrate spectra and image data (Purkis, 2005).

The many wavelengths that comprise a hyperspectral image introduce the possibility of expressing the spectral response (or signature) of a pixel as a spectrum. Data manipulation methods can then be applied that treat data as continuous spectra. These include the alternative image classification approaches such as spectral angle mapping and spectral unmixing, and derivative analysis. These emphasise unique shapes to the spectral curves associated with targets using mathematical derivatives and usually result in the production of new, derivative wavebands (Hedley, 2013). Spectrally continuous approaches yield maps that estimate the proportions of endmember coverages (land-cover types). These can be determined from decomposition of the entire spectral curve. Often, several maps can be produced in this way, each corresponding to a different land-cover type. Examples include the use of semi-analytical techniques to map coral, algae and sand in Kaneohe Bay, Hawaii, from AVIRIS satellite imagery (Goodman and Ustin, 2007) and the use of spectral unmixing to map live coral cover at the Al Wajh Bank, Red Sea, from AISA-Eagle hyperspectral data (Hamylton, 2011). This is a fundamentally different map product from one that nominates a single category of dominant ground-cover type for each individual pixel. From a statistical perspective, this is an important distinction because maps produced at a higher measurement level represent much more versatile datasets. These can be input into a greater range of models (Haining, 2003).

In terms of data structures, most maps are commonly provided in raster or vector format (see Chapter 2). Vector maps partition an area in such a way that every location falls into exactly one polygon and is assigned a value that is assumed to be that of the variable for all locations within the polygon. Polygons are therefore limited in their ability to represent heterogeneous data, as they permit no internal variation within the polygon area. Raster layers partition an area into regular grid cells, which are represented as homogeneous for all locations within the cell.

4.11 Fieldwork for Mapping Coastal Environments

Until the development of self-contained underwater breathing apparatus (SCUBA) in the 1940s, field-based observations of coastlines were largely restricted to mangroves, saltmarsh, dune–beach environments, intertidal reef flats and the upper 10 m of the water column. Upon accessing deeper environments, the diversity of seafloor communities became apparent. Marine biologists began to record discontinuous distributions of variously sized organisms living on, in and over highly three-dimensional underwater topographies. The combination of intensive sampling with underwater environments placed inherent limitations on the geographical scope of such field survey methods. This was due to the restricted time available for surveys conducted with the aid of SCUBA. Underwater surveys have historically tended to draw on quantitative plant ecology techniques, such as the use of the quadrat (e.g. Kershaw, 1957; Greig-Smith, 1961). Subsequent concerns over the effects of varying sample size and sampling area meant that quadrat-based estimates came to be usurped by line transect and plotless methods, also derived from terrestrial forest ecology (e.g. Loya, 1972). These remain prevalent today.

Fieldwork campaigns undertaken for the purpose of mapping are commonly referred to as 'ground truthing' or 'ground referencing' remotely sensed imagery. At the most basic level, these must collect spatially referenced records of land surface cover. Records may take the form of field notes, photographs, gridded or categorised entries along transects, video footage (above- and underwater) or notes annotated onto maps (see Figure 4.10). Field survey methods focusing on coastal vegetation such as seagrass and mangroves might also include associated characteristics such as leaf area index, shoot density or standing crop in spatially defined areas, such as quadrats (Mumby and Green, 2000).

All information collected must have accompanying spatial referencing information. This is commonly collected *in situ* with the mobile global positioning system (GPS). The horizontal accuracy associated with commercially available devices has improved over the last 20 years (approximately 4–8 m). Higher accuracies (locational error < 1 m) are possible using differential GPS technology to simultaneously model, and correct for, atmospheric influences on the satellite signal. With the help of navigational devices, GPS technology can be used to track the location of a person in the field and simultaneously display this on a computer screen. This is overlaid onto a spatially referenced map or satellite image in real time. Such an approach offers

Figure 4.10 The author collecting ground referencing information by lowering a video camera down from a boat. Working from a boat platform facilitates rapid access across large areas. Mounting a camera onto a 50 m length of cable allows deeper environments to be surveyed. GPS coordinates of all survey points are recorded to allow comparison with other image datasets. (A black and white version of this figure will appear in some formats. For the colour version, please refer to the plate section.)

the distinct advantage of enabling the fieldworker to visit sites deemed to be of interest on the basis of the remote sensing dataset (e.g. satellite image) itself. This is clearly advantageous, as this is the information that forms the foundation of the map itself.

Ground referencing information taken directly from the field is used to interpret the signal from a remote sensing image in terms relating to the physical properties of the seafloor target. Because remote sensing takes place at a distance from the object of interest, the successful application of classification algorithms to a satellite image requires the use of additional information on the nature of that environment. Field datasets can be compared with the satellite image in one of two ways. They can be employed within the classification procedure as a training dataset for a supervised classification algorithm. Alternatively, they can be utilised as an independent dataset for validation of the classified product. This is an important distinction, because records that are used for calibration play a role in the production of a map. Therefore, they should not subsequently be used to validate that same map product. Using the same record twice introduces a circular logic in which the accuracy of a map will be artificially inflated because the reliability of the final product is being assessed by comparison against a dataset that played a role in its initial production.

As virtually all image classification algorithms are spectrally based, the coupling of imaging spectroscopy with air- or spaceborne platforms provides a powerful tool for coastal mapping. Field spectroscopy involves the collection of information on the spectral characteristics of objects *in situ*, in the form of field spectra (Figure 4.11d). This is of practical value for planning broader-scale airborne remote sensing surveys because it helps to identify the optimal spectral bands and viewing configuration to collect imagery that will best distinguish between different coastal features. It also acts as a tool for the calibration or training of classification models that relate the spectral signatures of coastal features to different land-cover classes (Milton, 1987). Spectroscopy therefore offers a cost-effective and repeatable field survey method for the refinement of remote sensing approaches.

The spectral signature of land-cover features such as sand and seagrass is an inherent property of the features themselves. This remains constant regardless of location from which the signature is collected. A *Sargassum* seagrass signature collected in the Mediterranean Sea could equally be used to map *Sargassum* seagrass in the Pacific Ocean. This means that libraries of signatures can be collected that are of broad use for mapping coastal features using remote sensing. Examples of such libraries include the Ozcoasts Australian Shallow Waters Spectral Library of Seagrass and Macroalgae (GeoScience Australia, 2008) and the spectral library of biotic and abiotic features of Heron Island on the Great Barrier Reef (Roelfsema and Phinn, 2006). Developments in the availability of associated underwater housings and software for downloading and manipulating signatures have led to wider uptake of hand-held spectrometers, such as the GER500, Zeiss MMS-VIS and Ocean Optics models for coastal remote sensing. To integrate field-collected spectra into the analysis of remotely sensed imagery, the spectra need

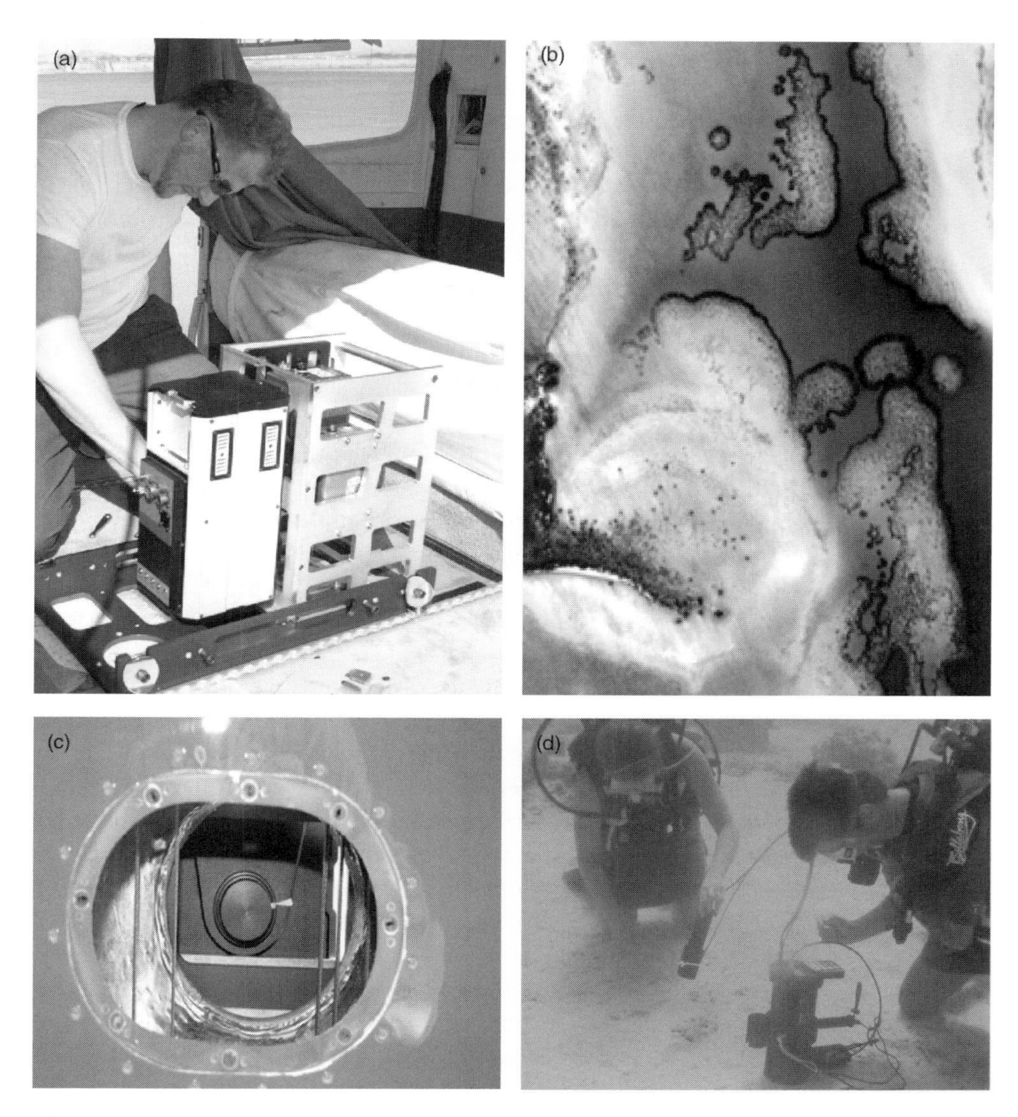

Figure 4.11 The use of remote sensing spectrometers. (a) An AISA-Eagle hyperspectral instrument mounted inside a Cessna seaplane. (b) Image of reef systems in the Red Sea collected by the airborne hyperspectral instrument in (a). (c) Sensor lens as seen through the base of the seaplane. (d) Hand-held spectrometer being operated underwater for the collection of sand spectra. (A black and white version of this figure will appear in some formats. For the colour version, please refer to the plate section.)

to be brought into a common frame of reference with the image wavebands. Either the image must be corrected for the effects of the atmosphere and water column using methods that preserve the signal-to-noise ratio, or the spectra must be 'forward' modelled by artificially adding a water column and atmospheric component to the reflectance signature (Kutser et al., 2006).

A fieldwork exercise to support map production from a remote sensing image must represent a range of potential seafloor characters or coastal sub-environments. As far as possible, information that is collected should be representative of each of the ground-cover types to be mapped. To construct a coastal map that distinguishes between possible saltmarsh, exposed cliff, sand and mud, field records must enable interpretation of the remotely sensed signal of each and every one of the ground-cover types from the satellite. Because both the satellite image and field data are spatially referenced, the geographic coordinates provide a basis for linking the field and image datasets. If a photograph of saltmarsh is plotted onto a satellite image, it is possible to extract the radiance from the corresponding satellite image pixel across each of the wavebands for this ground reference point. This can then be used to inform the classification algorithm. This tells the computer approximate radiance values for saltmarsh pixels across each waveband. Pixels in other parts of the image with similar radiance values across these wavebands will also be classified as saltmarsh. It is therefore useful to anticipate the number and nature of classes in the final map, such that a representative range of ground reference points can be collected to support the classification algorithm across the complete range of ground-cover types. Furthermore, it is necessary to account for the spatial referencing uncertainty associated with the technology in the sampling procedure. If the hand-held GPS unit has a spatial referencing accuracy of ±4 m, then it makes more sense to collect a reference point from the centre of a saltmarsh that is 20 m × 20 m than it would to collect one from the edge of a patch that is 5 m × 5 m. This would allow for a spatial margin of error associated with both the GPS and the locational information embedded within the satellite image. Because coastal environments are very dynamic, additional processes such as wave energy and water depth will give rise to variation in the reflectance characteristics of a given ground-cover type across environmental gradients. It will therefore also be useful to consider representativity in relation to environmental variability, e.g. by collecting ground reference points for sand across a range of water depths.

The geographical scope or area covered by a field campaign is determined by practical considerations such as time available, budget and accessibility of terrain. In the USA, most State and local government mapping agencies produce maps at scales ranging from 1 : 2400 to 1 : 24 000. The scale adopted depends on the intensity of development and length of coastline to be mapped. A descriptive hierarchy might be established in which larger-scale maps (1 : 600 to 1 : 10 000) are produced for areas that require immediate, concentrated management and regulation. Smaller-scale maps (e.g. 1 : 24 000) might suffice in areas less in need of rigorous management (Ellis, 1978). A feature-led approach to mapping is more likely to be dictated by the spatial scales of variation that determine the processes that underpin the features being mapped. For example, if the objective is to map kelp communities along a rocky intertidal foreshore, then the processes that determine their distribution will have an important influence on what geographical area is to be considered. These include the larval pool, physical transport processes, hydrodynamic circulation, substrate availability and biotic interactions of competition and predation (Velimirov and Griffiths, 1979; Sala and Graham, 2002). These will operate across a hierarchy of influences to determine a spatially variable realisation of kelp patterning. Thus, it is possible to be prescriptive by relating the overall geographical scope to the dimensions of relevant processes (see Section 4.4).

4.12 Mapping From Unmanned Aerial Vehicles (UAVs)

An unmanned aerial vehicle (UAV), or drone, is an airborne platform that is flown with no pilot on board and controlled from the ground. UAVs can fly autonomously based on pre-programmed flight plans, or be controlled remotely in real time (Figure 4.12). The potential for UAVs as platforms for collecting remotely sensed imagery above coastal environments has been recognised for a while. For example, flying of kites

Figure 4.12 The use of unmanned aerial vehicles, or drones, to map coastal environments. (a) Example of a pre-programmed flight survey path around southeastern Heron Island, Great Barrier Reef. (b) Mosaic of raw images derived from a UAV survey of southeastern Heron Island. (c) Simultaneous in-water collection of seafloor photographs during a UAV survey. (d) Operating a UAV using hand-held remote controls from the beach at Heron Island. Images used with permission from Dr Karen Joyce and Dr Chris Roelfsema. (A black and white version of this figure will appear in some formats. For the colour version, please refer to the plate section.)

to acquire aerial photography of reef islands (Scoffin, 1982) provides an early example of their use. UAVs or drones are unpiloted platforms that may take the form of kites, helicopters, blimps, balloons or a combination of these (Klemas, 2012). The adaptation of sensors, particularly reductions in the size and weight of hyperspectral cameras, LiDAR, synthetic aperture radar and thermal infrared sensors, has made them suitable for operation from a UAV platform (Klemas, 2015). Advantages of using UAVs as a survey platform include the ability to control survey deployment to coincide with favourable weather windows, frequency of coverage, a greater spatial resolution achieved from closer proximity to the Earth, flexibility in spectral resolution (particularly optimising band placement for the littoral zone) and the ability to adjust operational settings for maintenance and calibration throughout airborne surveys (Malthus and Mumby, 2003). Disadvantages include the relatively poor spatial coverage in comparison to satellite imagery as comparable areas need to be captured in several flights composed of multiple survey lines. They are also dependent on good weather, particularly low winds, as most UAV craft are designed to be light and therefore are affected by the slightest breeze (e.g. the popular DJI Phantom UAV model should not be flown in winds above 15 m s^{-1}). The costs of operating UAV platforms include all of the preparation (including training and licensing), integration with the sensor and ground control. Generally, they are not allowed to fly over populated areas or in air spaces used by other aircraft, although special permission can be sought (Peterson et al., 2003).

Some processes in coastal environments can only be observed with high spatial and temporal resolutions that satellite remote sensing cannot provide (i.e. sub-metre pixels with acquisition intervals of less than a few hours) (Klemas, 2012). These processes include tidal fluctuations, the influence of river plumes on water quality, and movements of sedimentary landforms such as beaches and spits arising from high-energy events such as storms. While field surveys provide *in situ* observations, they cannot sample instantaneously at such a broad scale in a bay or along a beach as airborne sensors could within a short period. To this end, UAVs can be used for a wide range of coastal environmental applications, such as wetland mapping, LiDAR bathymetry, flood surveillance, measuring wave run-up during periods of high swell, tracking oil spills, studies of coastal urban environments, and Arctic ice investigations (Casella et al., 2014; Klemas, 2015). Despite these applications and associated developments that have made UAV technology much more affordable and widespread for hobbyists over the last five years, operational licensing restrictions imposed on the use of drone platforms for commercial image acquisition have limited their uptake among the coastal management and research community. Kites have therefore continued to be used as a platform for detailed mapping.

Recent examples of the application of UAVs in coastal mapping include the use of kites to collect images for mapping vegetation cover and the reef systems around Durai Island of the Anambas Archipelago, Indonesia (Currier, 2015), and the construction of high-resolution, three-dimensional models of intertidal rocky shores using image feature detection and matching, structure-from-motion and photo-textured terrain surface reconstruction algorithms (Bryson et al., 2013). Drone platforms have been used to acquire multiple images over time to enable an assessment of how coastal features are changing and evolving. Examples include the monitoring of boulder movements in intertidal zones of beaches in northwestern Spain (Pérez-Alberti and Trenhaile, 2015), the detection of volumetric changes in sand dunes and dykes following storms in Oleron Island along the Atlantic coast of France (Guillot and Pouget, 2015), and the quantification of shoreline erosion by coupling UAV survey data with hydrodynamic and transport models (Rovere et al., 2014).

Technical advances anticipated in the near future will include the development of automated algorithms for programming UAVs to revisit the sites at regular time intervals at high spatial resolution (Pereira et al., 2009) and continual increases in the spectral specification of sensors, accompanied by reductions in their weight. The drive for cheaper, better and faster airborne imagery has placed a large emphasis on smaller, more cost-effective sensors, and recent designs for CCD cameras provide higher spectral resolution sensors that are substantially smaller, more compact and less costly than their predecessors (Peterson et al., 2003). In combination with relaxations in the licensing restrictions for their operation, this creates exciting new opportunities for enhancing and expanding their application in coastal environments.

4.13 Mapping in Three Dimensions: the Importance of Bathymetry

Bathymetry is related to many important processes in coastal environments as it controls ambient levels of light, photosynthesis in seagrass or kelp beds (Brey, 2009), sediment and nutrient loading, infill of estuaries and accretion of mangrove forests (Swales et al., 2007). Bathymetric variability determines associated seafloor topographic properties such as rugosity, slope, terrain ruggedness, aspect and bathymetric position index (Brown et al., 2011). These properties also determine the biological character of nearshore seafloor communities, and associated populations of fish and invertebrates (Arias-González et al., 2012; Harborne et al., 2012). Information on the bathymetric character of coastal shelves is therefore an important component of most coastal mapping exercises.

4.13.1 Light Detection and Ranging (LiDAR)

LiDAR technology employs active sensors that emit electromagnetic radiation in the visible and near-infrared wavelengths and record return signals as discrete points within the sensor swath (Figure 4.13). There are two types of airborne LiDAR instrument: those with a terrestrial focus and those with a bathymetric (underwater) focus. Terrestrial sensors collect topographic information for applications such as geomorphology, landscape ecology, coastal engineering and volumetric calculations. Bathymetric instruments are designed to penetrate the water column and simultaneously measure water depth and sensor elevation above the water surface. By using two lasers, one in the near-infrared and one in the blue–green section of the electromagnetic spectrum, topographic and bathymetric data can

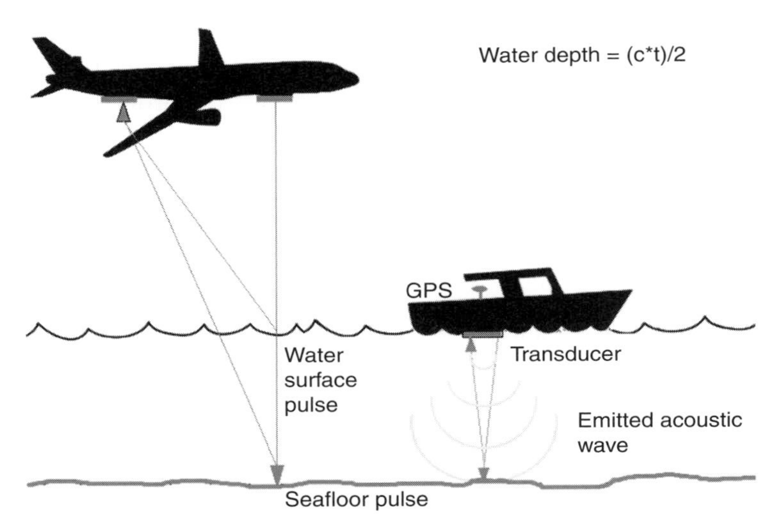

Figure 4.13 A schematic of the principles of LiDAR and sonar survey instrument operation. Water depth can be calculated from the multiple LiDAR pulses from the travel time difference between the water surface and seafloor pulses.

be collected simultaneously (Mather and Koch, 2011). Light pulses from the two different systems can be compared to establish water depths and shoreline elevations, and profile objects on the seafloor. The laser that operates in the green section of the electromagnetic spectrum is designed to facilitate seawater penetration. The lasers are usually mounted inside an aircraft that flies less than 1 km altitude (Figure 4.13).

LiDAR instruments measure the time taken by a light pulse to reach the ground and for a part of the scattered radiation to return to the sensor. Because the emitted electromagnetic energy travels at the speed of light, the calculation of distance, D (m), between the sensor and ground is simple (see (4.1)):

$$D = cT/2 \tag{4.1}$$

where c is the speed of light (2.998×10^8 m s^{-1}) and T is the time (s) between signal emittance and detection of the return. Because coastal LiDAR instruments measure the distance to the water surface and the distance to the bottom of the water body (Figure 4.14), the water depth is calculated as the difference between the two.

Surveys yield information in the form of a point grid of bathymetry measurements. The density of the point measurements will vary depending on the frequency of the sensor employed. These points can be subsequently interpolated to a continuous raster surface of elevation. Key differences among LiDAR sensors relate to the laser's wavelength, power, pulse duration and repetition rate, beam size and divergence angle, the specifics of the scanning mechanism, and the information recorded for each reflected pulse (Lefsky et al., 2002). Purkis and Brock (2013) define three types of LiDAR instruments as profilers, discrete return and waveform sensors. Profiling sensors record a single return at a given sampling density within the swath, while discrete return sensors record multiple returns (typically around five) per pulse of light emitted. Waveform instruments record the most information in the form of a digitised time-varying intensity profile of the full return pulse.

The objective of waveform analysis is to interpret the continuous form of pulses that correspond to vertically graduated surfaces from which the light may have been reflected. Additional information about the target can be inferred from variations in properties such as intensity. The intensity of the laser pulse returned after intercepting a morphologically complex surface, such as a mangrove vegetation canopy, will be a combination of reflected energy from surfaces at numerous distances from the sensor (see Figure 4.14).

Although typically used to map elevation, LiDAR data can also provide information on coastal geomorphology, rugosity, texture and bedform geometry (Purkis and Brock, 2013). Until recently, coastal geomorphologists have traditionally based studies of beach erosion and accretion, sediment transport and budgets, and island movement dynamics on repeated in-situ topographic profiling and shoreline mapping. The

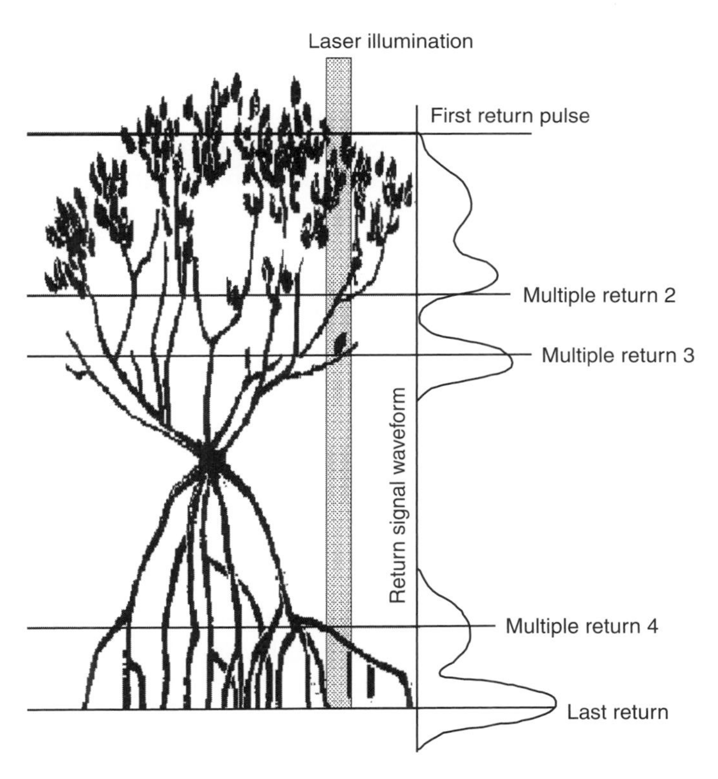

Figure 4.14 Illustration of the differences between waveform and discrete LiDAR returns from a mangrove shrub. The hypothetical return signal waveform in the centre shows peaks that correspond with the multiple returns from the vertically graduated canopy, branches, roots and ground surfaces.

development of LiDAR mapping has allowed analysis of beach and dune microtopography along broad swaths that span the land/water interface. Repeat surveys allow volumetric change analysis and the quantification of local sediment budgets (Liu et al., 2007).

4.13.2 Sonar, Single- and Multi-Beam Echosounder

Optical remote sensing technologies are limited to shallow, clear waters, whereas acoustic methods are more commonly used in deeper environments (i.e. in water depths exceeding 20 m). Acoustic methods for profiling underwater topography are based on measurements of the speed of sound in water. As with LiDAR, they are active sensors that record the travel time between emission and reception of a sound pulse. Sensors emitting sound pulses are either towed behind or mounted to the hull of a boat (Figure 4.13). Acoustic signals vary in strength, width and pulse orientation. They can be interrogated to provide further information on the water body, sediment surface or the seafloor sub-bottom (sediment or bedrock) interior. Sound waves also carry energy, although they have very different transmission properties from electromagnetic radiation (e.g. light). Energy is transported by waves that propagate in a periodic vibrational state (Benenson and Stöcker, 2002). This mechanical disturbance can propagate up to thousands of kilometres through water. The sound pulse itself takes the form of a longitudinal wave made up of localised regions of compression and rarefaction occurring along the direction of wave propagation (Riegl and Guarin, 2013).

The extent to which sound gets absorbed in seawater is a function of the frequency of the acoustic pulse. Sound-based instruments usually operate within the frequency range between 1 Hz and several hundred KHz (Tolstoy and Clay, 1966). Higher-frequency waves give rise to faster vibration of seawater molecules. These take more energy from the acoustic wave that is absorbed faster. It therefore travels comparatively short distances. The average velocity of sound in water is approximately 1500 m s^{-1}, although this varies with temperature, density, salinity, depth and environmental conditions (Jones, 1999). A sound–speed profile can be detailed by plotting speed as a function of depth, which is specific to a particular location and is commonly obtained by lowering a CTD (conductivity, temperature, depth) instrument through the water column.

Figure 4.15 provides a broad overview of how acoustic remote sensing works. One of the primary goals of sonar (sound navigation and ranging) technology is to capture signals that can be usefully differentiated from background noise. Background noise might come from biological or thermal underwater sources, or effects at the surface such as wind, waves, rain or shipping. Echosounder surveys are planned in such a way that the line orientation and swath width provide complete coverage of an area at the required spatial resolution. Often this takes the form of a series of parallel lines crossed by perpendicular 'tie' lines for validation. Transducers convert electrical impulses to sound – they generate and transmit the initial acoustic waves and receive reflected pulses. They can be a single unit, or in a linear, planar or volumetric array configuration. Calculations are performed to account for the various angles at which signals were detected and any tidal fluctuations that may have occurred during the survey period. The signal is then processed to enhance the signal-to-noise ratio before

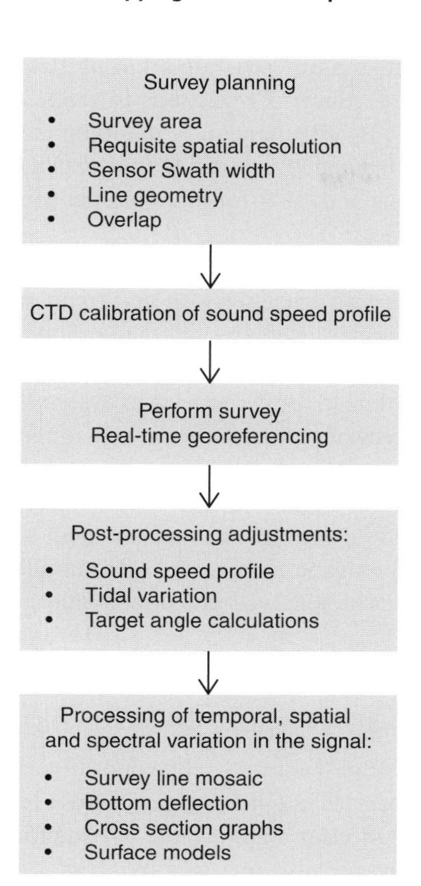

Figure 4.15 Schematic overview of the processes involved in conducting an acoustic seafloor survey. See text for a detailed description.

any output signals exceeding a user-defined threshold are recorded. Finally, bathymetric map products are generated, such as cross-sectional profiles, topographic three-dimensional bathymetric models of the surface area, seafloor mosaics of survey lines or volumetric representations (Riegl and Guarin, 2013).

Four instruments commonly used for acoustic bathymetric survey are single-beam echosounders, multi-beam echosounders, side-scan sonar and interferometric sonar. As with LiDAR, single-beam echosounders harness the time differential between the sent sound pulse and received echo, to measure distance from the seafloor. They are often set up with multiple transducers operating at different frequencies for shallow and deep work. Beam spreading occurs as sound pulses spread out in a conical shape after leaving the transducer. This causes measurement error for larger footprints associated with further away and smaller objects. Because single-beam echosounders yield a series of point measurements, like LiDAR datasets they are often interpolated to produce a continuous bathymetric representation of the seafloor (see Section 6.12).

Complete profiling of the seafloor can be achieved using swath methods, which use an arrangement of single-beam transducers to make up a wider swath. Multi-beam sonar systems use beam forming and steering to generate a fan of sounding pulses originating from a single centralised transducer. Corrections in range and bearing can be made to record off-track depths. By simultaneously mapping more than one location on the seafloor at a time, a swath can be generated that increases the area surveyed per line. This reduces the ship time and costs for mapping the equivalent area of seabed with a single-beam echosounder.

Ship-mounted or towed side-scan sonar instruments map seafloor topography. As with echosounders, they are configured with a beam width and frequency suited to the water depth. Scanners provide an underwater view of the seafloor from above by emitting two fan-shaped beams from a transducer that is mounted from a towed platform. By mounting transducers in linear arrays, side-scan systems are used as imaging devices to detect textural features of the seafloor, such as roughness associated with strong scatter. Instead of providing depth information, images are primarily represented as scaled backscatter coefficients (the ratio of the intensity of sound scattered per unit area and the intensity of the incident plan sound wave). Interferometric sonar systems combine acoustic bathymetry with simultaneous side-scan measurements to provide two co-registered, well-positioned images of bathymetry and backscatter.

4.13.3 Estimating Bathymetry From Satellite Images

The last two decades have seen the development and refinement of procedures for estimating water column depth from optical satellite imagery across large, shallow (water depth < 30 m) platforms through clear water. Table 4.4 summarises developments in estimation techniques, which are also reviewed by Dekker et al. (2011). Methodologies can be broadly subdivided into empirical approaches (Clarke, 1966; Lyzenga, 1981; Philpot, 1989) and optimisation approaches (Lee et al., 1998; Lee et al., 1999; Goodman and Ustin, 2007; Klonowski et al., 2007). Empirical approaches assume that light becomes attenuated exponentially with depth through ocean surface waters. They manipulate the varying extent to which attenuation occurs at different wavelengths to derive estimates of water depth. The manner in which light becomes attenuated is captured mathematically in a 'coefficient of attenuation'. This is dependent on local water quality. It can be derived empirically by comparing some independently collected *in situ* depth data with a combination of different wavebands in a satellite image. Once these coefficients have been estimated, they are assumed to be constant across the entire image.

Optimisation approaches deconstruct measured spectra from satellite images into components relating to water depth, bottom type and water quality. Instead of empirically deriving the coefficients of attenuation, a more constrained system is adopted that defines physical ranges for the spectral attenuation of the water column based on knowledge of the nature of typical water constituents. The optimisation method allows variation in water optical properties from pixel to pixel across the image. A range of properties, including seafloor cover and water depth, are deduced for every pixel from the best optimisation model that fits within user-specified constraints.

Table 4.4 Developments in the derivation of shallow water bathymetry from remote sensing imagery.

Bathymetric mapping development	Reference
Range of depths to which light penetrates through the water column at different wavelengths employed to map 'depth of penetration' zones for a given water quality	Jupp (1988)
The effects of absorption and scattering in the water column are addressed through pairwise correction of wavebands to derive a single depth-invariant index from the ratio of the radiance of two independent bands	Lyzenga (1978, 1981)
A relationship between the ratio of two water penetrating waveband pairs and water depth is established. Linearised (natural logarithm transformed) reflectance values of two image bands are plotted against each other to determine the extent to which the ratio of the two (i.e. gradient) varied with depth, independently of bottom albedo	Stumpf et al. (2003)
A semi-analytical model developed for shallow water remote sensing based on radiative transfer	Lee et al. (1998)
An inversion optimisation approach developed to simultaneously derive water depth and water column properties from hyperspectral data	Lee et al. (1999)
A semi-analytical model incorporating the assumption of constant water optical properties across the scene	Adler-Golden et al. (2005)
A semi-analytical model that includes a quantitative comparison of model-derived depth with high-resolution multi-beam acoustic bathymetry data	McIntyre et al. (2006)
Linear unmixing of the benthic cover incorporated into semi-analytical approaches	Giardino et al. (2012), Goodman and Ustin (2007), Klonowski et al. (2007)
Look-up table approach for matching image spectra to a library of spectra corresponding to different combinations of depth, bottom type, and water properties	Louchard et al. (2003), Mobley (1994), Lesser and Mobley (2007)

4.14 Mapping the Low Isles, Great Barrier Reef, Australia

Low Isles is a low wooded island in the middle of the Great Barrier Reef that is bordered by a windward shingle ridge on a horseshoe-shaped coral reef. This provides protection from storms and cyclones. A sand cay has developed near the leeward reef edge, and the relatively high reef platform has been colonised by mangrove forest (Steers, 1929). On a global scale, Low Isles is believed to be 'unique in the study of islands' (Fairbridge and Teichert, 1948, p. 67) because of the extensive history of mapping activity that has taken place there. This rich history is due in part to the fact that Low Isles provided an easily accessible base for the 1928–29 *British Museum Great Barrier Reef Expedition* and the 1973 *Royal Society and Universities of Queensland Expedition* to the northern Great Barrier Reef. The collection of maps that have been produced of this island provides insight into a range of approaches historically applied to mapping reef islands (summarised in Table 4.1), as well as more recent technologies. Figure 4.16 illustrates six different maps that have been produced of Low Isles using different fieldwork techniques.

Figure 4.16(a) is a 'physiographical sketch map' (Steers, 1929) outlining the major features of the low wooded island. This was produced for the purpose of discussing island formation by E.C. Marchant during the 1928–29 *Great Barrier Reef Expedition*. The map was made using a plane-table made from a drawing board mounted on a camera tripod with an alidade sighting instrument, ruler and compass (Debenham, 1936, 1956; Speak, 2008). The island was sketched by placing the table at a series of locations that afforded views of landscape characteristics, such as a lighthouse, beach rock or mangrove stands. The table was set in relation to a reference point and multiple 'rays' were recorded (lines marked along the sight rule) from set positions to mark both the direction and distance of an object of interest. This method compared favourably to others for speed for a given order of accuracy and dependence on the assistance of other workers (Debenham, 1936). Difficulties included the reliance on correct orientation of the table at each sighting station and the limited vertical range of sighting alidades. Such a basic set-up commonly produced a map that lacked correspondence between three or more ray intersections for a given point of interest. This raised questions about the accuracy of this technique, which was supplemented with a theodolite to enable more accurate triangulation of finer detail (Debenham, 1937).

Figure 4.16(b) was produced by M.A. Spender, also during the 1928–29 *Great Barrier Reef Expedition*. Theodolite triangulation methods were used alongside the first aerial photographs of Low Isles, flown in September 1928 at a scale of 1 : 2400. This map is the earliest known record of ground observations being cross-referenced against aerial photography for the purpose of mapping coral reefs, and this revolutionary technology is doubtless responsible for the remarkable level of detail incorporated into this map. The use of aerial photography for detailed reef mapping was not widely adopted until later (Steers, 1945). A theodolite measures angles at high precision from a mounted telescope which moves along both a horizontal and vertical plane. Precise measurements of angles can be made along these planes when the telescope is pointed towards features of interest (Ganci and Shortis, 1996). Triangulation uses a framework of points and lines to impose order on haphazard features, such as meandering coastlines and beach ridges. These are removed from the finished map product. The curve of a coastline is sketched by superimposing sighted lines over natural features. One practical advantage of this approach is that the position of an emergent coral boulder can be accurately established across an inaccessible reef flat by sighting the boulder from either end of a fixed baseline and measuring angles towards it. Subsequent plotting of the points to scale, or use of trigonometry, can determine the exact boulder location. This simple process of triangulation is fundamental to field mapping. It is through

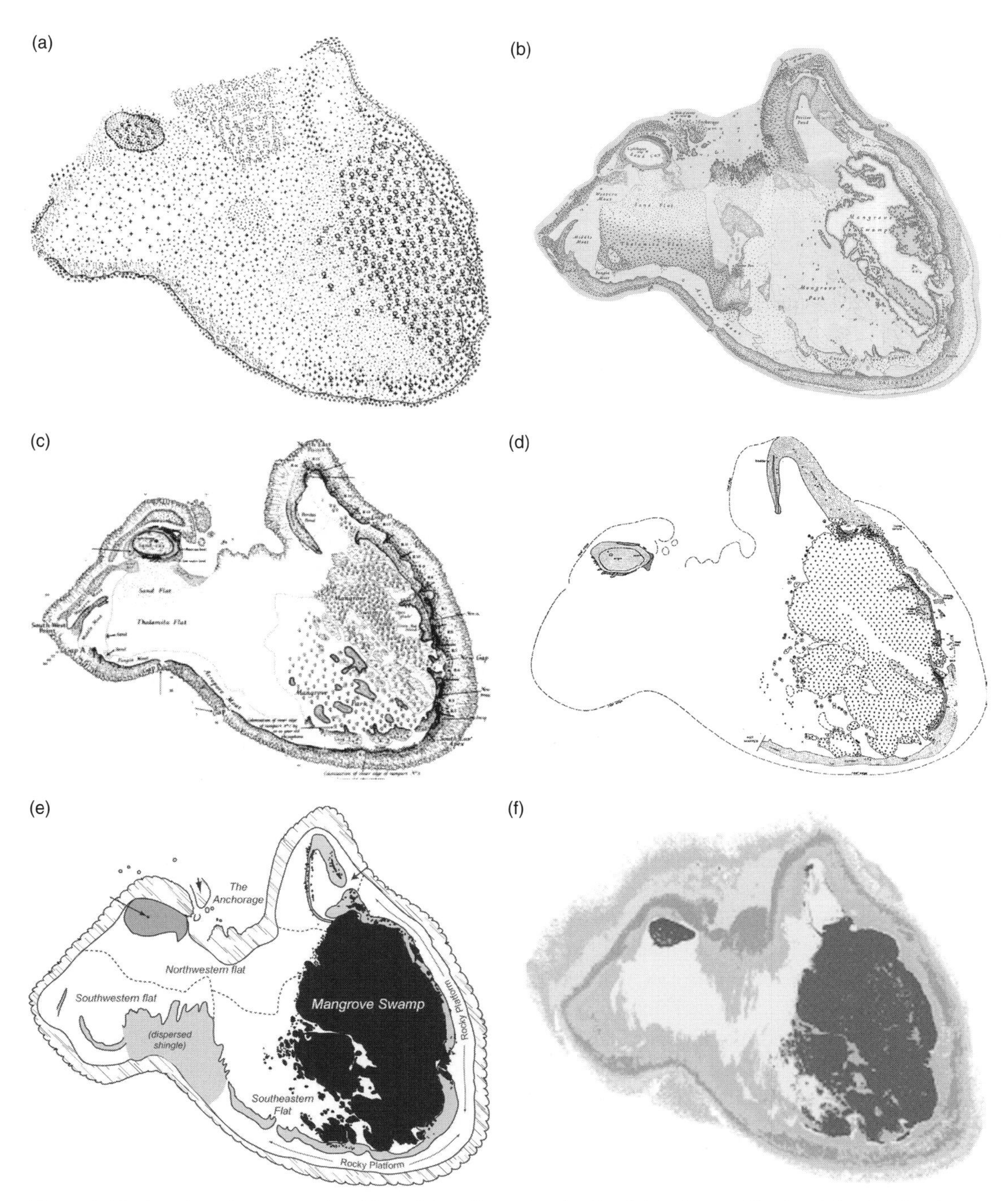

Figure 4.16 A series of maps made of Low Isles, Great Barrier Reef: (a) plane-table sketch by E. Marchant (Steers, 1929); (b) theodolite triangulation plane-table sketch by M. Spender (Steers, 1929); (c) aerial photograph trace by Fairbridge and Teichert (1948); (d) compass-traverse survey by D. Stoddart (Stoddart et al., 1978); (e) aerial photograph sketch by Frank and Jell (2006); and (f) satellite image classification by the author (Hamylton, 2014, unpublished). (A black and white version of this figure will appear in some formats. For the colour version, please refer to the plate section.)

the build-up of substantial triangulation networks that large-scale land surveying is conducted.

The advent of aerial photography revolutionised coastal mapping. Its application to reef islands made use of extensive aerial surveys that were undertaken by the military in the Second World War (Teichert, 1995). Low-angle, oblique aerial photographs charted beach landings and located small Pacific islands. Although these images were acquired for military purposes, they began to be adopted widely for scientific studies that tracked morphological changes to shorelines, the structure and orientation of platform reefs and the patterns of surface waves as they became refracted around islands and headlands (Steers, 1945). For the first time, synoptic information on land surface cover could be combined with features plotted from triangulation networks to generate maps documenting both continuous information on the nature of ground surface cover and the boundaries delineating zones of different cover types. Continuous zones of shingles, boulders, sand and mud could be identified across a wide and shallow reef flat, while features such as the outer boundary of the reef perimeter, channels and shingle ramparts could be fixed by triangulation (Figure 4.16c and e). Detail was transferred from the aerial photograph to the map by tracing discernible boundaries by hand from the air photograph to the draft map once the scales had been aligned.

Remote coastal environments were often mapped using a technique known as a compass traverse survey (Figure 4.16d). The direction and distance from a known position to an unknown location are measured during a traverse, which proceeds in a loop. The final measurement is an extension from the end point back to the beginning, such that a closure error can be calculated (Debenham, 1937). Field notes list landforms, bearings and distances obtained through calibration of the pacing distance. Features can be plotted either directly with a protractor or through computation onto a rectangular coordinate grid established by two axes. Once a network depicting the major landform features has been established, additional detail, such as beach sediment composition and vegetation around the island margins, could be added in. This information might have been recorded during the initial traverse, or collected later through a series of transects normal to the shore (Stoddart, 1962, p. 129, Appendix 'Surveying').

The most recent mapping initiative at Low Isles developed a high-resolution digital map of the entire low wooded island site, encompassing the submerged outer reef periphery, shingle ramparts and mangrove swamp. This was achieved by applying a supervised maximum likelihood image classification to a multispectral WorldView-2 satellite image of the island (spatial resolution 2.4 m). Image analysis was supported by the collection of a ground referencing dataset covering the complete island system (Figures 4.16f and 4.17). The ground referencing dataset comprised a series of point locations at which a snapshot (approx. 30 s) of video footage was taken. This was subsequently viewed and the benthic seafloor character was interpreted.

Figure 4.17 illustrates a more contemporary approach to storing and sharing information on the character of the seafloor around Low Isles. Information from a ground referencing exercise has been loaded into an online Web viewer and hosted via the open-source Google My Maps application. One advantage of using this approach is that multiple users can interact with the information. They can pan around

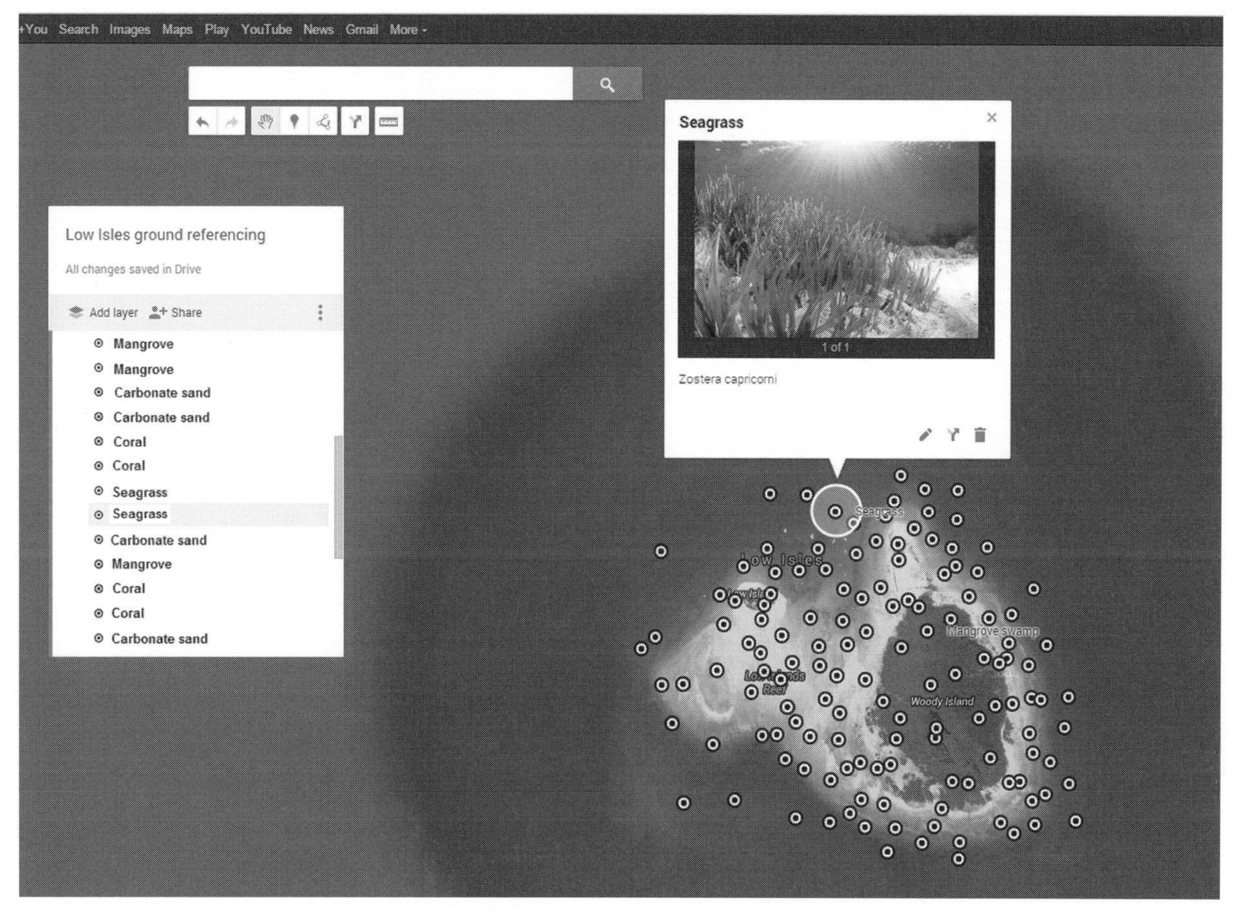

Figure 4.17 A Google My Maps project developed to host ground referencing data points collected at Low Isles on the Great Barrier Reef, illustrating associated photo records of seafloor cover and field notes. (A black and white version of this figure will appear in some formats. For the colour version, please refer to the plate section.)

the map, zoom in and interrogate points of interest. For each of the 135 ground referencing points, it is possible to access photographs and further notes, as shown in Figure 4.17. The selected point shows the species of seagrass found in that location, along with a photo of that species. Ground referencing points are overlaid onto a base map satellite image hosted by Google.

It is important to distinguish between the combination of satellite imagery and field data used to generate the map illustrated in Figure 4.16(f), and the provision of spatially referenced information without necessarily generating a map, shown in Figure 4.17. The two often occur simultaneously because digital maps can be posted online to be interactively interrogated and downloaded by users. To make a map from a remote sensing image, it is necessary to first acquire a full satellite image dataset, often through a commercial image supplier. This allows the user to access data underlying the picture, i.e. the full collection of wavebands that comprise the image. In turn, they are able to perform a classification (see Section 4.10.2). This is not the case when spatial information is overlaid onto an image base map.

The satellite image classification shown in Figure 4.16(f) was achieved by using half of the ground referencing points to train a maximum likelihood classifier. This was then able to statistically distinguish the ground-cover types associated with each pixel, on the basis of eight bands, across an entire Worldview-2 satellite image of Low Isles. The remaining half of the ground referencing points was then used to validate the output map. The resulting digital raster map could then be uploaded to an online Web atlas, or spatial data infrastructure to maximise usage by coastal management authorities (see Section 9.7).

4.15 Summary of Key Points: Chapter 4

- Digital maps are valuable datasets that can be explored for patterns and interrogated for clues as to how and why the features are arranged in space. Maps can be used as evidence to inform scientific debate and management decision-making.
- When defining a geographical area to be mapped for spatial analysis, it is useful to do so with respect to the spatial dimensions of the processes or features being analysed.
- It is important to understand how maps were historically produced, as they represent baseline information against which assessments of coastal change can be carried out. Historical mapping methods include theodolite triangulation, plane-table survey, compass traverse, tacheometry, optical rangefinder and horizontal sextant methods.
- Remote sensing represents a key modern source of raster-based spatial information. It is derived from a sensor that is not in close proximity to the object of interest.
- Optical remote sensing is commonly used to map shallow water coastal environments because light in the visible portion of the electromagnetic spectrum penetrates clear water. Coastal mapping varies markedly in terms of descriptive and spatial resolution, and accuracy.

4.15 Summary of Key Points: Chapter 4

- Remote sensing images hold information in three domains or 'spaces' (image space, spectral space and feature space). Each of these domains provides useful information for spatial analysis.
- Remote sensing images need to be pre-processed to account for the effects of the atmosphere, water surface and water column before they can be interpreted in terms of seafloor features to make coastal maps.
- Conducting fieldwork in coastal environments can be challenging because of weather, tides and the need to work underwater.
- LiDAR is a form of active remote sensing that uses pulses of light to map above- and underwater topography from a plane.
- Sonar is another form of active remote sensing that uses pulses of sound underwater to map topography from a range of different sensors from a boat hull or towed platform.
- Bathymetry (water depths) can be estimated from satellite images and field data.

5 Monitoring Coastal Environments

5.1 Monitoring Change Over Time

Shorelines are continuously adjusting in response to winds, waves, tides, sediment supply, changes in relative sea level and human activities. These processes change the position of the shoreline over a variety of timescales. These range from the daily and seasonal effects of winds, waves and tides to longer-term changes in sea level over centuries to thousands of years. Shoreline changes are not constant, and frequently switch from negative (erosion) to positive (accretion) and vice versa.

According to the Population Division of the Department of Economic and Social Affairs (DESA) of the United Nations, the world population reached 7.2 billion in mid-2013. This is projected to increase by almost 1 billion people within the next 12 years, reaching 8.1 billion in 2025, and to further increase to 10.9 billion by 2100 (DESA, 2013). It is estimated that, in 2010, 620 million people globally lived within the coastal zone, and that, by 2060, this will have grown to approximately 1400 million (Neumann et al., 2015). Of the 16 megacities with a population exceeding 10 million people, 12 are located on the coast (McGranahan et al., 2007). Such population densities invariably lead to an increased presence of human settlements, industrial activities, tourism, agriculture, aquaculture, fishing and pollution. The findings of the most recent Intergovernmental Panel on Climate Change (IPCC) report indicate that the Earth's climate is changing at unprecedented rates and that increased atmospheric concentrations of greenhouse gases and aerosols generated by human activity are leading to substantial environmental changes in the coastal zone. These include alterations to the frequency and intensity of regional extreme weather events such as storms and cyclones, coastal erosion, sea-level rise, elevations in sea surface temperatures, and chemical changes such as acidification of ocean surface waters (Stocker, 2014).

At the heart of attempts to manage, mitigate or adapt to increasing levels of coastal change lies our ability to monitor them continuously. This monitoring may be necessary across large geographical areas. In some cases, change may be expected and considered desirable; for example, the rejuvenation of coastal dune vegetation following implementation of a conservation measure. In other situations, change might indicate a problem to be addressed; for example, inadvertent algal blooms as a result of pollution. Whether change is considered desirable or not, decision-makers need reliable information to make informed judgements regarding policy and actions in order to manage changes to the coastal environment.

Monitoring coastal environments involves the collection of continuous, systematic observational records over a period of time, so that a picture can be established of how those environments are changing. Historically, *in situ* observations have provided important data on the form of coastal environments. Since the widespread adoption of remote sensing instruments for mapping the coastal zone, this technology has emerged as an ideal tool for monitoring changes in coastal environments. This is because it offers several practical advantages over surveying changes directly in the field. In a survey of 60 managers and scientists throughout the world, the detection of change in coastal environments was found to be the most common application of remote sensing for coastal management (Green et al., 1996).

5.2 Observing Change Over Time

Change detection is the process of identifying differences in the state of an object or phenomenon by observing it at different times. Assessments of change across multiple images relating to different points in time are variously referred to as time-series analysis, change detection and multi-temporal image analysis (Goodman et al., 2013). Figure 5.1 provides a schematic overview of the general procedures involved in a change detection exercise. Raster-based approaches compare two images or maps to produce continuous change estimates. Vector-based approaches focus on specific features of interest, which are digitised as point, line or polygon features and subsequently overlaid onto features from another point in time. The choice of whether to employ a raster- or vector-based approach will depend in large part on the nature of the changes to be observed. Remotely sensed Earth observation data are ideally suited as a source of information for change detection exercises. This is because they are consistent in quality (resolution, completeness, accuracy and error; see Chapter 8 for a discussion of spatial data quality). They also provide repetitive coverage for a given area of the ground at short intervals.

To assess change in coastal environments over time, it is necessary to have a baseline dataset against which change can be measured. The earliest European records of many coastal features date back to the eighteenth century and are often associated with charts made during times of exploration and discovery. For example, in the late eighteenth century the cartographic expeditions of Captain James Cook in the *Endeavour* across the Pacific represent some of the earliest coastal maps for these islands. Early hydrographic charting was undertaken using a navigational technique known as 'dead reckoning' (Crone, 1966). This procedure estimates the position of landforms in relation to a previously determined position. The estimated position is advanced on the basis of known or estimated speeds travelled over a known elapsed time and bearing. Since the position of new landforms is calculated solely on the previously surveyed position of neighbouring landforms, any errors in the process are propagated along the coastline. Hydrographic surveyors accounted for this by accompanying these relative measurements with an absolute determination of position using celestial means. A sextant

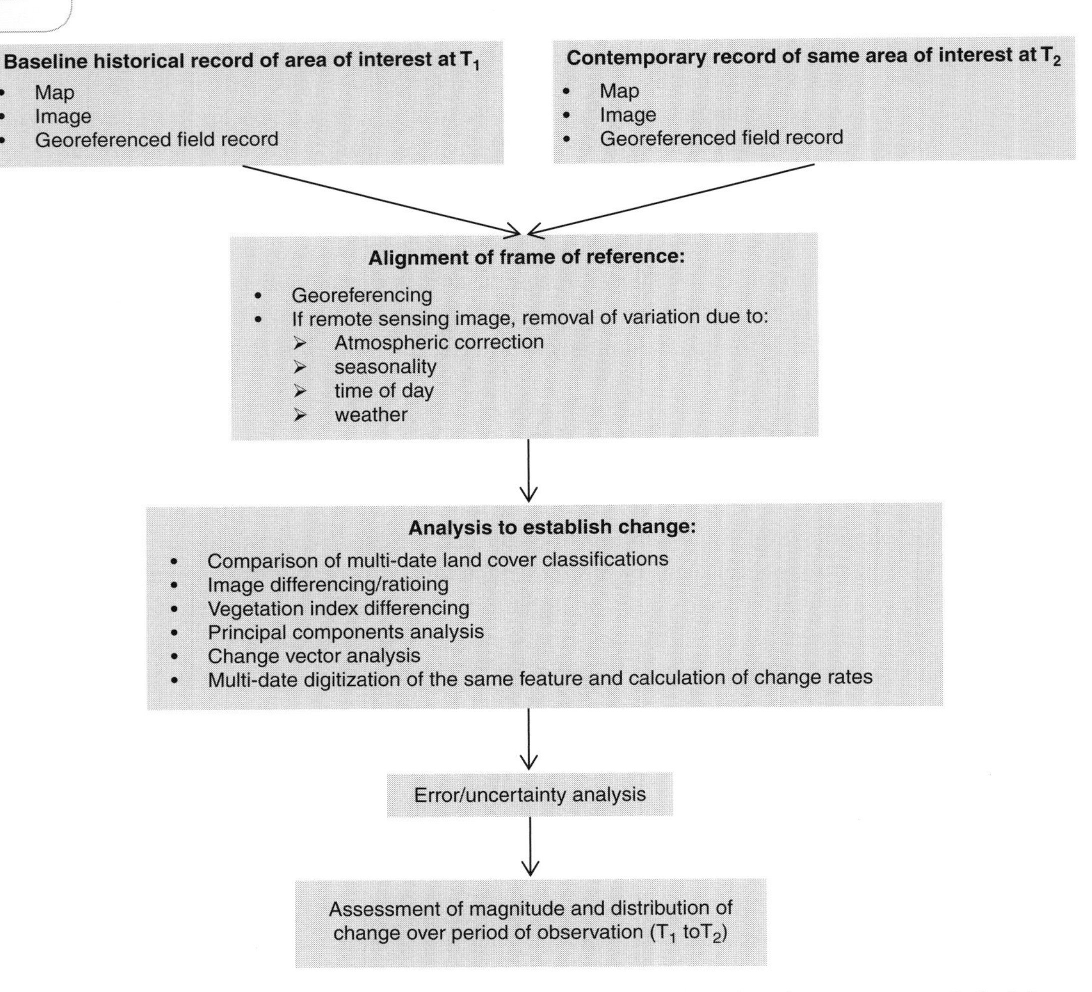

Figure 5.1 A schematic overview of the procedure for detecting change over a period of time (between T_1 and T_2) using multi-temporal spatially referenced datasets.

was used to track the movements of the stars and determine an accurate measure of longitude (Sobel, 2007).

Figure 5.2 is an extract of the east Australian coastline as mapped by James Cook and published shortly after the return of the *Endeavour*. The route that Captain Cook followed, which was annotated onto the map, is emphasised by the darker (blue) line. The black stars that were also placed into the detail of the map record where longitude was established by astronomical observations. Periodic measurements of absolute longitude were important given the use of dead reckoning and the potential introduction of inaccuracies through prevailing winds, the East Australian Current and the presence of the Great Barrier Reef system, which prevented the ship from carrying out sightings closer to the Queensland coastline.

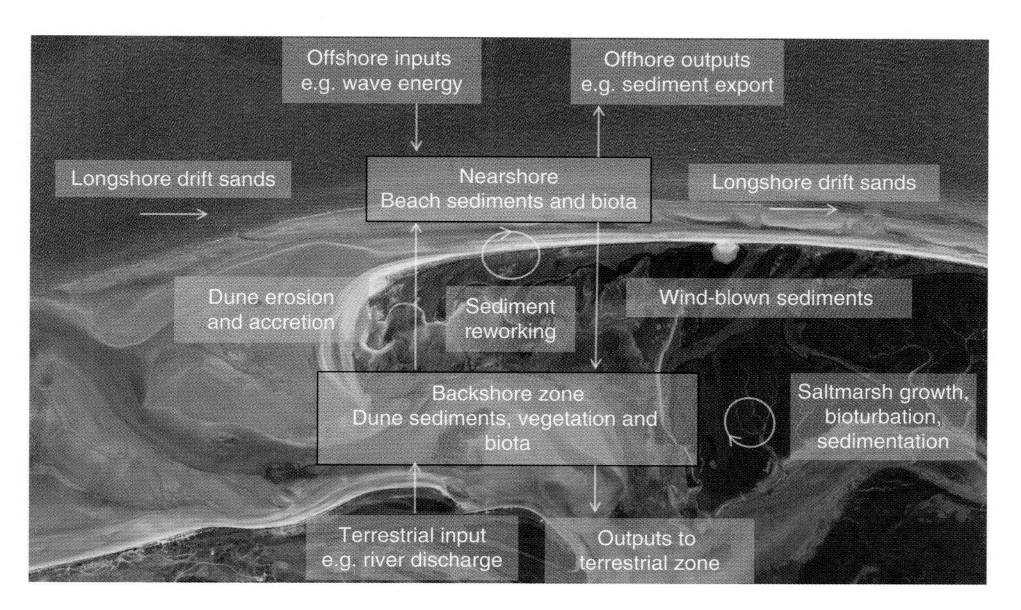

Figure 1.5 A simple coastal biogeomorphological system.

Figure 1.8 A conceptual model of sediment accretion on Bora Bora (French Polynesia), identifying the different components (sources, sinks) of the sedimentary system and processes linking them (production, breakdown, export, transport) around different sub-environments of the atoll lagoon (terrestrial, seagrass, coral, mangrove).

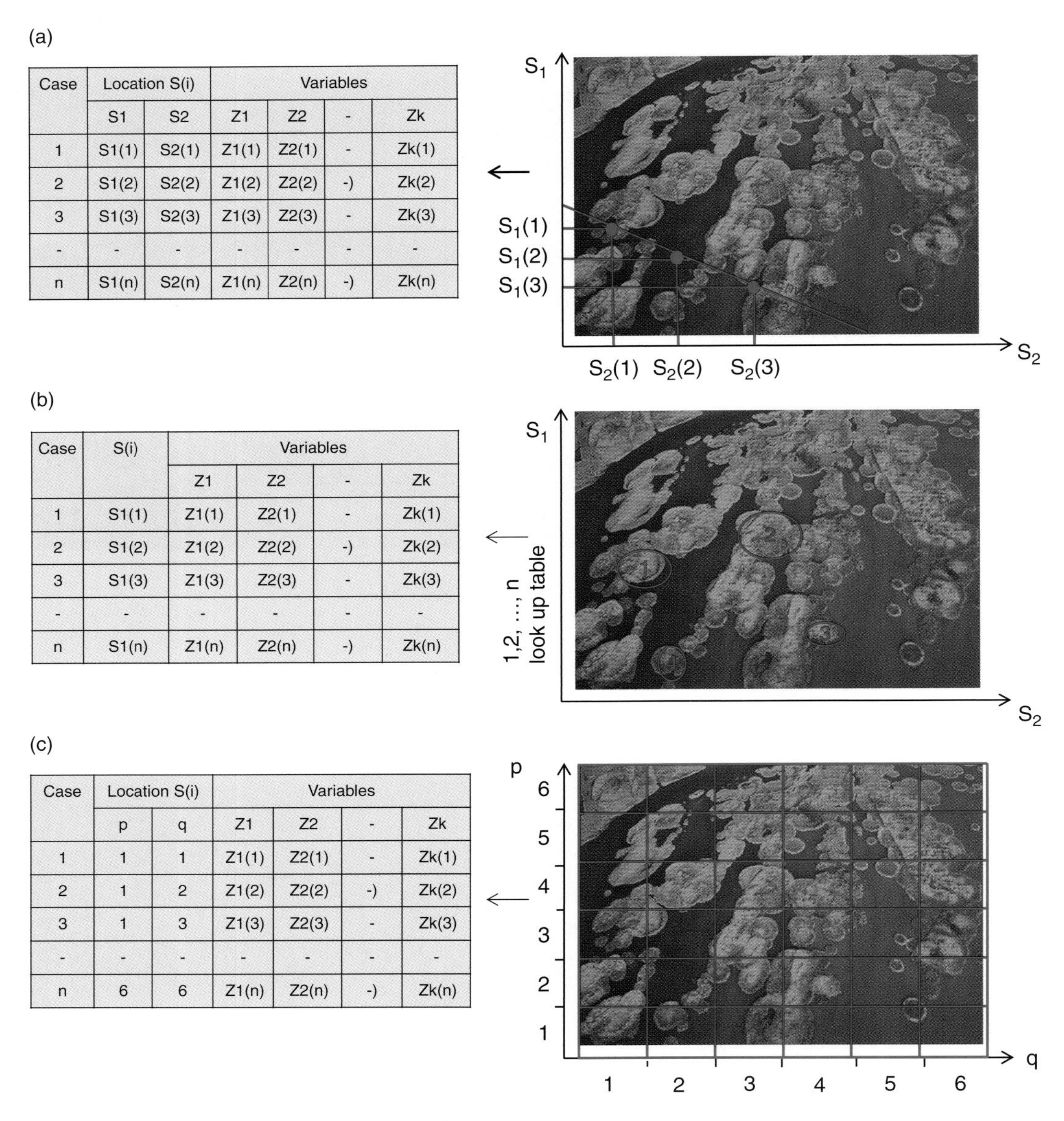

(a)

Case	Location S(i)		Variables			
	S1	S2	Z1	Z2	-	Zk
1	S1(1)	S2(1)	Z1(1)	Z2(1)	-	Zk(1)
2	S1(2)	S2(2)	Z1(2)	Z2(2)	-)	Zk(2)
3	S1(3)	S2(3)	Z1(3)	Z2(3)	-	Zk(3)
-	-	-	-	-	-	-
n	S1(n)	S2(n)	Z1(n)	Z2(n)	-)	Zk(n)

(b)

Case	S(i)	Variables			
		Z1	Z2	-	Zk
1	S1(1)	Z1(1)	Z2(1)	-	Zk(1)
2	S1(2)	Z1(2)	Z2(2)	-)	Zk(2)
3	S1(3)	Z1(3)	Z2(3)	-	Zk(3)
-	-	-	-	-	-
n	S1(n)	Z1(n)	Z2(n)	-)	Zk(n)

(c)

Case	Location S(i)		Variables			
	p	q	Z1	Z2	-	Zk
1	1	1	Z1(1)	Z2(1)	-	Zk(1)
2	1	2	Z1(2)	Z2(2)	-)	Zk(2)
3	1	3	Z1(3)	Z2(3)	-	Zk(3)
-	-	-	-	-	-	-
n	6	6	Z1(n)	Z2(n)	-)	Zk(n)

Figure 2.1 Assigning locations to spatial objects within a riverine marsh landscape: (a) points, (b) polygons and (c) pixels stored in the format of the spatial data matrix.

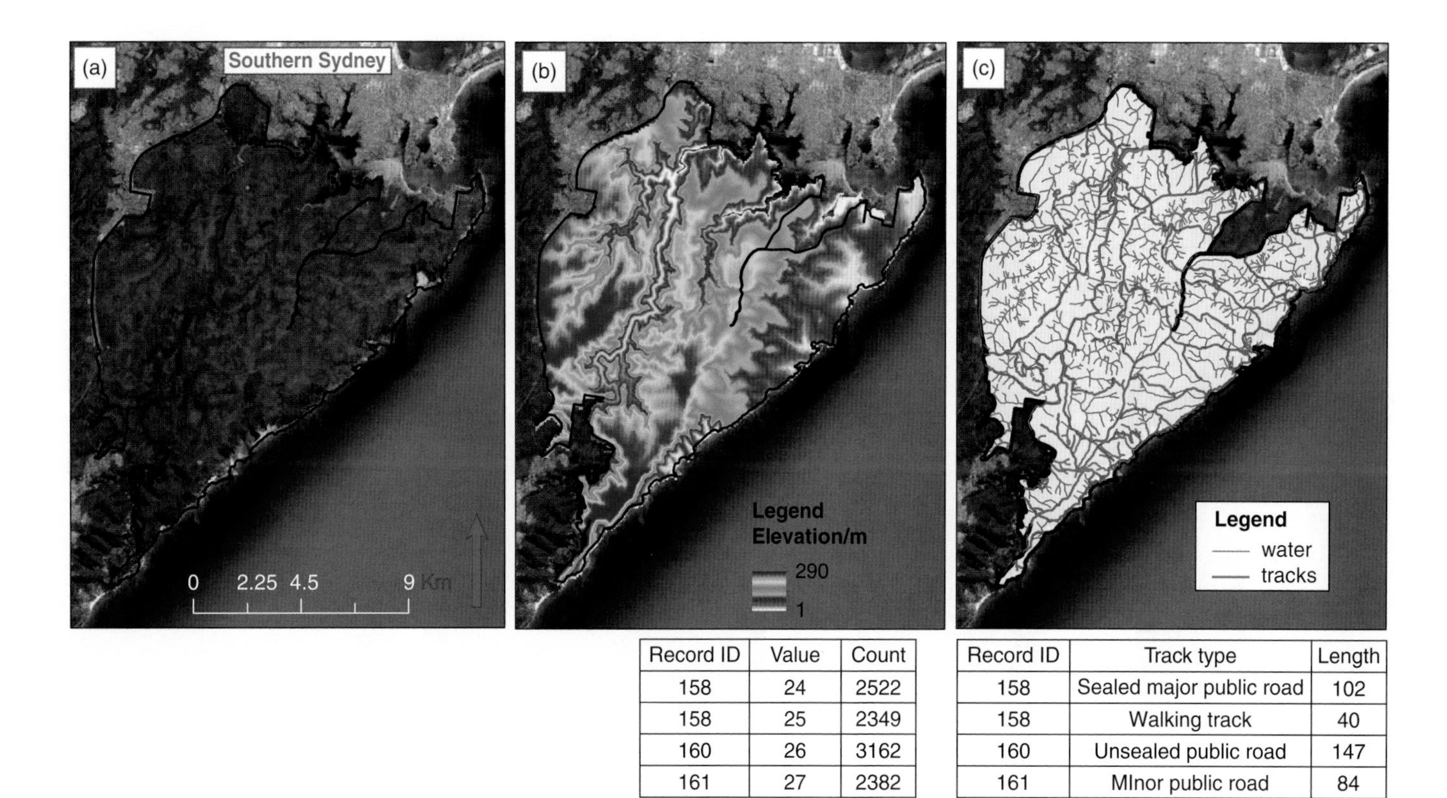

Record ID	Value	Count
158	24	2522
158	25	2349
160	26	3162
161	27	2382

Record ID	Track type	Length
158	Sealed major public road	102
158	Walking track	40
160	Unsealed public road	147
161	MInor public road	84

Figure 2.4 Different formats for the storage of spatial data: (a) satellite image of the Royal National Park, coastal New South Wales; (b) digital elevation model, providing a continuous raster surface of elevation above sea level; (c) polyline vector datasets depicting rivers and roads inside the Royal National Park. Tables beneath (b) and (c) provide illustrative examples of the attribute information associated with raster and vector datasets.

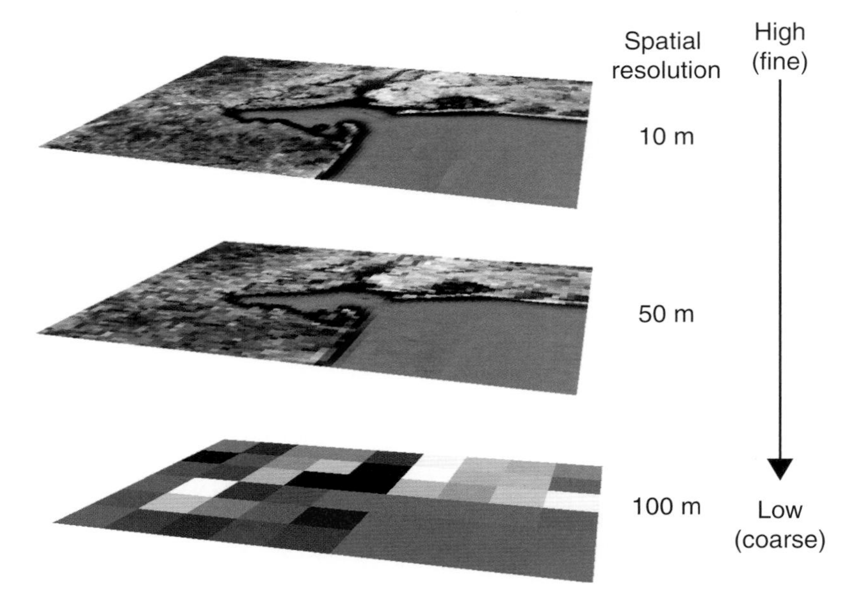

Figure 2.10 A series of raster images of varying spatial resolution showing New York Harbour.

Figure 2.11 Some different applications of GPS technology for the collection of spatially referenced data. (a) A snorkeler towing a GPS unit along in a floating waterproof bag while taking ground referencing photographs above a coral reef. (b) Research volunteers carrying GPS drifters into the surf zone as part of a study on rip currents. (Photo credit: Patrick Rynne.) (c) A drone collecting imagery from a GoPro above mangroves in Minnamurra River, New South Wales. (Photo credit: Mark Newsham.) (d) A coupled GPS monitor and side-scan sonar instrument display attached to a boat. (Photo credit: Frank Sargeant.).

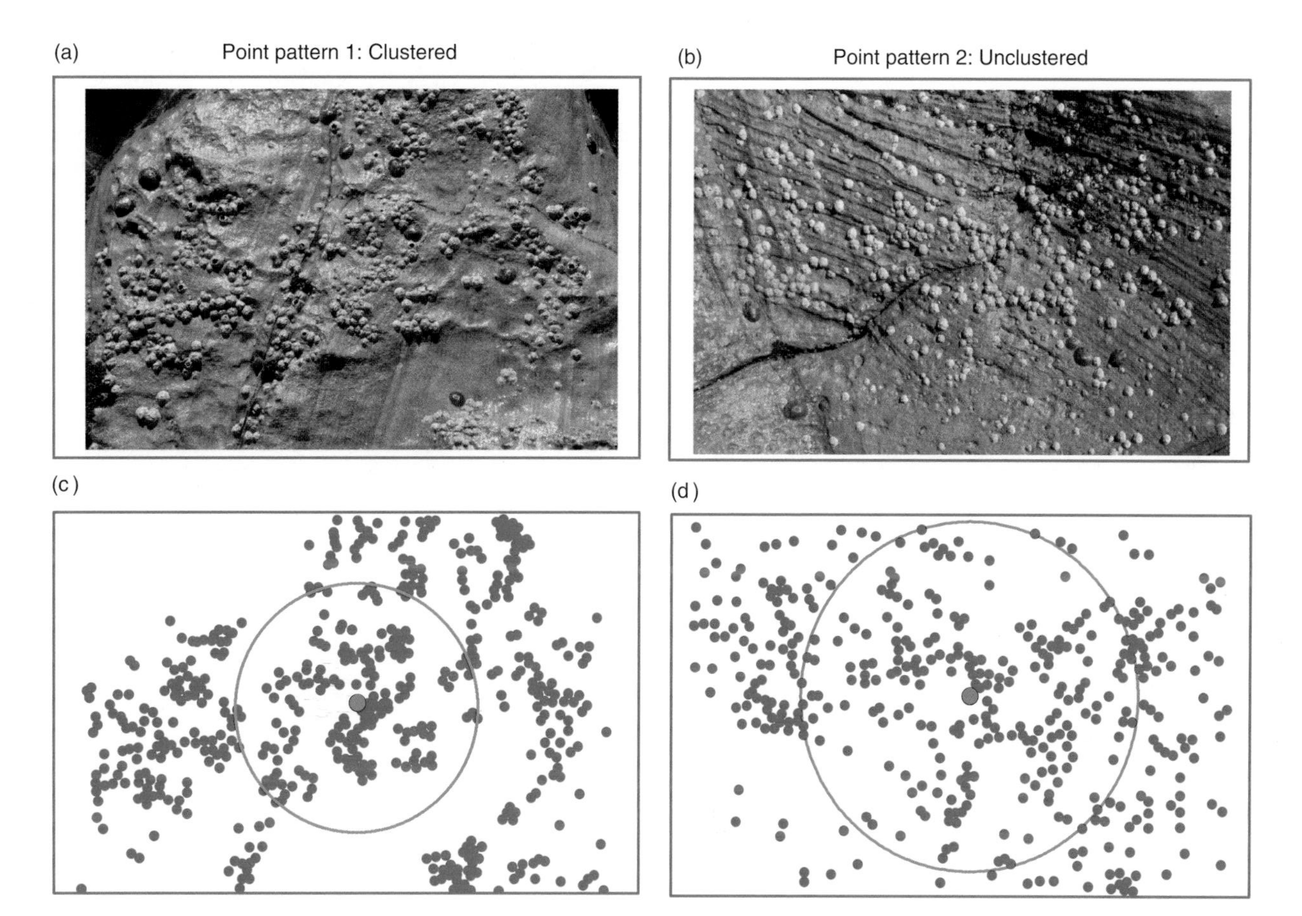

Figure 3.2 Barnacle patterns observed on a rock platform: (a) clustered; (b) unclustered; (c) mean centre and standard distance of the clustered pattern in (a); and (d) mean centre and standard distance of the unclustered pattern in (b) shown by the large point and circle, respectively.

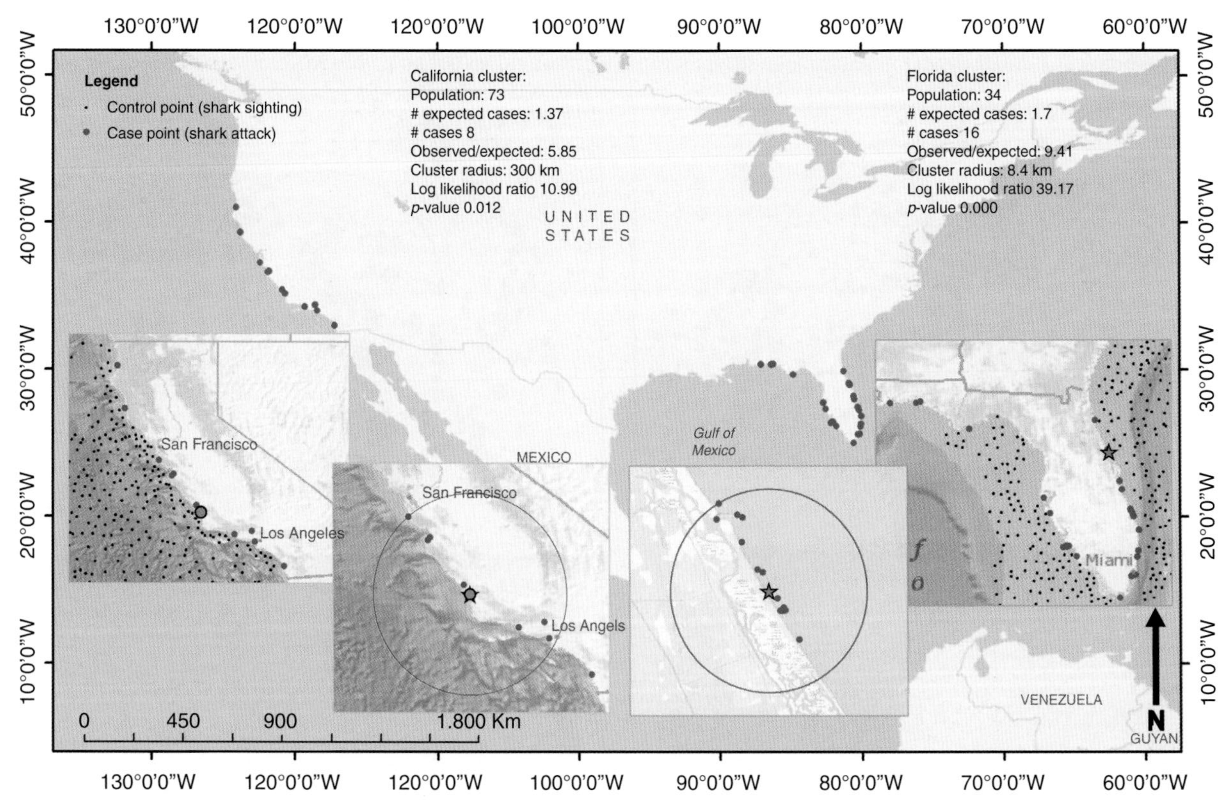

Figure 3.6 A map of shark attack records along the coasts of California and Florida, USA. Insets show the location of attacks in relation to known locations of sharks where attacks have not occurred, derived from tagging programmes. Red circles indicate statistically significant clusters or 'hotspots' of shark attacks, where it is statistically more likely that an attack will occur. Details of clusters are provided in the inset text.

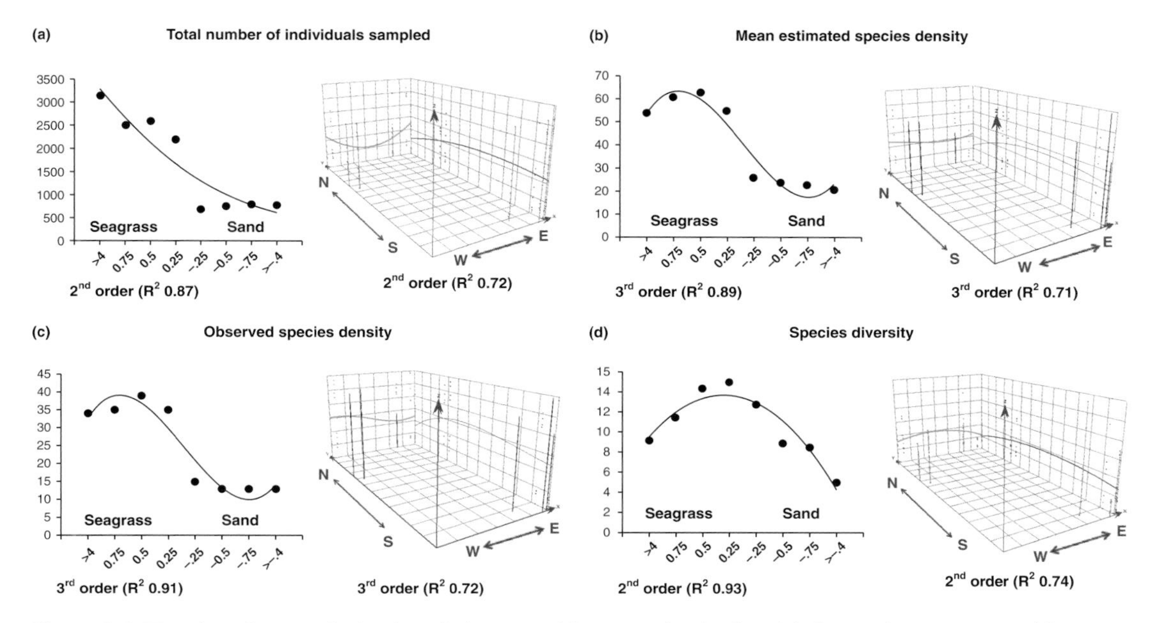

Figure 3.7 Trend surface analysis plot of the assemblage metrics for benthic invertebrate communities sampled across a sand/seagrass interface in Moreton Bay, Queensland. Mean values for each sample lattice are shown on the left-hand side as distances (m) from the seagrass/sand interface and values are plotted in their geographical (x, y) locations on the right-hand side with the z axis depicting the value of the metric. The x axis on the plots is a proxy for distance from the seagrass/bare sand interface. Green and blue lines indicate lines of best fit along the spatial (north to south, east to west) axes.

Figure 3.9 Worbarrow Bay, east of Lulworth Cove, Dorset, UK. Headland and bay coastline exposing strata from Portland stone to chalk. Features include (a) Iron Age hill-fort on chalk ridge, (b) recent landslides in chalk cliffs with talus extending below sea level, (c) coastal path at risk from cliff-top retreat, (d) stacks resulting from erosion of limestone steeply dipping strata which continue as a submerged ridge to the headland in the left foreground, (e) incised dry valley in chalk, (f) landslides in chalk, sands and clays, and (g) shingle beach with occasional landslides.

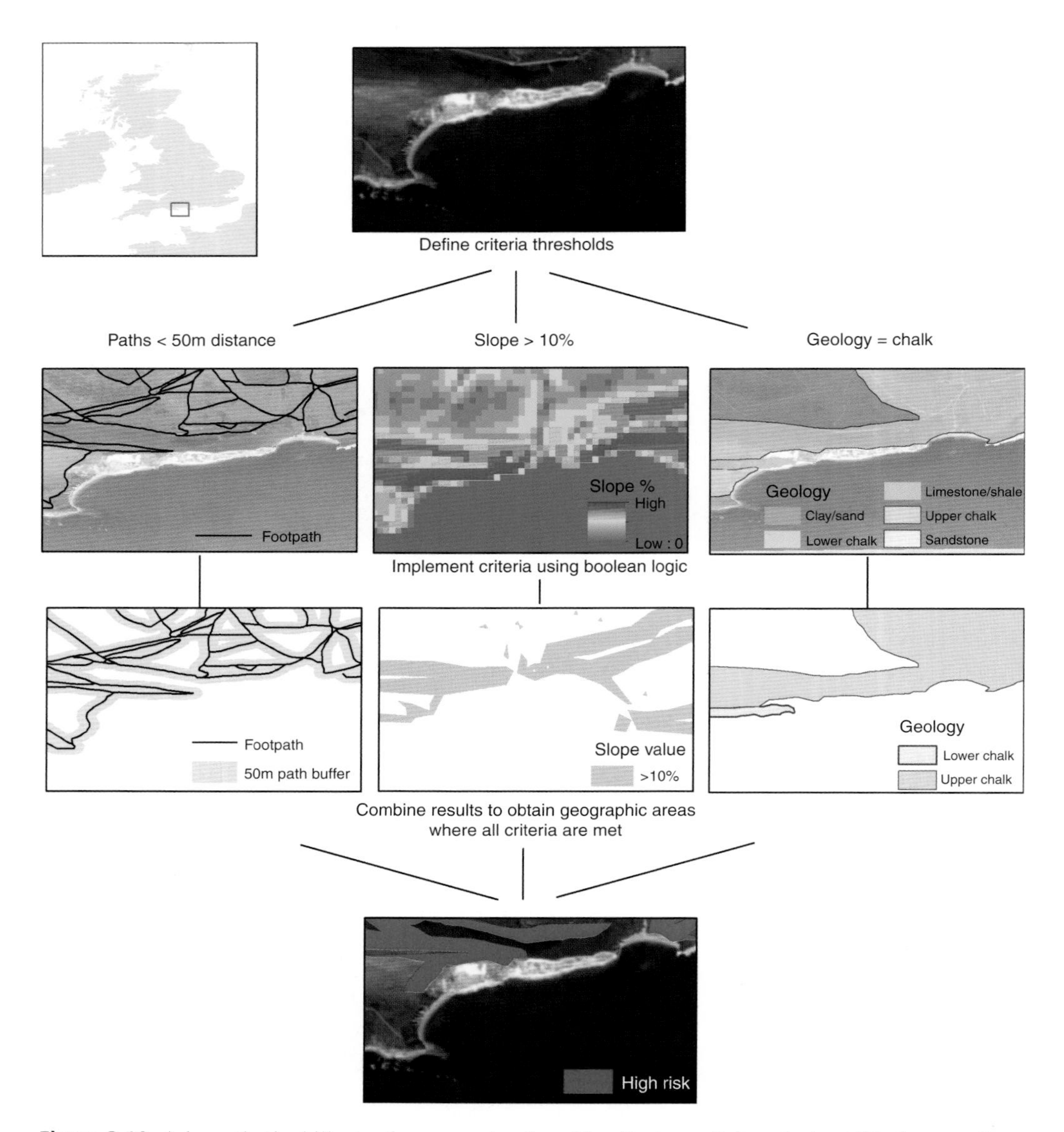

Figure 3.10 A hypothetical illustrative example of multi-criteria spatial analysis at Worbarrow Bay near Lulworth Cove, Dorset, along the coastline of southern England. The risk of injury from rock falls can be modelled based on distance to footpath, topographic slope and underlying geological character. Text boxes indicate key parts of the process of multi-criteria analysis.

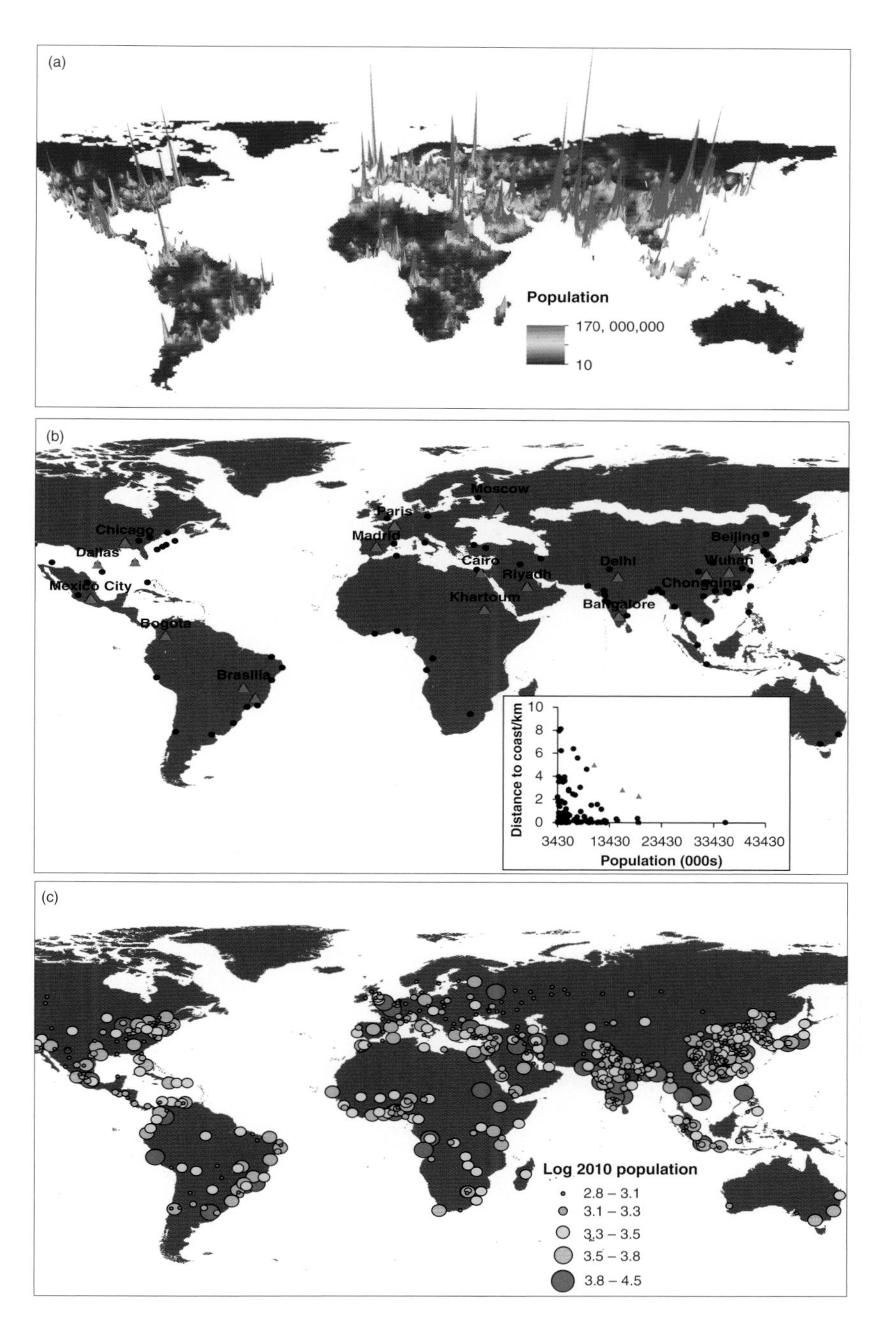

Figure 3.11 Some ESDA techniques. (a) Focusing: an oblique, three-dimensional view of world population centres, showing height above ground as a function of population (cell size, 10 km). (b) Linking: a scatterplot of distance to coast vs population (lower left), with outliers highlighted and plotted as triangles on the map; the identity of outlying cities is shown on the linked attribute table (right-hand side). (c) Arranging Views: a choropleth map of coastal cities with the symbology (size and colour of circles) linked to population as an example of an arranged view of a point dataset. (Data source: Natural Earth public domain dataset.)

Figure 4.2 World Ocean Floor Panorama (Bruce C. Heezen and Marie Tharp, 1977). Copyright by Marie Tharp 1977/2003. Reproduced by permission of Marie Tharp Maps LLC, 8 Edward Street, Sparkill, New York, NY 10976.

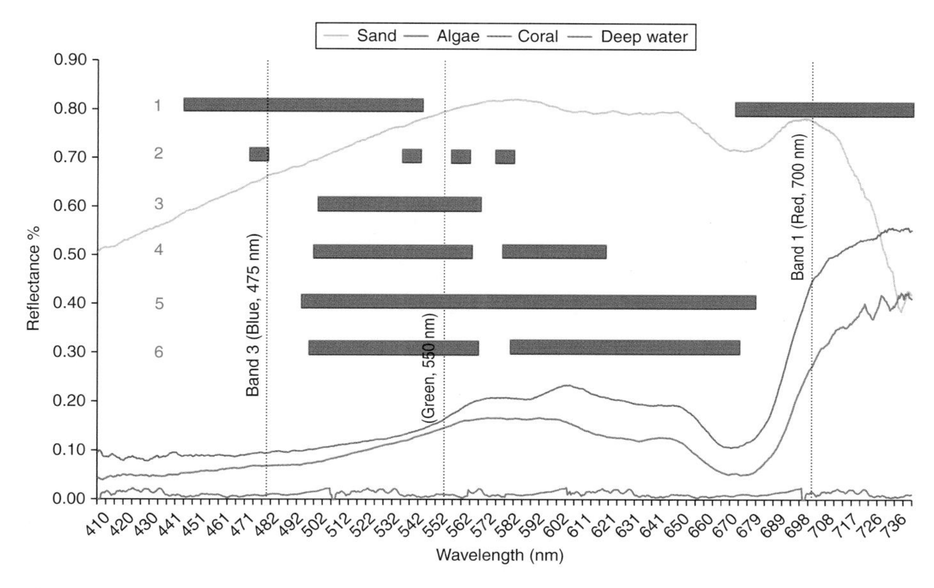

Figure 4.8 Spectral signatures of common coastal seafloor types (sand, algae, coral and deep water) collected with a GER1500 spectrometer from Aldabra Atoll, Seychelles, and associated published recommendations for the placement of wavebands within the electromagnetic spectrum for discriminating between seafloor types. Red numbers indicate the following publications: 1, Myers et al. (1999); 2, Hochberg and Atkinson (2000); 3, Wettle et al. (2003); 4, Holden and LeDrew (2000); 5, Kutser et al. (2000a, 2000b); 6, Holden and LeDrew (1999). The positions of the three different bandwidths of the satellite images employed for mapping in Section 4.10.3 are shown as vertical lines (after Joyce, 2005).

Figure 4.10 The author collecting ground referencing information by lowering a video camera down from a boat. Working from a boat platform facilitates rapid access across large areas. Mounting a camera onto a 50 m length of cable allows deeper environments to be surveyed. GPS coordinates of all survey points are recorded to allow comparison with other image datasets.

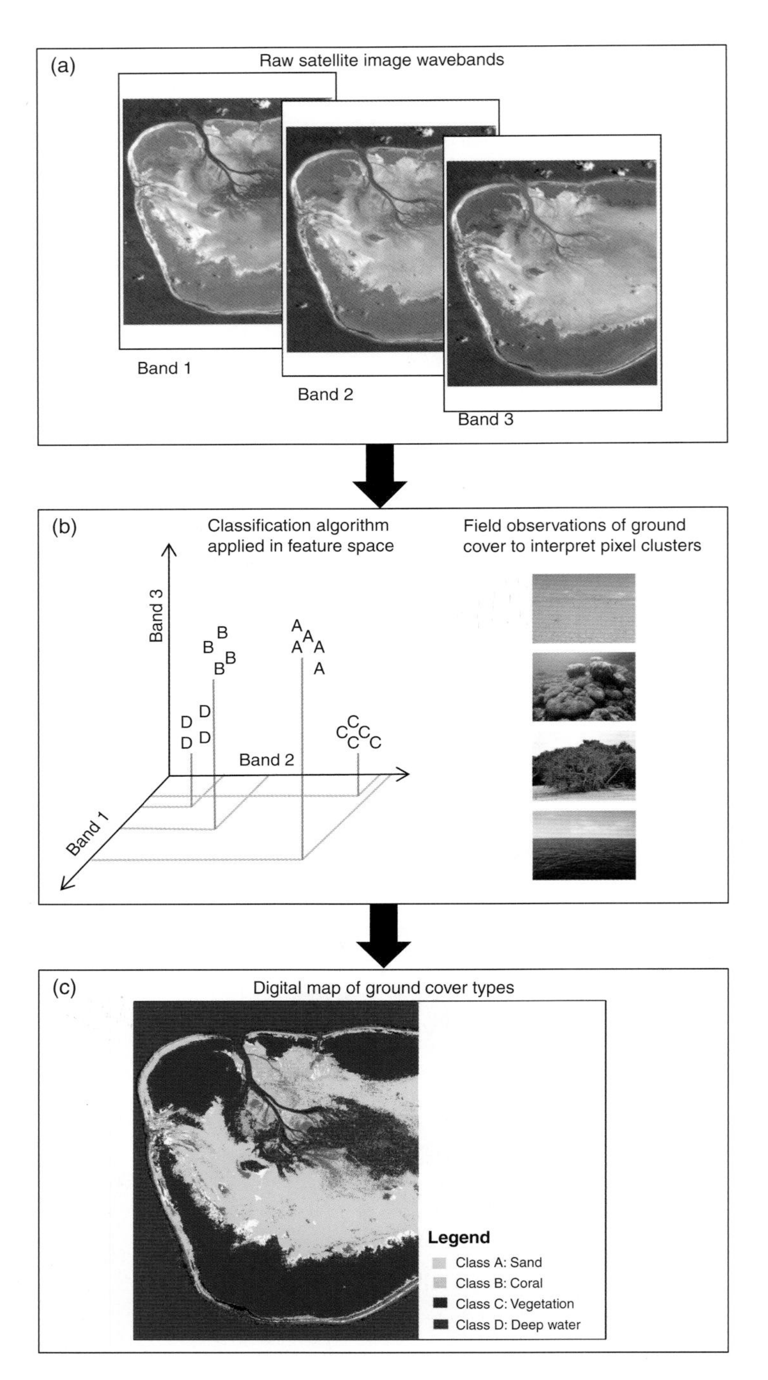

Figure 4.9 A flow diagram illustrating the application of a supervised classification algorithm to the raw bands of a satellite image of the western section of Aldabra Atoll to distinguish four ground-cover classes in a digital map: sand (class A), coral (class B), terrestrial vegetation (class C) and deep water (class D).

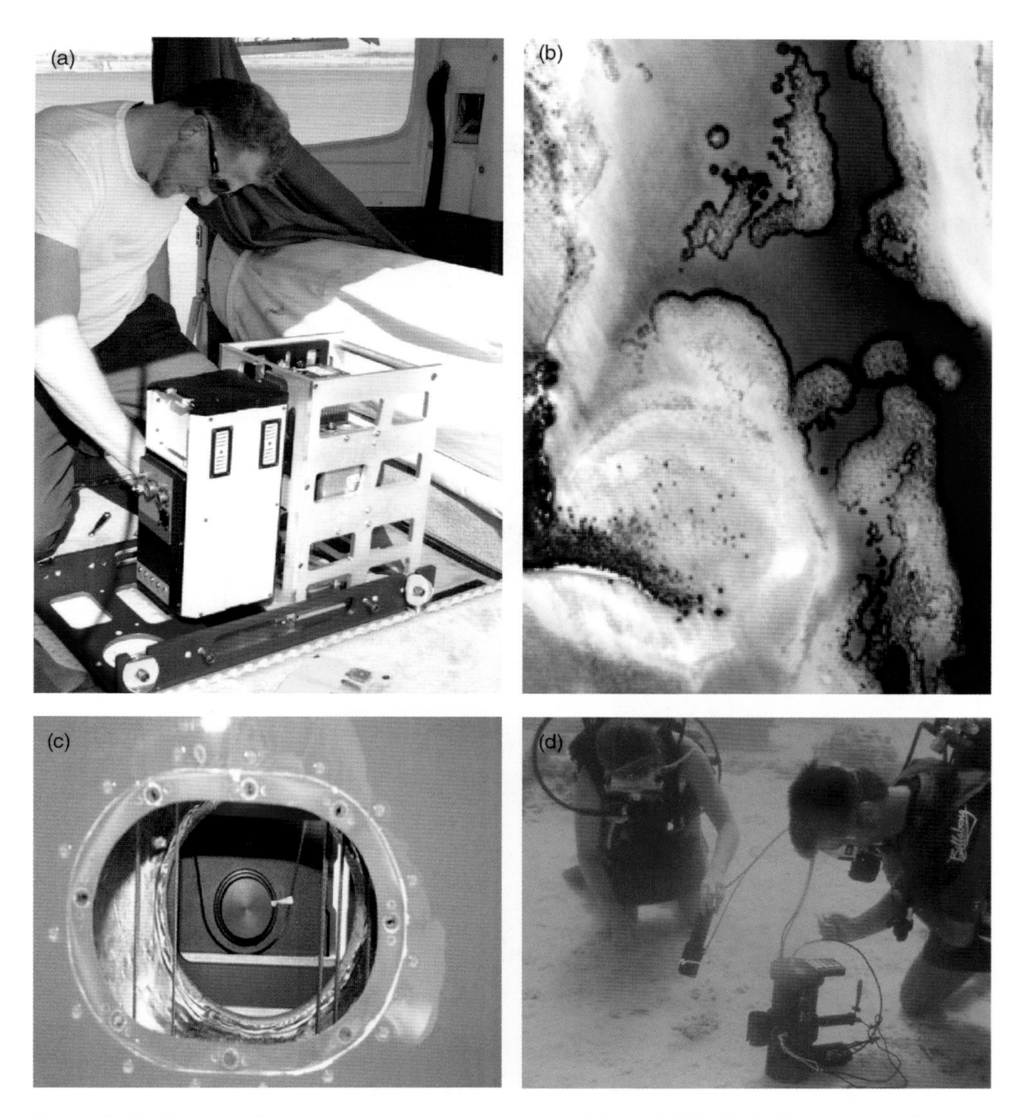

Figure 4.11 The use of remote sensing spectrometers. (a) An AISA-Eagle hyperspectral instrument mounted inside a Cessna seaplane. (b) Image of reef systems in the Red Sea collected by the airborne hyperspectral instrument in (a). (c) Sensor lens as seen through the base of the seaplane. (d) Hand-held spectrometer being operated underwater for the collection of sand spectra.

Figure 4.12 The use of unmanned aerial vehicles, or drones, to map coastal environments. (a) Example of a pre-programmed flight survey path around southeastern Heron Island, Great Barrier Reef. (b) Mosaic of raw images derived from a UAV survey of southeastern Heron Island. (c) Simultaneous in-water collection of seafloor photographs during a UAV survey. (d) Operating a UAV using hand-held remote controls from the beach at Heron Island. Images used with permission from Dr Karen Joyce and Dr Chris Roelfsema.

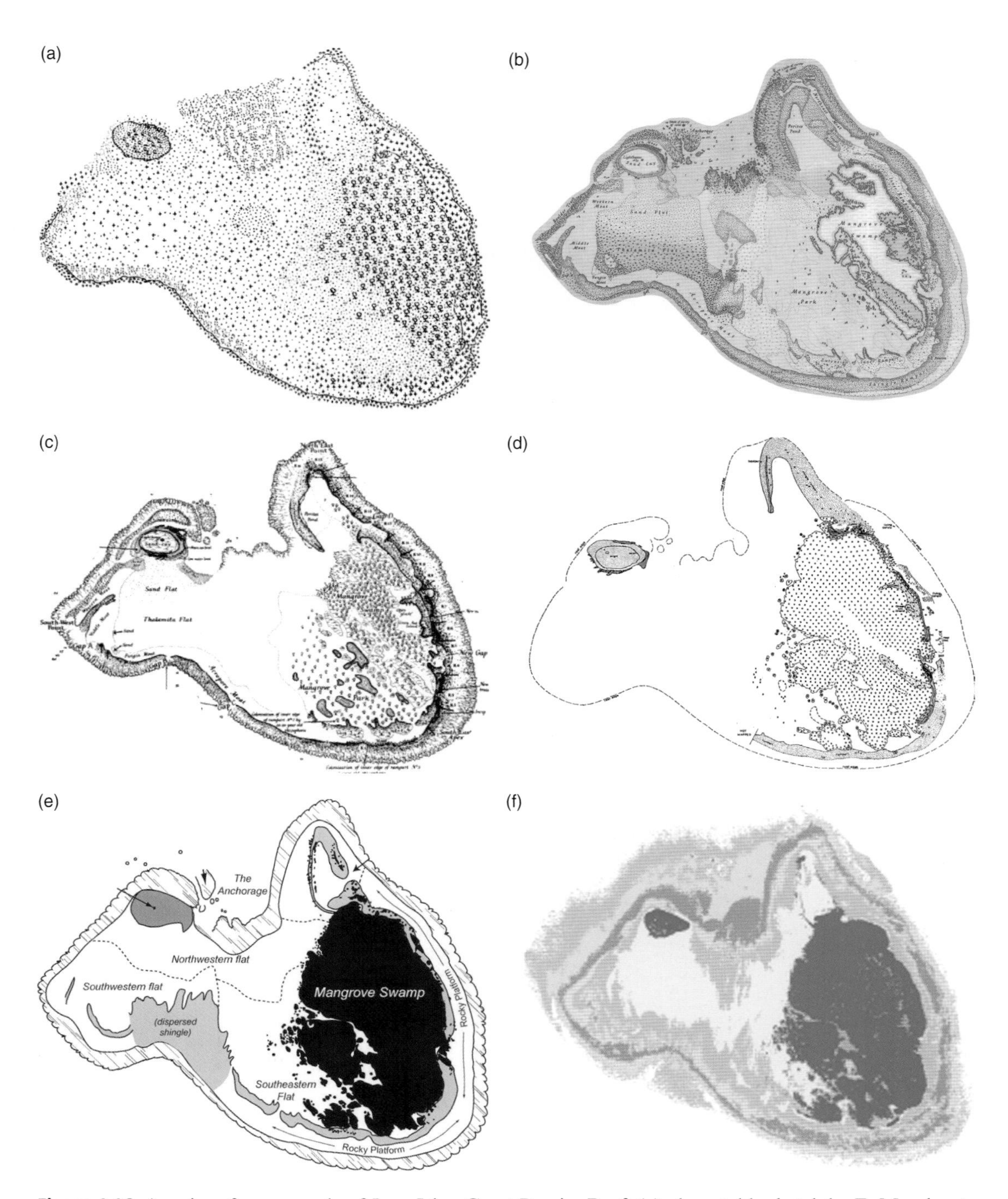

Figure 4.16 A series of maps made of Low Isles, Great Barrier Reef: (a) plane-table sketch by E. Marchant (Steers, 1929); (b) theodolite triangulation plane-table sketch by M. Spender (Steers, 1929); (c) aerial photograph trace by Fairbridge and Teichert (1948); (d) compass-traverse survey by D. Stoddart (Stoddart et al., 1978); (e) aerial photograph sketch by Frank and Jell (2006); and (f) satellite image classification by the author (Hamylton, 2014, unpublished).

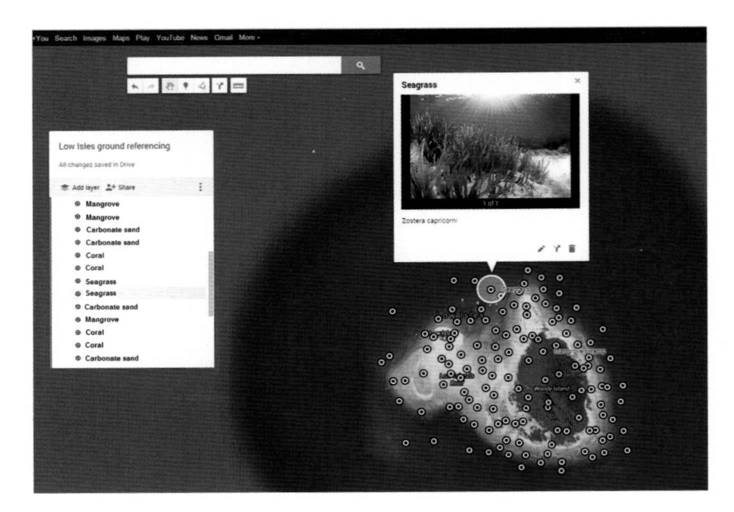

Figure 4.17 A Google My Maps project developed to host ground referencing data points collected at Low Isles on the Great Barrier Reef, illustrating associated photo records of seafloor cover and field notes.

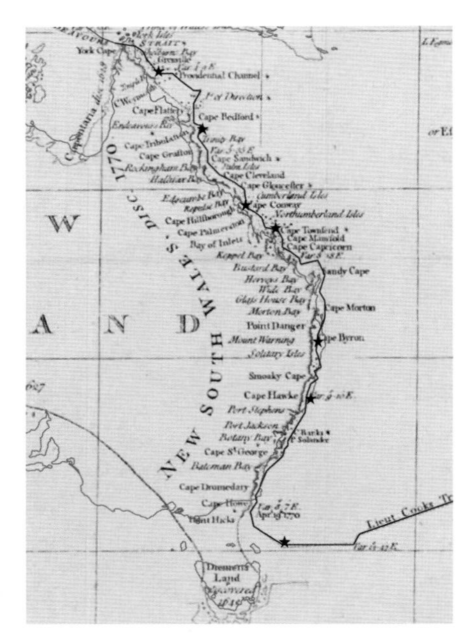

Figure 5.2 East Australian extract from the *Chart of part of the South Seas showing the tracks and discoveries made by His Majesty's ships: Endeavour, Lieutenant Cook, 1769*. (Accessed through Cambridge University Map Library, January 2014.) Features originally annotated onto the map are emphasised by the darker (blue) line (Cook's ship track) and black stars (points of astronomical determination of position). The overlaid lighter (red) line denotes the coastline of Australia as mapped using a georectified mosaic of Landsat TM satellite images (2000).

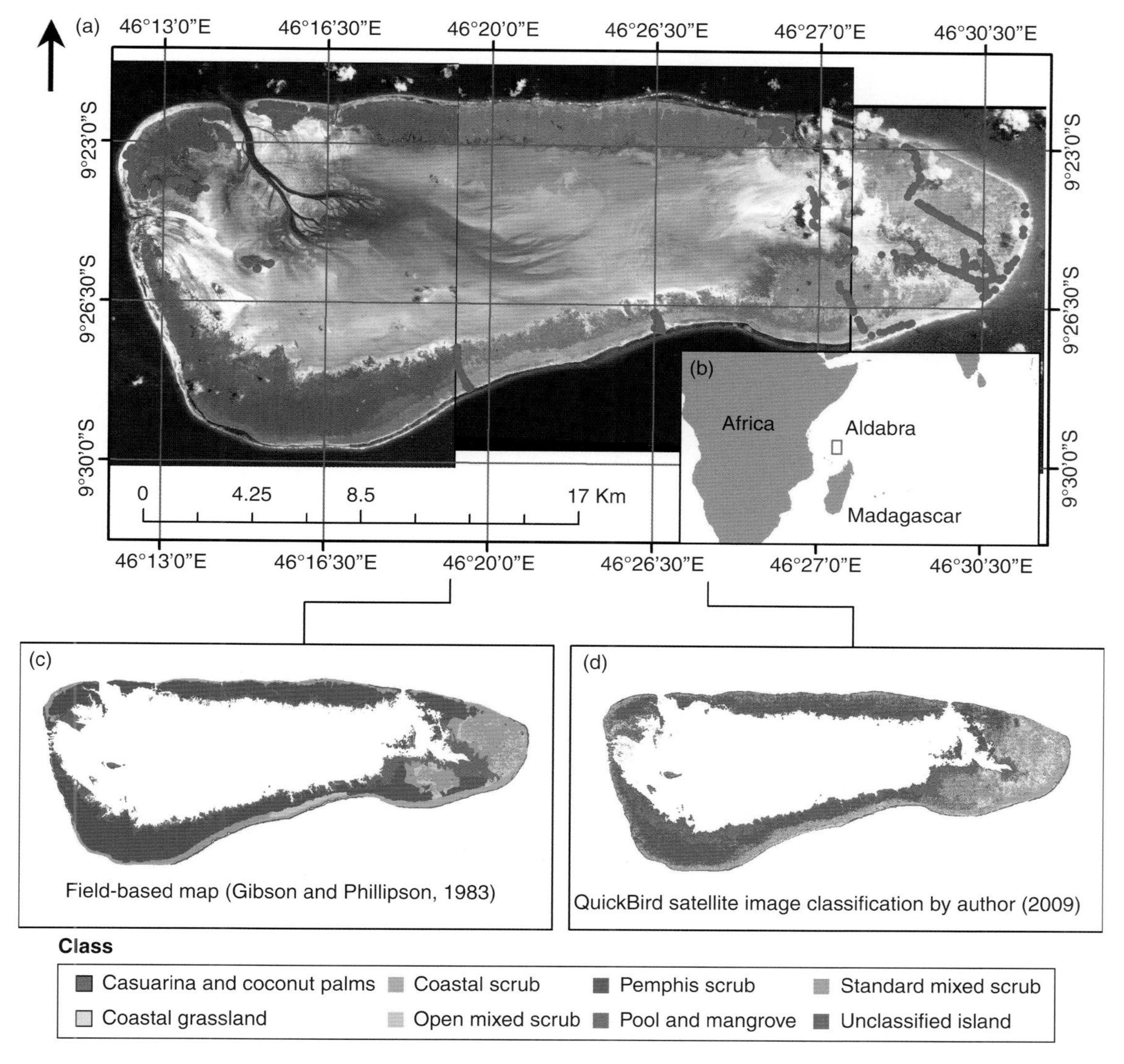

Figure 5.3 (a) Three QuickBird satellite images showing Aldabra Atoll. The (red) dots indicate field vegetation surveys undertaken in 2009. (b) The location of Aldabra Atoll in the west Indian Ocean in relation to the African continent and Madagascar. (c) Aldabra terrestrial vegetation map produced by Gibson and Phillipson (1983). (d) Vegetation mapped via classification of the QuickBird images by the author in 2009.

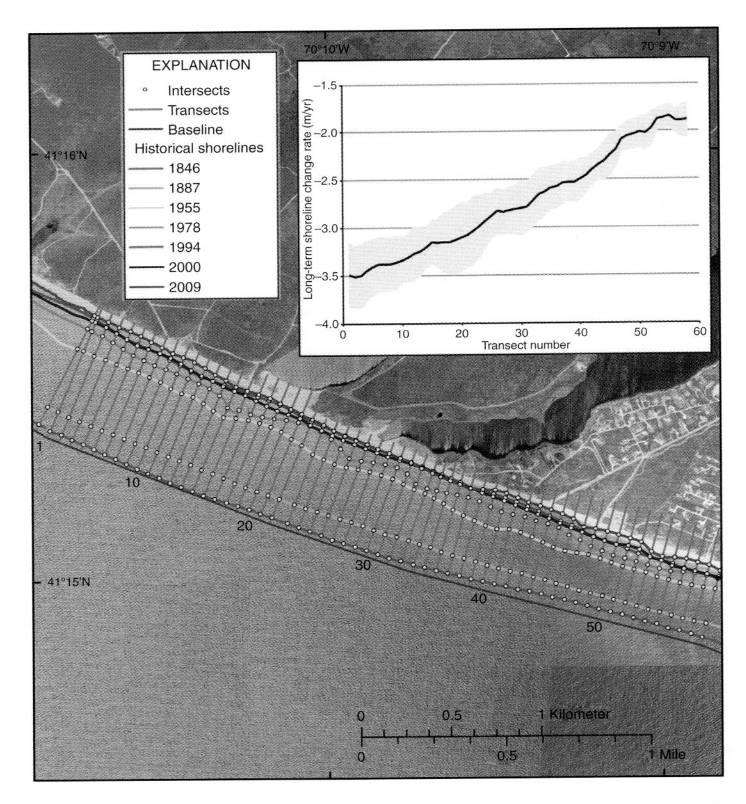

Figure 5.4 Change detection analysis of the Cisco Beach area, southern shore of Nantucket Island, Massachusetts. This was conducted using the US Geological Survey's Digital Shoreline Analysis System. Historical shorelines digitised from older aerial imagery are shown as the coloured lines corresponding to the date range 1846–1994. DSAS-generated transects are established at 100 m spacing with the graph showing the long-term rate of shoreline change based on simple linear regression. Grey shading indicates the 90% confidence interval for the calculated rates. (Courtesy E. Robert Thieler, USGS.)

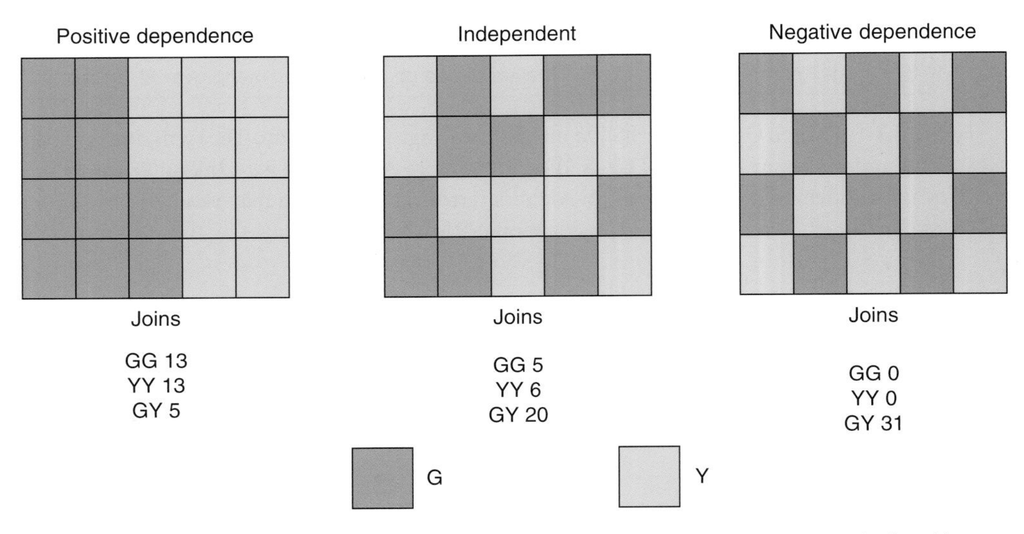

Figure 6.2 Join-counts for different patterns of mapped distributions of seagrass (depicted in green, G) on sand (depicted in yellow, Y). Adapted from Haining (2003).

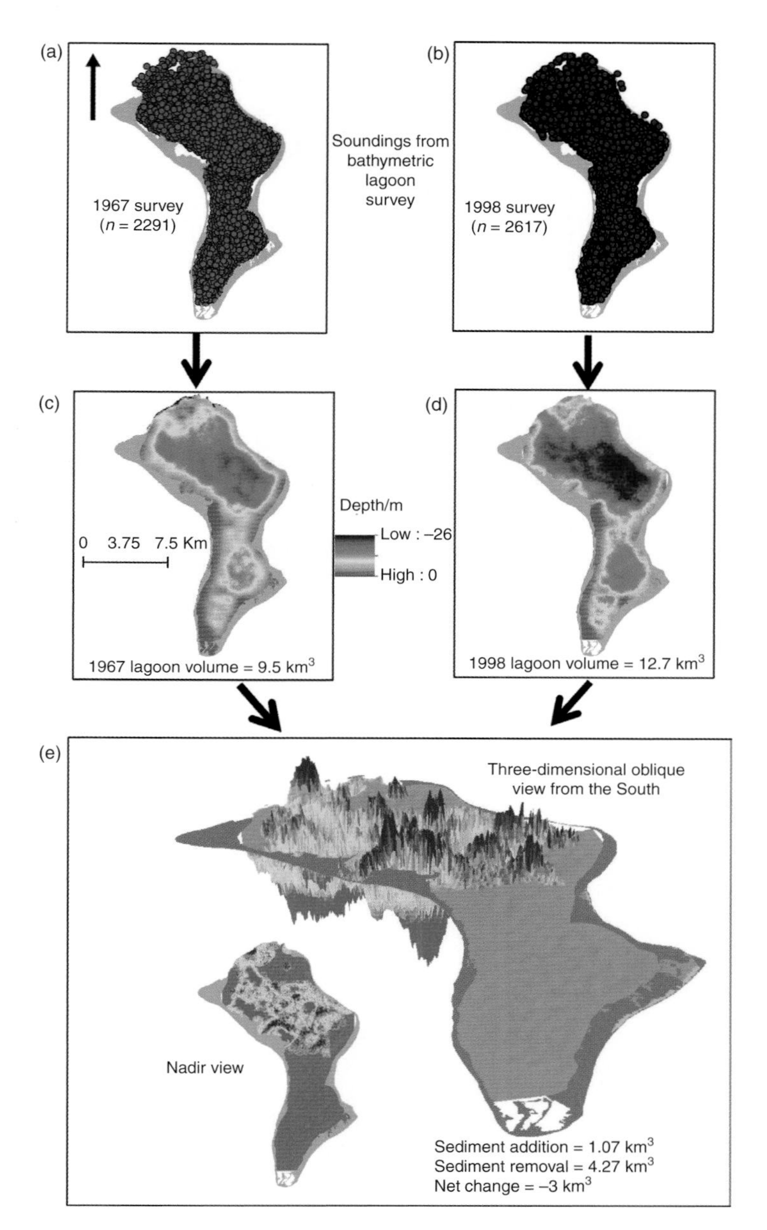

Figure 5.5 An analysis of lagoon volume change at Diego Garcia Atoll as a result of dredging and dumping of sediments during development of a military airbase in the 1970s. (a) Individual echosounder points from the 1967 survey. (b) Individual echosounder points from the 1998 survey. (c) DEM derived in 1967 by interpolating echosounder points. (d) DEM derived in 1998 by interpolating echosounder points. (e) Volume change map derived by subtracting the 1998 DEM from the 1967 DEM. Red (darker) indicates areas of sediment loss from dredging of boat channels; green (lighter) indicates localised areas of sediment gain due to the dumping of dredge spoil.

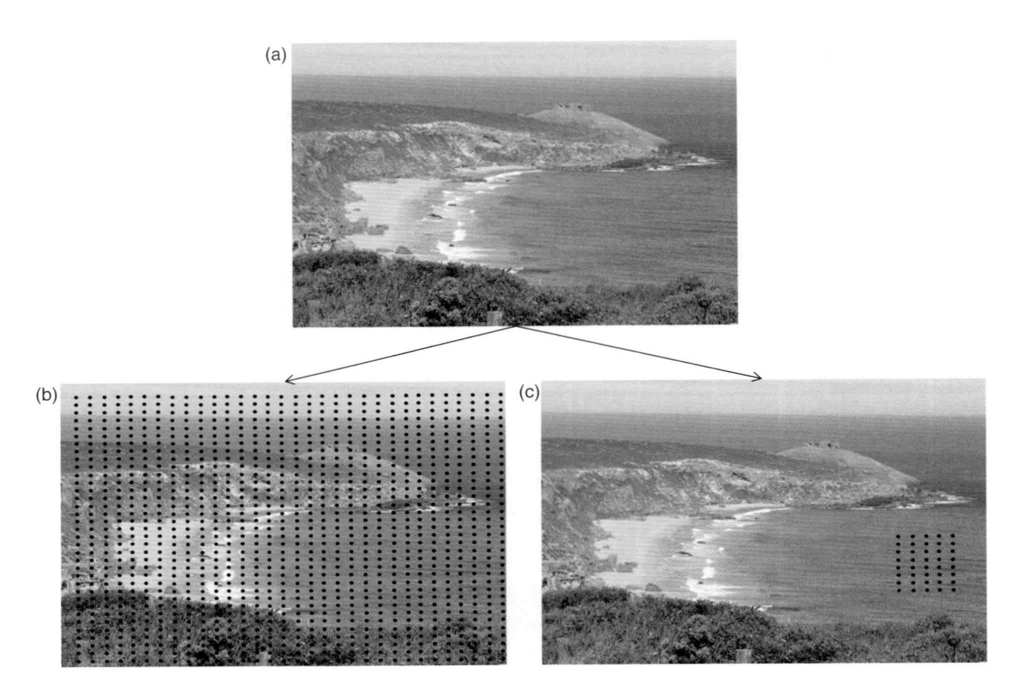

Figure 6.1 (a) A panoramic view across Weirs Cove to Remarkable Rocks, Kangaroo Island (South Australia). (b) Grid points covering the entire image to produce a global analysis of image properties. (c) Grid points covering a subsection of the image to produce a local analysis of image properties; see text for further detail.

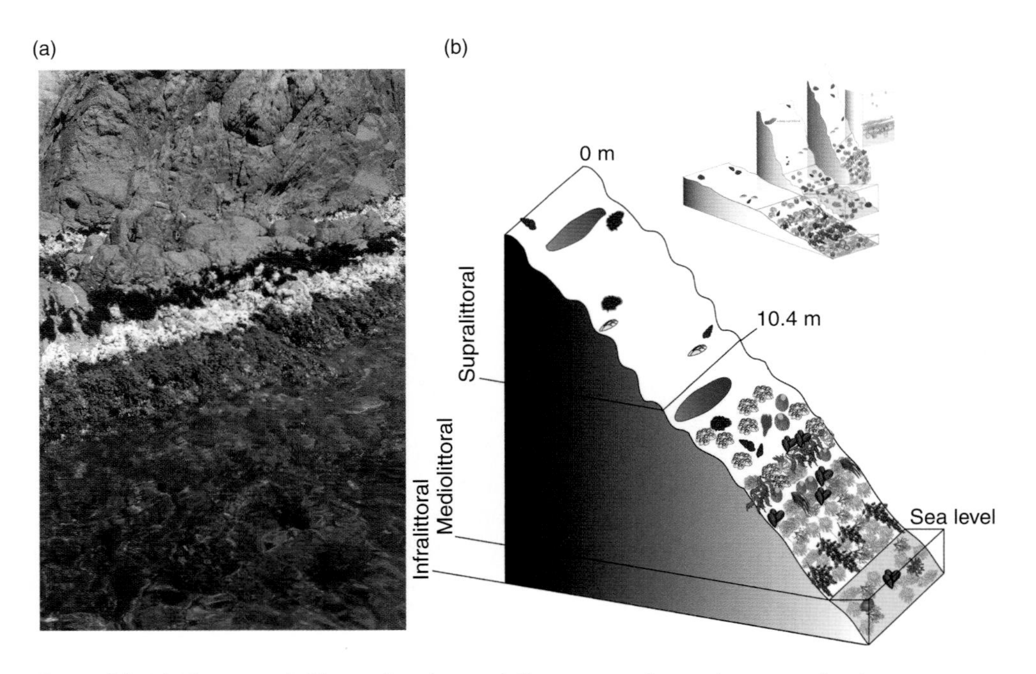

Figure 6.7 (a) Photograph illustrating the spatially structured zonation occurring in species inhabiting the littoral zone along a rocky shore. (b) Patterns of benthic community distribution vary along both horizontal and vertical gradients, depending on factors such as cliff slope (upper right). Reproduced with permission from Chappuis et al. (2014).

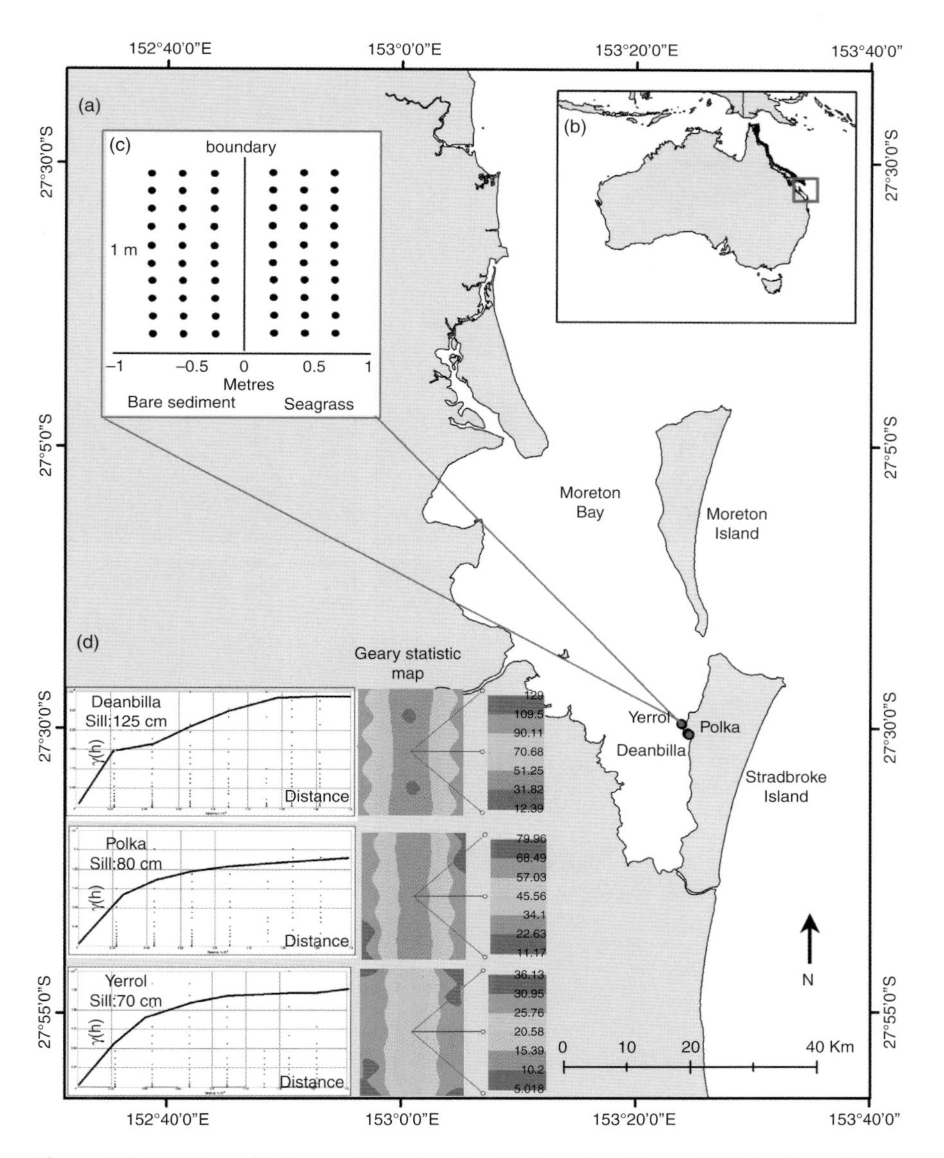

Figure 6.6 (a) Map of Moreton Bay showing the locations from which lattices of core samples of benthic invertebrates were extracted at Polka, Yerrol and Deanbilla. (b) The location of Moreton Bay adjacent to Brisbane along the Queensland coastline, Australia. (c) The configuration of the lattice of sediment cores sampled across seagrass boundaries from Moreton Bay. (d) A series of semivariogram plots for counts of benthic invertebrates sampled from Deanbilla, Polka and Yerrol sites. Coloured squares to the right of semivariogram plots indicate values of the Geary statistic, mapped across the lattice configuration for each sample site. High values of the statistic, shown in red, indicate positive spatial autocorrelation within the bare sediment or seagrass habitat. Low values of the statistic, shown in blue, indicate weaker positive autocorrelation across the sand/seagrass interface.

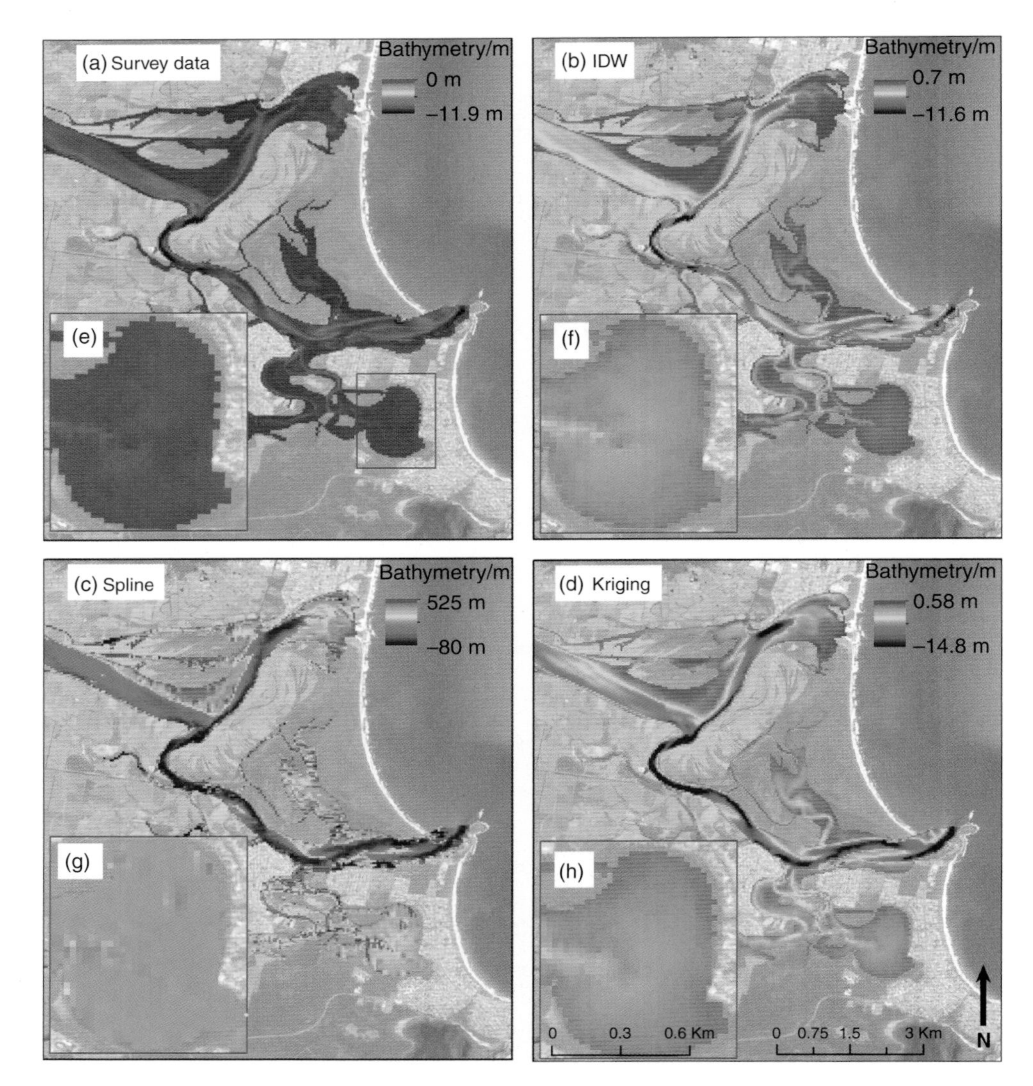

Figure 6.13 A series of interpolations on bathymetric data collected from the Shoalhaven Estuary, New South Wales: (a) raw bathymetric surveyed point data; (b) inverse distance weighting interpolation; (c) spline interpolation; (d) ordinary kriging interpolation; (e)–(h) larger-scale representations of the Curley's Bay area to the south (indicated by the red box in panel (a)). To facilitate comparison, the same colour scheme has been applied to panels (e)–(h).

Figure 7.2 Some examples of French coastal landscapes within which patterns can be observed: (a) Archipel des Lavezzi; (b) Delta de la Leyre, Gironde; (c) Baie des Veys, Manche; (d) Baie de Somme; (e) Grand Cul-de-Sac, Guadaloupe; (f) Marais D'Olonne, Vendée. (Photo credit: Frédéric Larrey.)

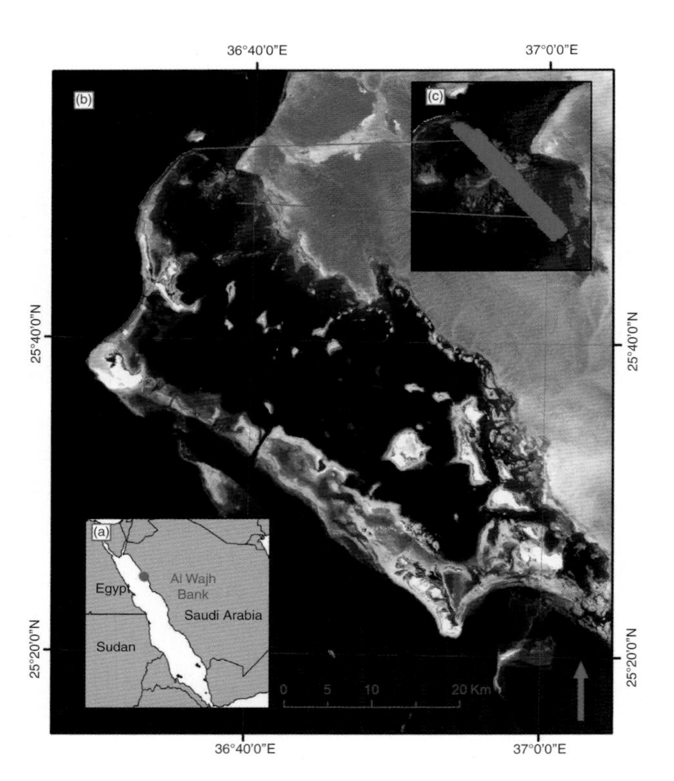

Figure 7.3 (a) Regional setting of the Al Wajh Bank (Saudi Arabia) within the Red Sea. (b) The Al Wajh Bank, a north–south aligned semi-enclosed reef complex. (c) The location of the study area covering a strip of water at the northern end of the Al Wajh lagoon.

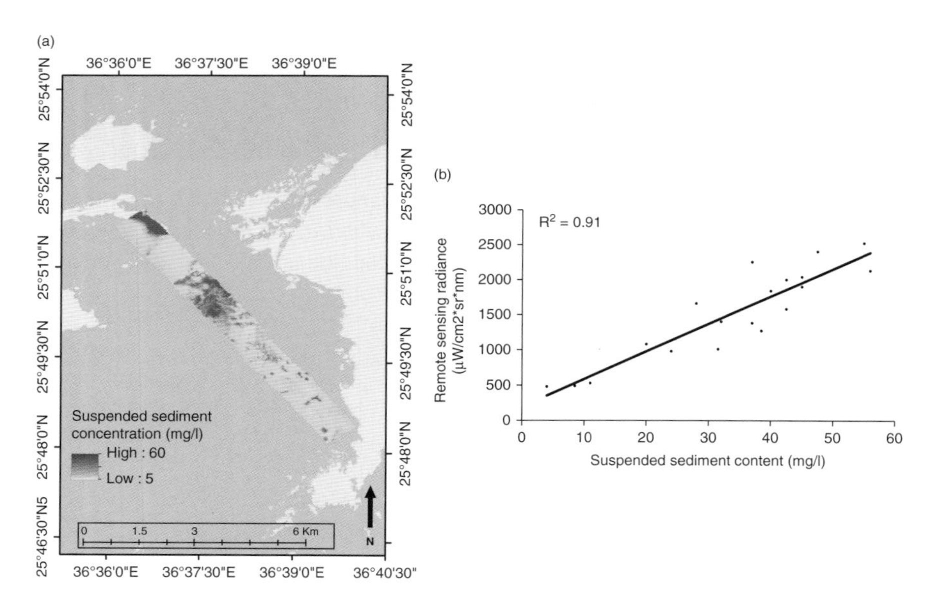

Figure 7.5 (a) Map of suspended sediment concentration across the reef systems inside the Al Wajh Bank, Saudi Arabia, as modelled from remote sensing radiance of the surface waters. (b) Empirical relationship between suspended sediment content of 19 water samples taken from the study area and the associated surface water radiance.

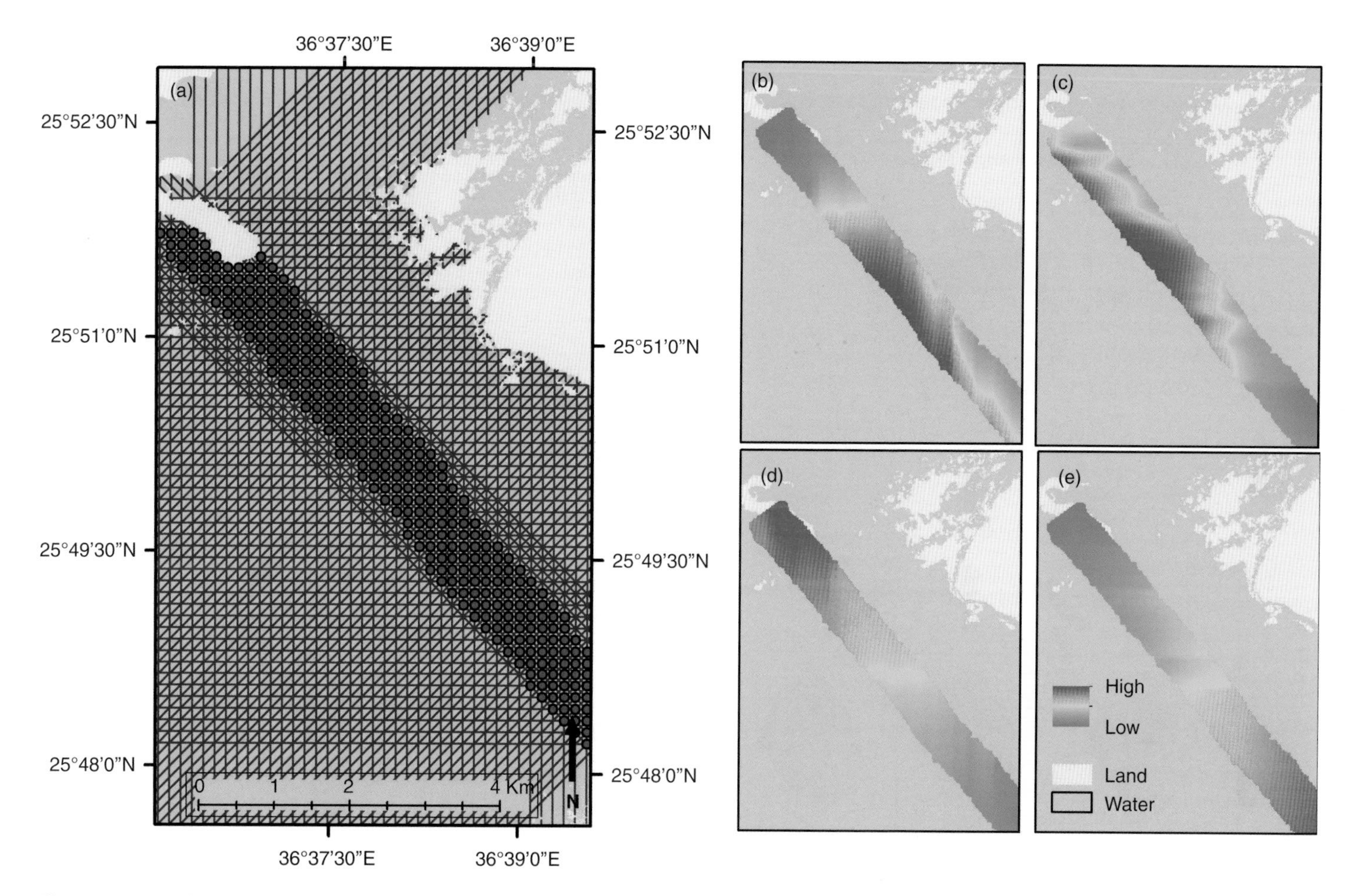

Figure 7.4 A series of maps illustrating high and low fetch distances from each cardinal direction in relation to land surrounding the study area, with emergent land features depicted in orange. (a) Fetch lines in all cardinal and subcardinal directions emanating from a grid of points covering the study area. Relative fetch values within the study area (red indicates high fetch, green indicates low fetch) for fetch directions (b) north, (c) east, (d) south and (e) west. Note the gap in land masses to the north, resulting in a non-fetch-limited area with respect to this direction in the centre of the study area.

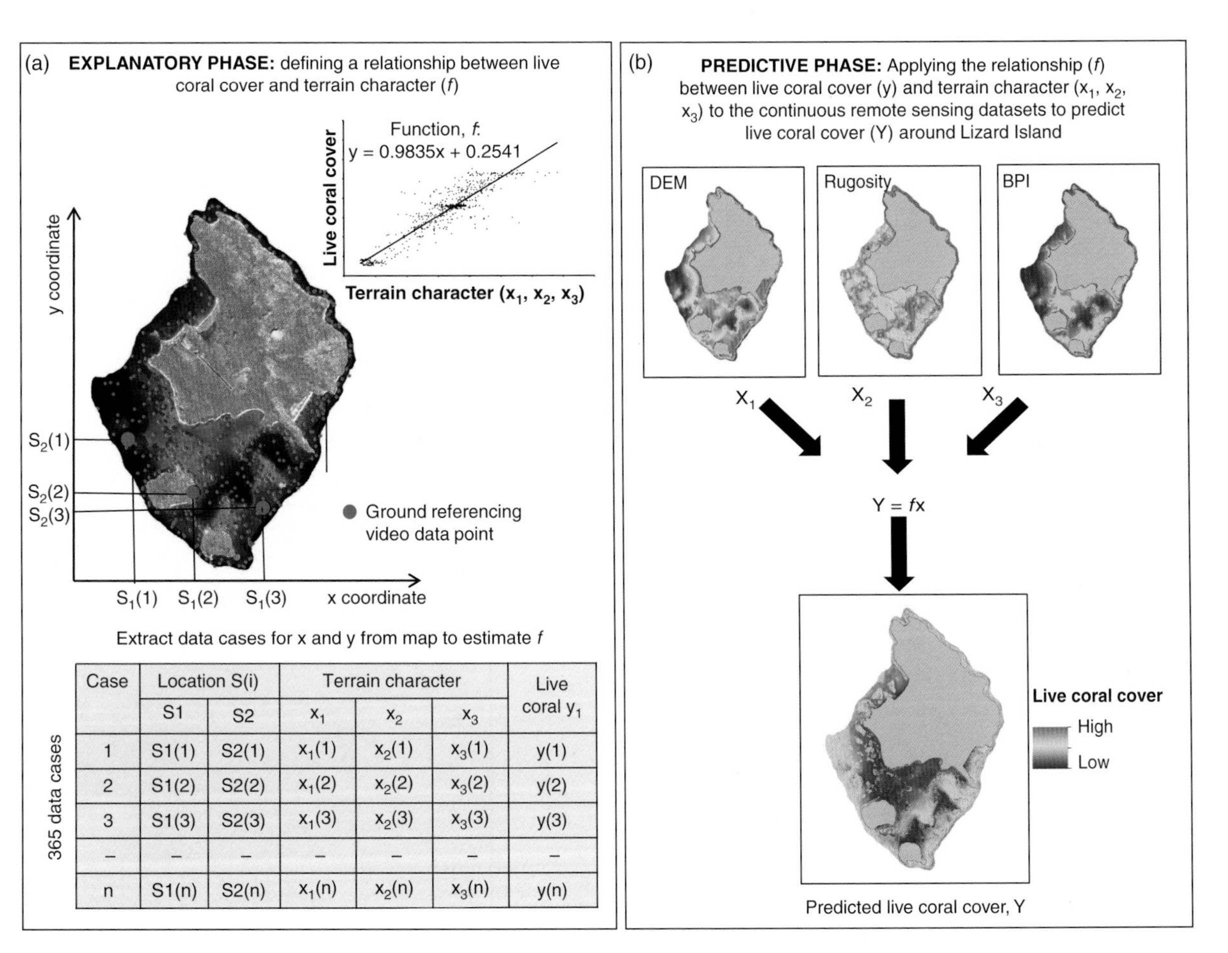

Case	Location S(i)		Terrain character			Live coral y_1
	S1	S2	x_1	x_2	x_3	
1	S1(1)	S2(1)	$x_1(1)$	$x_2(1)$	$x_3(1)$	y(1)
2	S1(2)	S2(2)	$x_1(2)$	$x_2(2)$	$x_3(2)$	y(2)
3	S1(3)	S2(3)	$x_1(3)$	$x_2(3)$	$x_3(3)$	y(3)
–	–	–	–	–	–	–
n	S1(n)	S2(n)	$x_1(n)$	$x_2(n)$	$x_3(n)$	y(n)

Figure 7.9 A schematic overview of the steps taken to develop an explanatory and predictive model of live coral cover around Lizard Island.

Figure 7.7 Aerial photograph looking north over the three islands that make up the Lizard Island Group, northern Great Barrier Reef. Lizard Island is in the background, with Palfrey Island (left) and South Island (right) in the foreground.

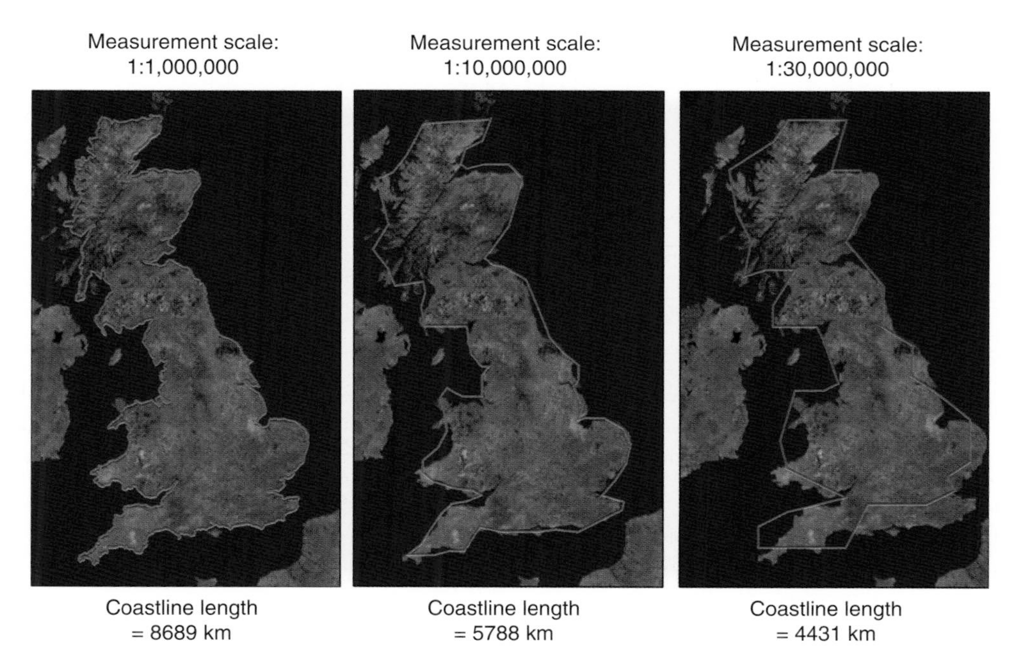

Figure 8.2 Three digitisations of the UK mainland coastline carried out across a range of scales. The coastline was digitised in order to estimate its length from the resulting vectors. The resulting length estimate increases as a function of the scale at which the digitisation was carried out.

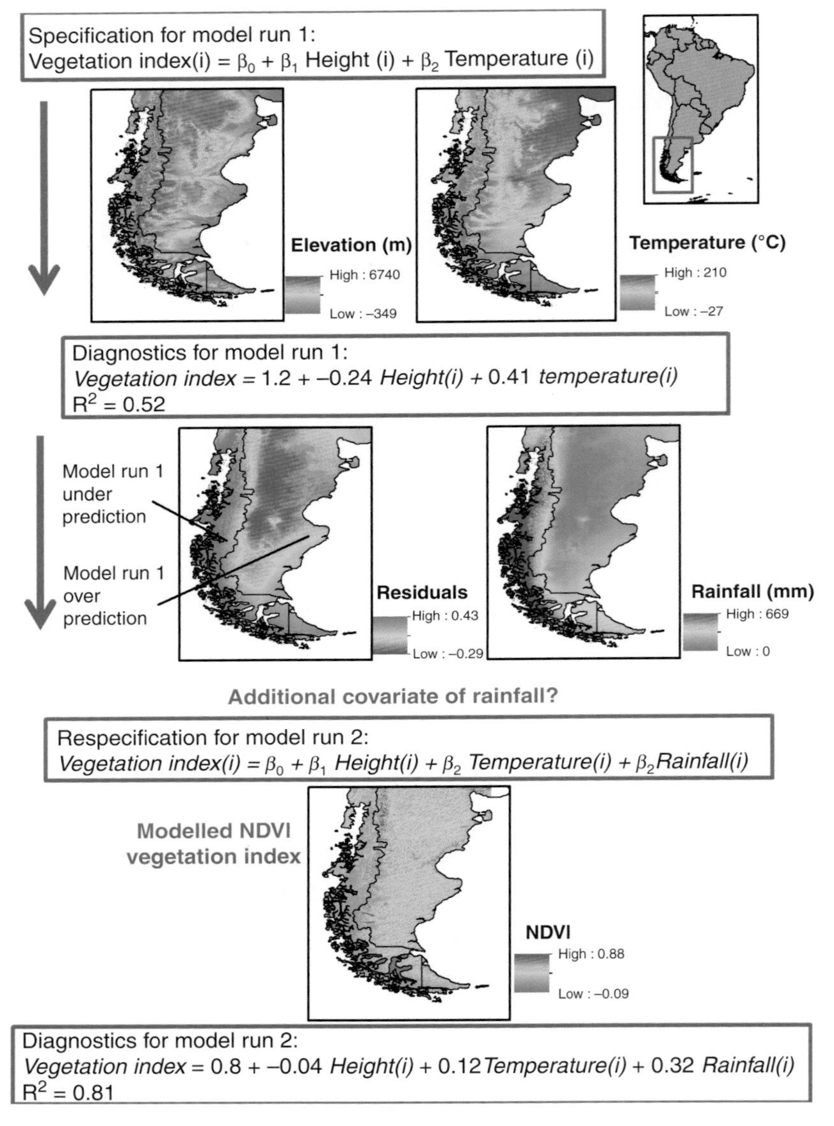

Figure 8.5 Modelling the vegetation of Patagonia across a gradient traversing the coast, a mountain range and the Argentine steppe-like plains. An illustration of model specification and re-specification using mapped residuals to indicate the potential identity of a missing covariate.

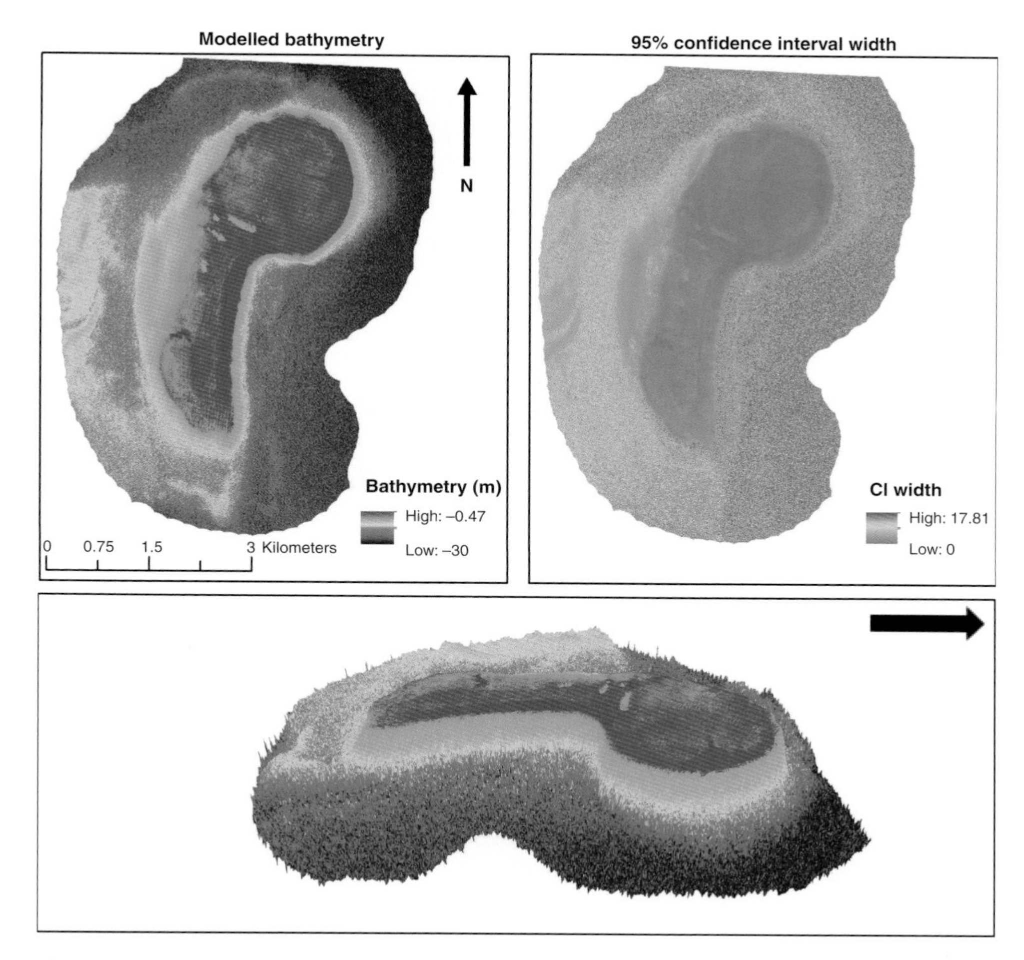

Figure 8.6 Modelled estimates of bathymetry (water depth) over Sykes reef platform (Great Barrier Reef). Sykes reef is a submerged platform extending from a shallow shelf in less than 1 m water depth to a deeper base in approximately 30 m water depth. The 95% confidence interval width illustrates the widening of confidence intervals associated with an increased spread of data points and higher levels of uncertainty in deeper areas. The lower image shows an oblique three-dimensional view of the reef platform bathymetry model, as seen from the eastern aspect.

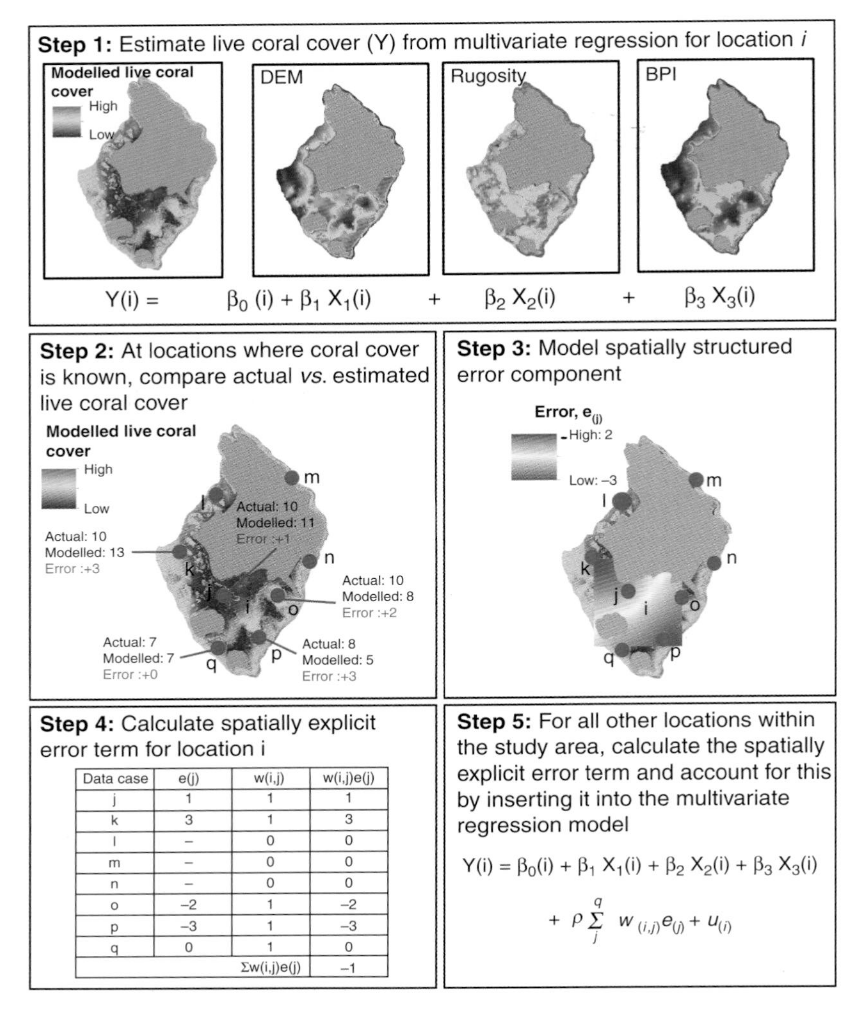

Figure 8.7 The steps taken to develop a spatial error model of live coral cover around Lizard Island (see text for further information).

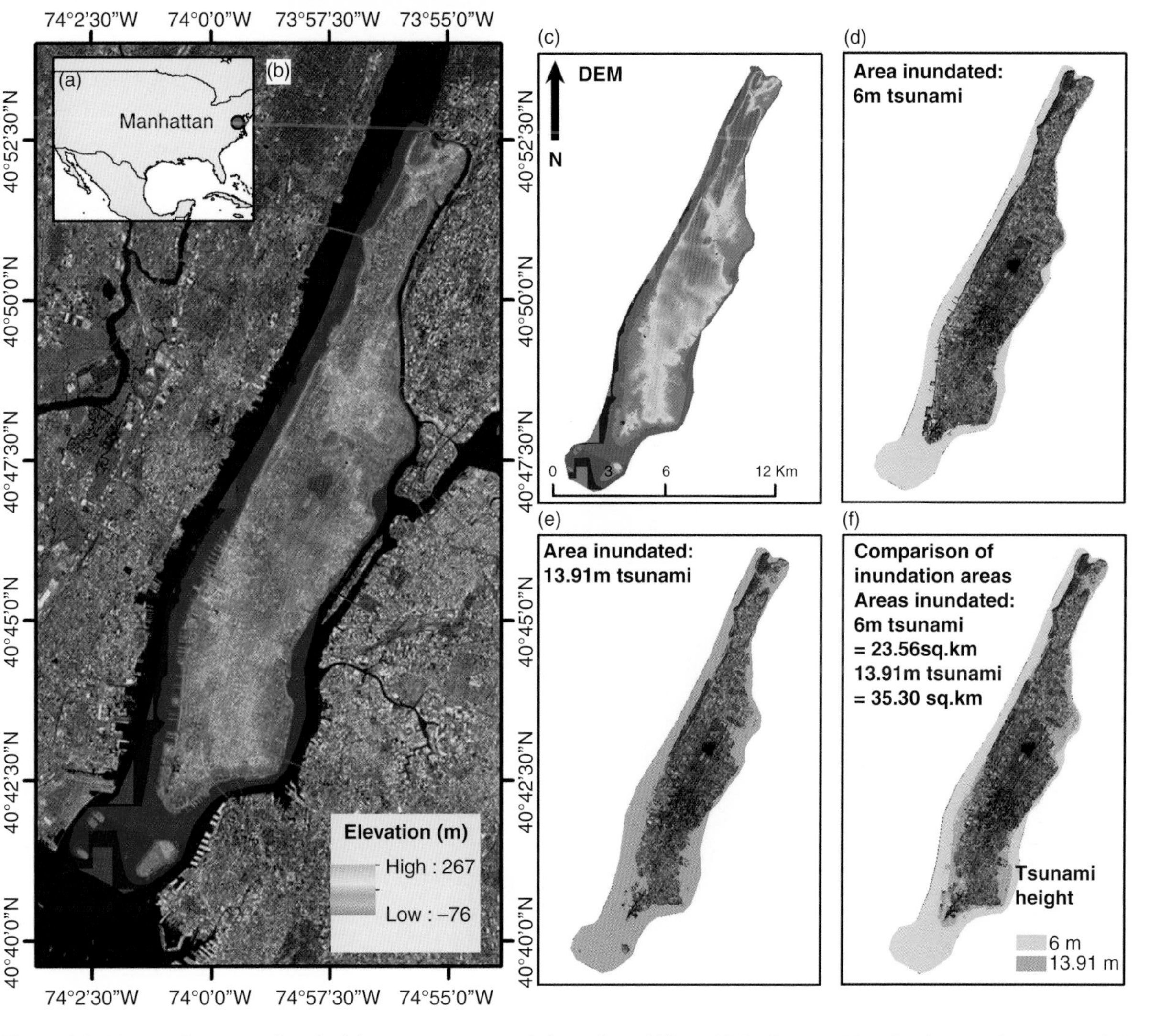

Figure 8.8 Uncertainty associated with an assessment of the vulnerability of Manhattan, New York, to a 6 m tsunami: (a) location of Manhattan on the North American continent; (b) DEM overlaid onto a satellite image of Manhattan; (c) DEM of Manhattan; (d) area inundated by a 6 m tsunami indicated in light blue; (e) area inundated by a 13.91 m tsunami indicated in dark blue (the upper 95% confidence limit); (f) area inundated by a 6 m tsunami (light blue) overlaid onto the area inundated by a 13.91 m tsunami for comparison of assessment outputs, allowing for uncertainty associated with the DEM.

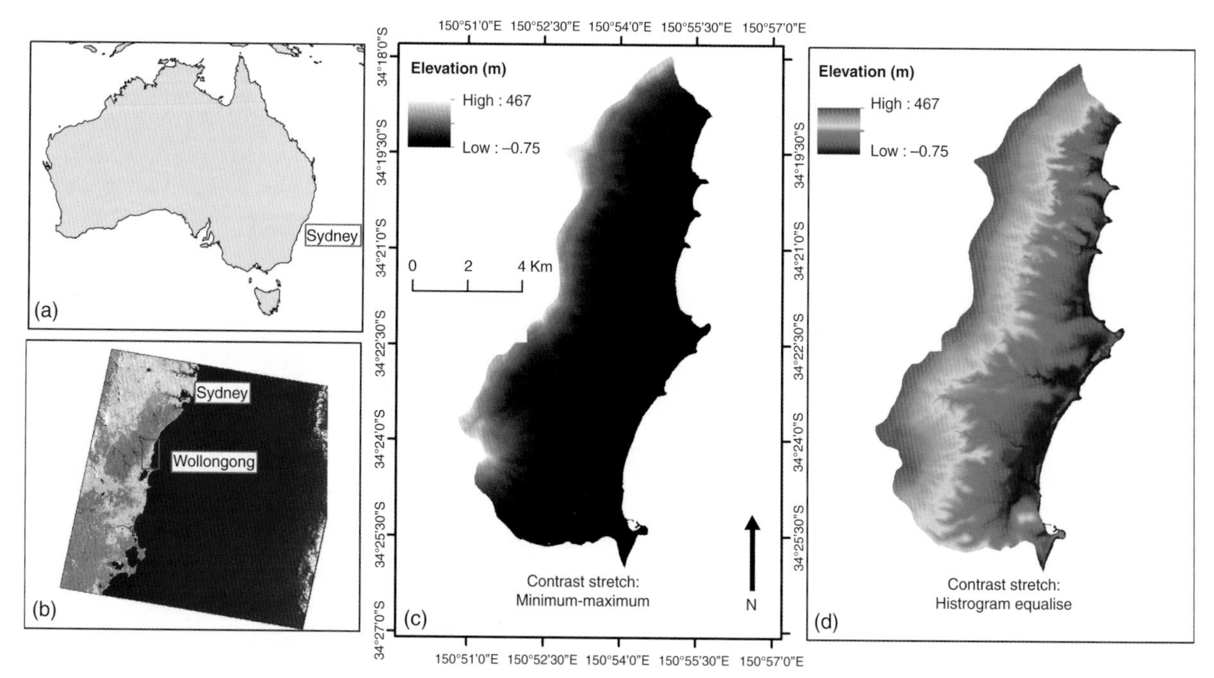

Figure 9.1 (a) and (b) The locations of Sydney and Wollongong on the eastern coast of Australia and New South Wales, respectively. (c) A DEM of the coast of New South Wales, where the city of Wollongong is located. (d) The same DEM with a 'histogram equalise' contrast stretch and a different colour scheme applied. Panels (c) and (d) illustrate the different ways in which information can be presented and, by extension, perceived as a result of applying different colour schemes and image contrasts.

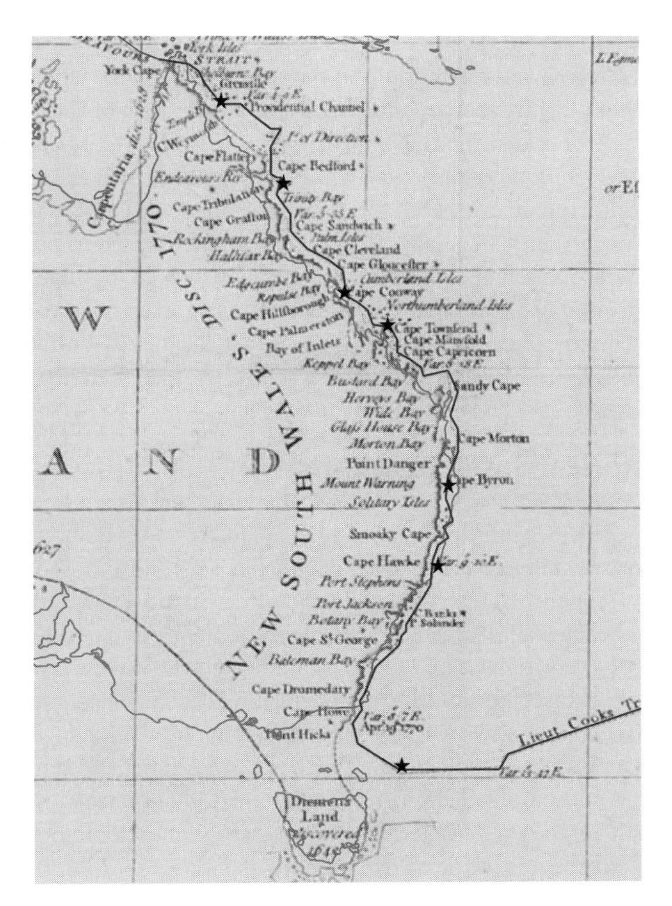

Figure 5.2 East Australian extract from the *Chart of part of the South Seas showing the tracks and discoveries made by His Majesty's ships: Endeavour, Lieutenant Cook, 1769*. (Accessed through Cambridge University Map Library, January 2014.) Features originally annotated onto the map are emphasised by the darker (blue) line (Cook's ship track) and black stars (points of astronomical determination of position). The overlaid lighter (red) line denotes the coastline of Australia as mapped using a georectified mosaic of Landsat TM satellite images (2000). (A black and white version of this figure will appear in some formats. For the colour version, please refer to the plate section.)

From Figure 5.2, it is possible to compare the coastline as mapped by Cook, which can be seen as the shaded line, with a more recent interpretation of that same coastline mapped from satellite technology, shown as the lighter (red) line. Both maps were projected on a flat surface in the same way. The distance between these two versions of the coastline reaches 80 km in places, illustrating the extent to which technologies have developed over this time. Moreover, the example illustrates the uncertainty inherent in older maps. These often represent the only information on the historical distribution and form of coastal features. As well as the dated technology of the time, they are subject to human error. Indeed, during the *Endeavour* voyage, the ship's astronomer,

Charles Green, became ill while performing his operations on the voyage and died. This led to criticism of astronomical readings upon the ship's return. How, then, is it possible to distinguish whether the offset between the two maps along much of the east Australian coastline in Figure 5.2 is because the coastline has genuinely moved, or because of uncertainty inherent in the techniques employed to map the coastline? This example highlights an important and unfortunate characteristic of older spatially referenced datasets; often the high degree of uncertainty associated with the technologies used to produce historic maps precludes their use as reliable data sources against which change can be measured. In many cases the ground distance corresponding to the overall error from the technique used to generate the map exceeds the change measured. This makes it impossible to reliably attribute any difference between the maps to anything other than instrument error.

5.3 Sources of Historical Baseline Information

Aerial photography provides information at a high spatial resolution and thematic accuracy. The interval between image acquisitions is typically long (i.e. more than five years). This may be too infrequent to detect the periodic or cyclic movements of coastal features such as spits, beaches and vegetation. Because of their comparably long deployment time, high rate of return visits and low cost, multispectral satellite sensors therefore provide more reliable datasets for monitoring change in coastal environments. Table 5.1 summarises the properties of some satellite sensors that are used in change detection.

The Landsat sensor is the longest-running Earth observation sensor. It has a high temporal resolution and comparably low cost – indeed most images are now freely downloadable. As a result of these characteristics, it is now the most commonly used type of image for change detection studies. Given that the spatial resolution of the Landsat sensor is 30 m, these images should only be used to detect large changes, generally of the order of magnitude of hundreds of metres or more. This maximises the magnitude of change (i.e. the 'signal') relative to the error associated with any change detection analysis applied (i.e. the 'noise'). By maximising the signal-to-noise ratio in this way, it is more likely that any change detected will be reliably identified as real change.

There are some notable practical constraints associated with using remote sensing for any form of change detection. These are summarised by Green et al. (1996) as:

1. the cost of obtaining multiple sets of comparable satellite imagery;
2. the need to standardise image data through a series of processing routines to ensure the validity of comparisons drawn between images;
3. the disjuncture between the availability of satellite images and the frequency with which field datasets can be collected; and
4. the spatial resolution and associated magnitude of change that can be detected on the imagery.

As a result of these constraints, care should be taken with the interpretation of images for change assessment. It may be advisable to apply multiple analyses directed towards the quantification of change. In this way, a multifaceted overall picture of observed changes can be obtained.

Table 5.1 Temporal attributes and characteristics (deployment years and revisit time) of multispectral satellite sensors.

Resolution	Deployment years	Sensor	Spatial resolution (m)	No. bands	Revisit time (days)	Approx. price ($US/50 km^2)
Low	1997–2011	SeaWiFS	1130	8	1	0
	1999–ongoing	Terra MODIS	250–1000	2–9	1–2	0
Medium	1972–2012	Landsat MSS	80	4	16	0
	1982–2012	Landsat TM	30	4	16	0
	1999–2003	Landsat ETM+	30	4	16	0
	2013–ongoing	Landsat OLI	30	5	16	0
	1986–96	SPOT HRV	20	3	26	1680
	1998–ongoing	SPOT HRVIR	20	3	26	2660
	2002–ongoing	SPOT HRG	10	3	26	1428
	1999–ongoing	Terra ASTER	15	3	16	120
	2007–11	ALOS AVNIR2	10	4	46	500
High	1999–ongoing	Ikonos	4	4	3	1000
	2001–ongoing	QuickBird	2.5	4	1–3.5	700
	2008–ongoing	GeoEye-1	2	4	3	625
	2009–ongoing	WorldView-2	2	7	1–4	1450

5.4 Techniques for Measuring Change Over Time

Techniques for the measurement of change over time using remote sensing images can usefully be subdivided into those that draw on the image, spectral and feature spaces of the imagery (see Section 4.9 for a detailed description of these image domains). These three 'spaces' refer to different domains or properties of the image from which information can be derived that is useful in the change detection process. A range of simple algebraic techniques can be applied across these domains to estimate the magnitude of change between two different digital images. These all assume that images have been independently registered to a common map projection. As far as possible, the pixels being compared can therefore be said to correspond to the same area on the ground. Table 5.2 summarises these techniques in relation to the different domains of a remote sensing image on which they draw.

Methods that operate in *image space* retain the location of each pixel within the raster grid of an image, without subdividing or manipulating the individual wavebands. Often these use contextual operators to compare pixels over time in the context of their neighbours within the grid environment (Section 3.5). Change detection approaches that operate in *spectral space* usually calculate and subsequently compare spectral indices designed to capture some feature of interest. These may represent the amount of coastal vegetation, including mangroves, saltmarsh, seagrass (Heumann, 2011) or live coral cover (Joyce and Phinn, 2013). Methods that operate in *feature*

Table 5.2 A summary of some different types of change detection methods operating in the image, spectral and feature domains of remote sensing images.

Image space	Spectral space	Feature space
Univariate image differencing	Principal components analysis	Post-classification comparison
Image regression	Change vector analysis	Change matrix
Image ratioing		
Vegetation index differencing		

space carry out a mapping exercise to completion at different dates. The finalised maps are then compared to ascertain change between these dates.

Many studies have sought to compare several of the change detection approaches listed in Table 5.2 within a coastal setting (Michalek et al., 1993; Macleod and Congalton, 1998; Muttitanon and Tripathi, 2005; Shalaby and Tateishi, 2007; Rodgers et al., 2009). Such comparisons yield variable findings, and no particular method emerges as being consistently more reliable than the others. It is clear that the performance will depend to a large extent on the nature of the change being observed, and how this is paired with an appropriate analytical method for detection.

5.4.1 Univariate Image Differencing

Univariate image differencing represents the most simple and common change detection technique for raster datasets. This procedure calculates the difference between the radiance values of two images corresponding to different times of acquisition at either end of the period over which change is to be detected. The basic premise for this approach is that changes in radiance values between two remote sensing scenes equate to changes in the corresponding area of land or coastline. This simple procedure can be applied iteratively across multiple bands of two multi-date images to produce an overall change image. This will have extreme positive and negative values in areas where changes have occurred, and values approaching zero where the land-cover character has remained relatively constant (Singh, 1989). It may be useful to define a threshold outside of which statistically significant change can be said to have been observed. For example, the combination of image differencing techniques with image histogram thresholding was applied to Landsat TM and ETM+ images to track reductions in the area of the hypersaline Lake Urmia (Iran) between 1998 and 2006 (Alesheikh et al., 2007).

5.4.2 Image Regression

The image regression technique also seeks to interpret changes in radiance values between two remote sensing scenes in terms of changes to features on the ground by plotting pixels from the before and after image against each other. A linear function is defined using least squares regression. Pixels where no changes are apparent should theoretically fall on the line that describes this function. Changes are estimated by calculating a 'difference image' as the difference between the predicted radiance value obtained from the regression line of best fit, and the actual value recorded by the 'after' image. The estimated magnitude of change is therefore the residual from this regression technique. As with the image differencing method, a threshold is then established to differentiate between areas of change. Berberoglu and Atkin (2009) applied image regression techniques in the Cukurova Delta in Turkey to assess the extent of erosion arising because of sand dune removal. Landsat TM images of the study area were used and the regression was performed on a band index that enhanced the difference between spectral features and suppressed topographic and shade effects (Lu et al., 2005). The technique was sensitive to changes in image radiance. Successful detection of erosion therefore depended on the selection of an appropriate band pair for input into the regression.

5.4.3 Image Ratios

Image ratio approaches compare two co-registered images from different dates by calculating the ratio of bands. Data are compared on a pixel-by-pixel basis and any deviation from the value of 1 (i.e. the same value in each image) is indicative of change. In the aforementioned study of the Cukurova Delta, image ratios were also calculated using log transformed bands. These were found to be a less reliable indicator of land-cover change as the complexity of the changes present was not captured (Berberoglu and Akin, 2009).

5.4.4 Vegetation Index Differencing

Indices are often developed to emphasise specific aspects of ground cover by virtue of their spectral reflectance curves. For example, the 'normalised difference vegetation index' (NDVI) indicates the amount of vegetation present in a pixel. It calculates the difference between reflectance of bands in the near-infrared (high-reflectance) and the red (low-reflectance) regions of the electromagnetic spectrum. Reflectance values on either side of the steep 'red edge' that characterises a vegetation reflectance signature are differenced to provide an index, for which high values indicate pixels with greater vegetation content than low values (Rodgers et al., 2009). Because high values of the index equate to the presence of more vegetation, simple comparisons can be drawn between values of the index on different dates to assess vegetation fluctuation. For example, to subtract an index calculated for one (earlier) date from an index calculated for another (later) date would yield positive values for vegetation gain and negative values for vegetation loss. In this way, a map of the spatial distribution of vegetation change can be established. In the Ban Don Bay area of the Gulf of Thailand, this index was calculated from multiple Landsat 5 images. It clearly identified land-use changes from agricultural activity to increased urban growth, including the development of shrimp farms across the period 1990–99 (Muttitanon and

Tripathi, 2005). The NDVI was also used to assess the sensitivity of coastal vegetation in the Weeks Bay Reserve, Alabama, USA, to the tropical cyclone Hurricane Katrina (Rodgers et al., 2009). Values of this index calculated from Landsat 5 images across a range of wetland vegetation types indicated variable recovery following the cyclone. Large decreases in estuarine emergent wetland vegetation were detected. These arose because of increased salinity introduced by the storm surge and regional drought conditions following the cyclone.

5.4.5 Principal Components Analysis

The multivariate analysis technique of principal components analysis (PCA) reduces the number of spectral bands into a single component that accounts for the most variance in relation to the original multispectral bands. This can be applied to the before and after images separately, before they are subsequently compared. Alternatively, it is possible to group all the bands from the before and after images into a single image dataset. Principal components analysis of this multi-temporal dataset will then identify areas of change as major components of variation in the radiance values of the two scenes (Singh, 1989). In the northeastern USA, changes in the coverage of eelgrass (*Zostera marina*) as a result of disease and pollution were mapped using PCA. Four bands of a 1990 Landsat TM image were compared with the same bands of a 1992 image. Standardised and unstandardised principal components analyses were subsequently performed on the resultant image and a threshold value was defined to separate changed and unchanged pixels of eelgrass density. The standardised approach was found to be more accurate (overall accuracy of 72%), although this was outperformed by a simpler image differencing technique (Macleod and Congalton, 1998).

5.4.6 Change Vector Analysis

Change vector analysis describes the direction and magnitude of change between two images from a vector. A threshold is applied to determine whether the magnitude of change is significant. As with the band ratio technique, this proceeds by plotting corresponding bands against each other and determining the position of pixels in relation to the theoretical line of equivalency or 'no change'. This approach can plot three bands against each other and interpret directions of change as particular transition types (e.g. vegetation regrowth, vegetation clearance or urbanisation). It was used near the village of Buen Hombre in the Dominican Republic, Caribbean, to document changes to intertidal mangroves, shallow water seagrasses, macroalgae and coral reefs (Michalek et al., 1993). Prior to change vector processing, two Landsat TM images were geocoded, registered and radiometrically normalised before transitions between the land-cover classes were assessed through coding of the vectors generated.

5.4.7 Post-Classification Comparison

Post-classification comparison of maps requires the complete production of two independent maps for comparison. It is necessary for the maps to be standardised in the classification scheme adopted for the land-cover features represented. This may restrict the choices available for thematic content to be included in contemporary mapping exercises where the objective is to compare a recent output map product with a historical map. A key advantage of finalising the preparation of both 'before' and 'after' maps before comparing them is that, for maps derived from remotely sensed imagery, any atmospheric and sensor differences between the datasets are inherently normalised. Assessments of change will usually take the form of a 'change map'. The accuracy of this change map will correspond to the product of the accuracies of each individual classified map (Stow et al., 1980). Alternatively, a 'change matrix' can be constructed to compare the maps (Section 5.5.3). Along the northwestern Mediterranean coastal zone of Egypt, maximum likelihood classification and post-classification matrices were tabulated to assess changes between different land-cover classes relating to agriculture and tourism development. Improvements in accuracy were achieved by integrating the automated classification algorithm with visual interpretation, resulting in a procedure that accurately detected increases in urban settlements and agricultural land, with associated losses of natural vegetation (Shalaby and Tateishi, 2007).

5.5 Constructing Change Matrices From Vegetation Maps of Aldabra Atoll

This case study describes the construction of a change detection matrix to compare two maps of the vegetation communities of the UNESCO World Heritage Site of Aldabra Atoll. This atoll in the western Indian Ocean supports the world's largest population of giant tortoises (*Dipsochelys dussumieri*). The islands that rim the Aldabran lagoon are colonised by a series of plant associations that are extremely rare worldwide. The vegetation supports the tortoise populations, which graze on 'tortoise turf', a mix of grasses, sedges, herbs and palatable shrubs and trees (Merton et al., 1976).

The exercise draws on an early vegetation map of Aldabra Atoll constructed as part of the Royal Society of London's Aldabra Research Programme and subsequently published in 1983 (Gibson and Phillipson, 1983). This was compared to a map made from satellite images of the atoll in 2009. This represents a common data availability scenario for a change detection exercise. A published historic map prepared from a combination of field survey and aerial photography is compared to a more contemporary digital map derived from classifying remotely sensed satellite imagery. By comparing two complete map products, the need to standardise between image properties (time of day for image acquisition, weather conditions, etc.) is overcome.

5.5.1 Aerial Photography and Ground Survey Procedures Applied in the 1970s to Produce Map 1

For the construction of the first map, field surveys were undertaken from 27 land areas of 1 km × 1 km that were randomly selected across the atoll. Within each of these areas, a belt transect was established. The width of this transect varied between 4 and 20 m, depending on the density of vegetation cover. Each transect was divided into 50 m lengths, along which both woody and ground vegetation were recorded. A cluster analysis was then used to generate a classification of 26 vegetation types to be mapped. Vegetation belt transects were superimposed onto aerial photography acquired by the UK Government's Department of Overseas Surveys in 1960 at a scale of 1 : 25 000. This procedure established boundaries between vegetation types that could be identified from both data sources. Extensive ground surveys throughout the atoll were then used to interpolate these boundaries between transects. Final boundaries were traced onto the aerial photography and a pull-out map was published at a scale of 1 : 1500 in Gibson and Phillipson (1983).

To draw a comparison, the Gibson and Phillipson map was scanned and georeferenced using information from the grid of coordinates overlaid onto the published map. Vegetation boundaries were then digitised at a scale of 1 : 1500. Digitising error was calculated as the standard deviation in map polygon positions from repeated digitisation of the same section of the vegetation map using the method of Ford (2011). Digitising error was found to be 0.59 m on the basis of 10 repeated digitisations.

5.5.2 Satellite Image Processing and Classification Procedures Applied in 2001 to Produce Map 2

In February 2009, the earlier Gibson and Phillipson transects were revisited by teams from the *Cambridge Coastal Research Unit* and the *Seychelles Islands Foundation*. Vegetation communities were surveyed against categories designed to enable comparison against the earlier classification scheme (see Table 5.3). In addition to this, an independent ground referencing dataset of 266 records was collected from around the islands. Each of these records represented a location point that was visited in the field where the type of land-cover vegetation was recorded in line with the recent classification scheme. This dataset was employed for accuracy assessment of the recent vegetation map.

Three QuickBird satellite images covering the case study area with a spatial resolution of 1 m were acquired on 4 and 6 February and 6 March 2009 (see Figure 5.3a for coverage of the three images). Each of these images comprised four wavebands: one in each of the red, green and blue sections of the visible electromagnetic spectrum (2.4 m spatial resolution) and one panchromatic band, spanning the red and near-infrared wavelengths (0.6 m resolution). As a pre-processing step, each image was atmospherically corrected using empirical line methods (Karpouzli and Malthus, 2003). All areas submerged by the ocean were removed to leave an image that only covered emergent island areas. An unsupervised classification was performed using a maximum likelihood parametric rule on the three pre-processed satellite images of the islands. The

Table 5.3 Comparison of the scheme employed for classification of the QuickBird Images and production of the Gibson and Phillipson (1983) map.

Satellite image classification (tier 1)	Gibson and Phillipson (1983) classification (tier 2)
Standard mixed scrub	Mixed scrub: standard Dune Jean-Louis mixed scrub 'Malabar', 'Gionnet' and intermediate types mixed scrub Picard mixed scrub
Open mixed scrub	Mixed scrub: open Groves *Cyperus* park-woodland Takamaka open
Pemphis scrub (*Pemphis acidula*)	*Pemphis* scrub: grade from local mixed type to *Pemphis*, open and closed Fern pocket *Pemphis* mixed scrub, open and closed
Casuarina and coconut palms	*Casuarina*, coconut (*Cocos nucifera*) and takamaka (*Calophylum inophyllum*)
Coastal grassland	*Sporobolus*/*Sclerodactylon* grassland and wooded coastal fringe Coastal
Coastal scrub	*Scaevola* or other wooded vegetation, open coastal mixed scrub
Pool and mangrove	*Thespesia populnea*/*Lumnitzera* pool vegetation *Avicennia* and mudflats Mixed mangrove woodland (closed canopy) Tidal pools with mangrove or *Pemphis* fringe Including species: *Rhizophora mucronata*, *Bruguiera gymnorhiza*, *Lumnitzera racemosa*, *Avicennia marina*, *Ceriops tagal*

classified output was a single thematic layer composed of 30 classes. These were subsequently interpreted with respect to a ground referencing dataset from recent terrestrial survey records from February 2009. Accuracy assessment proceeded by comparing the output classification to an independent ground referencing dataset of 266 records of vegetation cover. Overall accuracy was defined as the proportion of ground reference points that were assigned the same class as the map layer (Congalton, 1991). Of the 266 points used for the accuracy assessment, 220 were found to be classified correctly, yielding an overall accuracy of 82.7%.

5.5.3 Construction of a Change Matrix

Before a systematic comparison of the vegetation could be undertaken, it was necessary to group zones of similar vegetation type with respect to a simplified scheme. This was achieved using a two-tiered hierarchical scheme in which multiple classes of

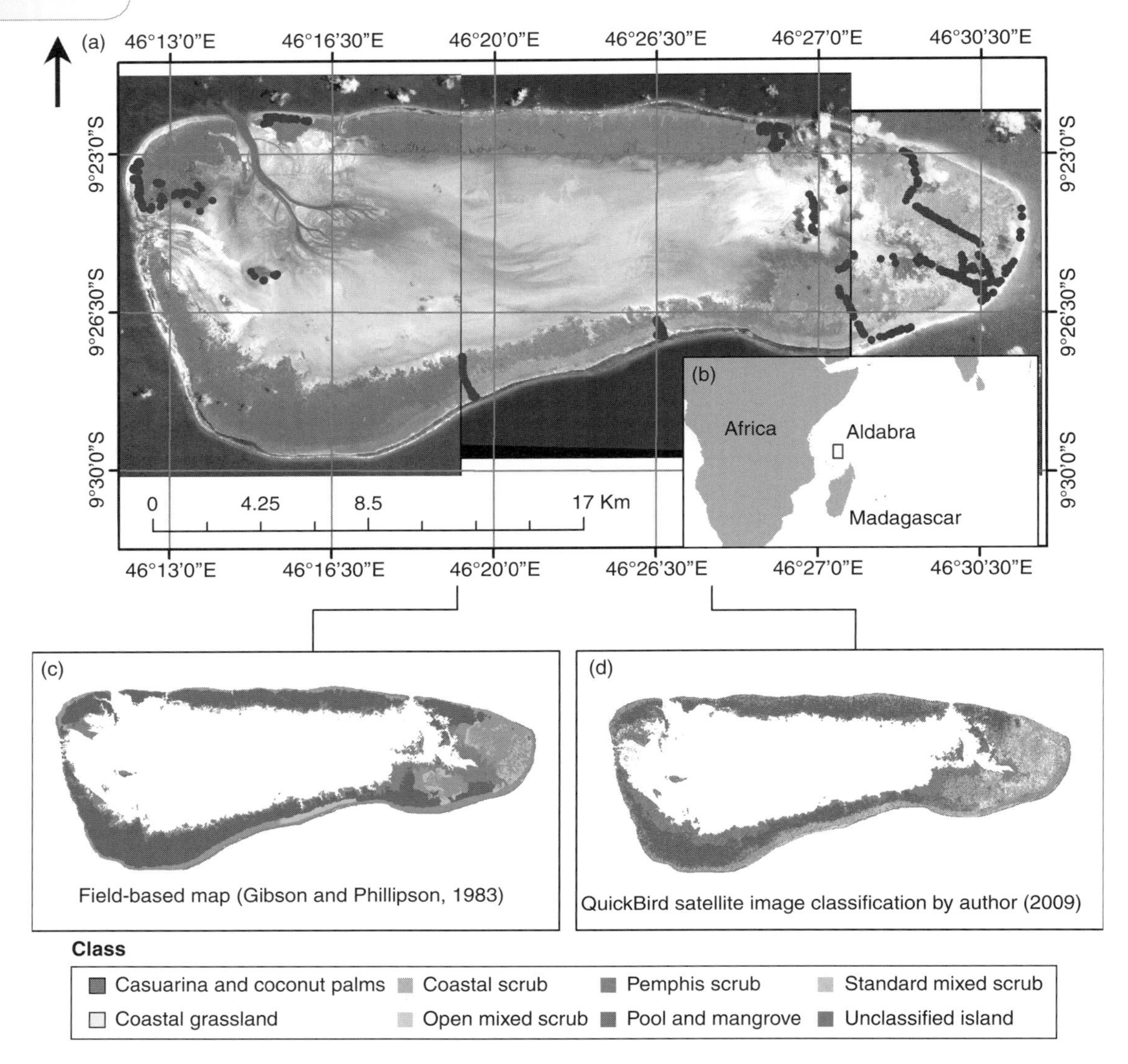

Figure 5.3 (a) Three QuickBird satellite images showing Aldabra Atoll. The (red) dots indicate field vegetation surveys undertaken in 2009. (b) The location of Aldabra Atoll in the west Indian Ocean in relation to the African continent and Madagascar. (c) Aldabra terrestrial vegetation map produced by Gibson and Phillipson (1983). (d) Vegetation mapped via classification of the QuickBird images by the author in 2009. (A black and white version of this figure will appear in some formats. For the colour version, please refer to the plate section.)

the more detailed Gibson and Phillipson map (tier 2) were grouped into a higher-level (tier 1) class for comparison (Table 5.3).

A change matrix was constructed to assess in detail the differences between the tier 1 classes depicted in the two maps (Table 5.4). Classes were coded numerically using successive powers of 2 to ensure every possible land-cover transition could be identified by a unique value, and the difference between the 2009 and 1983 digital files generated from the maps was then calculated for input into the change matrix.

At the scale of the overall atoll, 32% of the vegetation had changed. The most commonly occurring transition was '*Pemphis* scrub' to 'standard mixed scrub', and other common changes were bidirectional transitions between '*Pemphis* scrub' and 'pool and mangrove'. The accuracy of the change detection exercise was estimated to be 81.7%, based on the product of the accuracies associated with the individual maps.

5.6 Monitoring Shoreline Changes From Aerial Photographs

Geomorphic changes along shorelines, such as beaches, spits, cays and dune ridges, are driven largely by processes that serve to build up or break down sedimentary landforms. These might include the delivery of terrigenous sediments from inland rivers (Willis and Griggs, 2003), or offshore sediments through the action of waves (Stoddart and Steers, 1977) and cyclones (Scoffin, 1993; Hubbard, 1997). Human-induced changes to sediment dynamics may occur because of coastal developments (Flood, 1979). Notwithstanding some degree of natural coastal adjustment, the balance of sediment movement onto, away or along sections of shoreline depends on the sum impact of multiple local environmental processes superimposed onto one another. These processes often occur across timescales that are amenable to detection through the visual interpretation of aerial imagery. Features can be observed within the imagery and manually digitised for different snapshots of time. Their positions can then be compared to measure changes or shoreline movements.

Multi-temporal archives of historical aerial photography, satellite images and maps have been used to observe horizontal advancing or retreating island shorelines through comparison of their outlines (see Table 5.5). To manually extract accurate distance measurements from air photos, it is necessary to use georectified, 'controlled' mosaics that have been corrected for the tilt and magnification of aerial photographs, and to which orthorectifications have been applied to account for the variable effects on perspective across a field of view through the camera lens (Crone, 1966). These often select a visually discernible feature, such as the beach toe, vegetation edge, the high water line, the low water line, or the wet/dry line as a proxy for consistent definition of the shoreline (Boak and Turner, 2005). Features are visually interpreted and digitised from photographic archival records to build up a vector dataset representing the 'shoreline' at different times. The shoreline position can also be measured directly in the field by global positioning system (GPS) surveys, or extracted from three-dimensional light detection and ranging (LiDAR) surveys. Where multiple shoreline

Table 5.4 Vegetation change matrix calculated by comparison of two vegetation maps constructed for Aldabra Atoll. The earlier map was constructed from a combination of aerial photography and field observations by Gibson and Phillipson (1983). The later map was constructed from a combination of satellite imagery and field observations by the author in 2009. Numbers indicate land areas (units: km^2) of vegetation compared. Grey boxed numbers along the diagonal axis indicate areas mapped as the same vegetation type by each exercise.

QuickBird (2009)	**Gibson and Phillipson (1983)**						
	Standard mixed scrub	**Open mixed scrub**	***Pemphis* scrub**	***Casuarina* and coconut palms**	**Coastal grassland**	**Coastal scrub**	**Pool and mangrove**
Standard mixed scrub	17.3	3.8	9.6	0.1	0.2	2.5	2.0
Open mixed scrub	3.9	3.7	2.4	0.03	0.2	1.5	2.1
Pemphis scrub	2.4	0.3	57.7	0.1	0.0	0.6	4.9
Casuarina and coconut palms	0.2	0.0	0.2	0.3	0.0	0.1	0.0
Coastal grassland	0.2	0.0	0.1	0.0	1.3	1.1	0.0
Coastal scrub	1.9	0.7	2.2	0.1	0.1	4.4	0.2
Pool and mangrove	0.2	0.1	5.4	0.0	0.0	0.0	18.5
Total 1983	26.1	8.6	77.6	0.6	1.9	10.2	27.7

Table 5.5 A summary of reef island change assessments that use multi-temporal remote sensing.

Location	Summary	Reference
Central Pacific	Regional analysis of planimetric island adjustments across 27 atolls in the central Pacific in response to sea-level rise	Webb and Kench (2010)
Funafuti Atoll, Pacific	Assessment of vulnerability to flooding and inundation at Fongafale islet on Funafuti Atoll, including construction of a digital elevation model from stereo pairs of aerial photographs (1984) and reconstruction of land-use and settlement patterns	Yamano et al. (2006, 2007)
Hawaii	Use of ortho-photomosaics in the analysis of past shoreline change on Maui, Hawaii	Fletcher et al. (2003)
Kume Island, Japan	Use of aerial photography to assess the impact of typhoons on elongated sand cays in Japan	Hasegawa (1990)
Chagos Islands or British Indian Ocean Territory	Use of maps and satellite imagery in the assessment of atoll rim shoreline change dynamics at Diego Garcia	Hamylton and East (2012)
Marshall Islands, Pacific	Assessment of changes on an urbanised atoll, Majuro, Marshall Islands, in relation to sea-level rise	Ford (2011)
Palmyra Atoll, Central Pacific	Evaluation of sediment redistribution patterns	Collen et al. (2009)
Assateague Island, Maryland/Virginia, USA	Use of an automated method for estimating shoreline changes arising from storms from airborne topographic LiDAR data	Stockdon et al. (2002)
North Carolina, USA	Evaluation of airborne topographic LiDAR for quantifying beach change against ground-based GPS surveys	Sallenger et al. (2003)
Raine Island, Great Barrier Reef, Australia	Use of historical topographic surveys, a simple numerical model and GIS techniques to reconstruct a 40-year (1967–2007) shoreline history of Raine Island	Dawson and Smithers (2010)
Tarawa Atoll, Kiribati	Short-term reef–island area and shoreline change over 30 years determined by comparing 1968 and 1998 aerial photography	Biribo (2012)

positions are required, it will be necessary to bring data from different sources into a comparable frame of reference before estimating changes. For example, at Heron Island on the southern Great Barrier Reef, Flood (1974) used aerial photography to observe reef island shoreline adjustments to seasonal wind patterns. Southeasterly winds (between February and August) drove the reef island leeward, to the northwest part of the reef, and shoreline adjustments in the opposite direction were observed in response to northwesterly winds (between September and January). Elsewhere on the Great Barrier Reef, analysis of aerial photography has revealed consistent minor oscillatory motions in oblongate spit orientations at Heron island (Flood, 1974), Erskine Island (Flood and Heatwole, 1986), Low Isles (Stoddart et al., 1978) and Gannet Cay (Flood and Heatwole, 1986).

5.6.1 The Digital Shoreline Analysis System (DSAS)

The widespread use of aerial photographic records to analyse horizontal shoreline change is apparent from the range of software developed for this purpose, including the Digital Shoreline Analysis System (DSAS) (Thieler and Danforth, 1994) and the Analysing Boundaries Using R (AMBUR) (Jackson, 2010). These interrogate boundary datasets derived from multiple digitisations of the shoreline.

The DSAS application is a free extension to Esri ArcGIS developed by the US Geological Survey. This establishes a series of transects adjacent to the shoreline along which a statistical dataset of repeated change measurements can be built up. Shoreline vectors are manually digitised using a proxy feature to consistently represent the shoreline from available multi-temporal images, each of which correspond to a given point in time. Shore-perpendicular measurement transects are cast at user-defined intervals (e.g. every 10 m) off a baseline. This may correspond to a user-drawn position or the oldest available shoreline position. Measurement transects are established across the baseline in such a way that they intersect the shoreline vectors at points from which location and time information can be extracted to calculate rates of change. The intersect points between transects and vegetation lines are recorded and used in further analysis.

For a time series of shoreline vector data, rate-of-change statistics can be computed using a semi-automated approach. Statistical metrics are calculated via comparison of the baseline and each intersection point along a transect (see Nantucket example in Section 5.7). A range of change statistics can be calculated. These include:

- the shoreline change envelope (the distance between the shoreline farthest from and closest to the baseline at each transect),
- net shoreline movement (distance between the oldest and youngest shorelines for which information is available),
- end point rate (overall distance of shoreline movement divided by time elapsed between the oldest and most recent shoreline), and
- linear or least squares regression rates (obtained from the slope of a plot of the shoreline date against the distance from the baseline).

Calculations of shoreline rates of change are only as reliable as the shoreline maps used to generate the data. This depends on errors associated with both the analysis procedure (i.e. the calculation of change statistics) and representation errors when compiling each shoreline position (Thieler and Danforth, 1994; Moore, 2000). Rates of apparent observed shoreline movement are accompanied by estimates of uncertainty, such as the standard error, confidence intervals and goodness-of-fit diagnostics (R^2). Overall uncertainty associated with representing the shoreline itself should incorporate both positional and measurement error arising from aerial photo distortion or reprojection of older historical map sources (Crowell et al., 1991). A weighting can be used to emphasise change statistics that have been calculated from points with a lower uncertainty.

Limitations associated with the multi-temporal analysis of aerial photographs to monitor island changes over time include the differing resolution or quality of images (Anders and Byrnes, 1991), poor condition of negatives and the fact that historical images were seldom acquired for this purpose. Associated flight angles and paths, exposure, coverage and elevation are seldom optimal for comparing shoreline positions (Webb and Kench, 2010). Nevertheless, in many geographic areas they offer the best opportunity for determining coastal change over time.

5.7 Shoreline Changes on Nantucket Island, Massachusetts, USA

The Massachusetts coastline supports tourism, shipping and commercial fishing industries, all of which make an important contribution to the local economy (Thieler et al., 2013). In 1989, the Massachusetts Office of Coastal Zone Management initiated the Shoreline Change Project to assess shoreline erosion and accretion around Nantucket Island between the mid-1800s and 2009. Historical shoreline positions were delineated from a range of information sources. These included the National Oceanic and Atmospheric Administration's topographic map sheets (T-sheets) and aerial photographs from the Office of Coastal Zone Management and the US Geological Survey. Estimates of positional uncertainty incorporated errors from georeferencing, digitising, topographic surveys underpinning the T-sheets, aerial photograph acquisition and rectification and positioning of the high water level at the time of survey (Crowell et al., 1991).

Seven shoreline images of Nantucket Island were acquired between 1846 and 2009. Change statistics were calculated along a series of transects spaced 100 m apart running perpendicular to the user-defined baseline. These were established in such a way that they intersected the subsequent shorelines so that measurements could be calculated for the area immediately seaward of the current shoreline (Figure 5.4). Long-term rates of shoreline change were determined by fitting a least squares regression line to all shoreline positions from the oldest (mid-1800s) to the most recent (2007, 2008 or 2009). The rate of change is the slope of the regression line. The linear regression method for determining rates of shoreline change assumes a linear trend of change between the oldest and most recent shoreline dates. This was an appropriate approximation for this section of shoreline, which consistently eroded over the period of measurement.

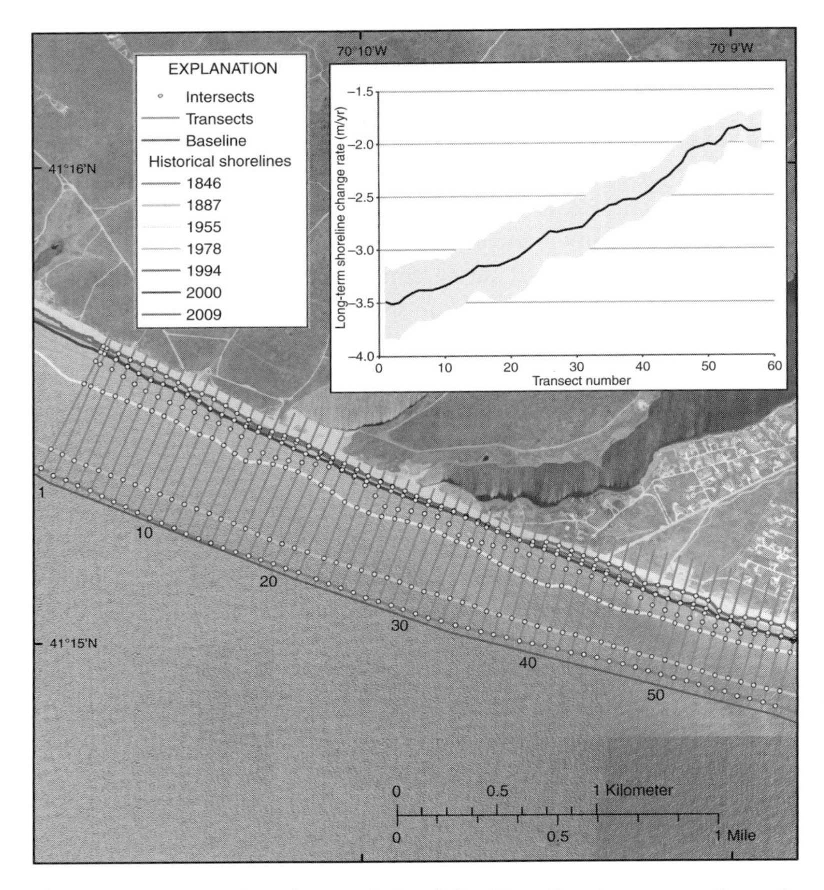

Figure 5.4 Change detection analysis of the Cisco Beach area, southern shore of Nantucket Island, Massachusetts. This was conducted using the US Geological Survey's Digital Shoreline Analysis System. Historical shorelines digitised from older aerial imagery are shown as the coloured lines corresponding to the date range 1846–1994. DSAS-generated transects are established at 100 m spacing with the graph showing the long-term rate of shoreline change based on simple linear regression. Grey shading indicates the 90% confidence interval for the calculated rates. (Courtesy E. Robert Thieler, USGS.) (A black and white version of this figure will appear in some formats. For the colour version, please refer to the plate section.)

Shoreline changes were variable around the wider Nantucket Island region, with erosion on the Atlantic Ocean-facing southern shore and accretion at the end of barrier spits (Thieler et al., 2013). Long-term rates of erosion on the southern shore showed an along-shore trend of decreasing rates going from west to east with a maximum rate of -3.5 ± 0.3 m yr^{-1}. Potential modes of shoreline behaviour included unidirectional shoreline change, fluctuating shoreline position, constrained shoreline position and anthropogenic shoreline relocation. Measurements were characteristic of unidirectional shoreline changes of moderate rates for the region. These reflected the exposed, open setting of the study area relative to the wider Atlantic Ocean (Thieler et al., 2013). For the Office of Coastal Zone Management, this identified geographical areas in the wider Massachusetts State administrative region that incorporates Boston, Cape Cod and Martha's Vineyard, where coastal management and protection resources were most needed. These could then be focal areas for management interventions such as seawalls, riprap revetments, groins, breakwaters and beach nourishment.

5.8 Assessing Volumetric Changes From Digital Elevation Models

5.8.1 Techniques and Tools for Assessing Volumetric Change on Coastlines

Combining information in the horizontal and vertical planes allows island volumetric changes to be assessed in three dimensions. Although it is possible to collect elevation information for coastal topography studies from LiDAR, the application of this technology to beaches, spits and islands is uncommon. This is because of the expense of these surveys. A cheaper, alternative approach is to derive surface elevation measurements from aerial photographs using a digital photogrammetric stereo-plotter (Yamano et al., 2006; Leon, 2010). Alternatively, oblique photographs can be taken from different angles and scanned, and specialised software, such as Agisoft Photoscan, used to generate three-dimensional terrain models (Agisoft, 2012). Photogrammetry can be applied to both contemporary and historical aerial photographs to generate digital elevation models from which it was possible to elucidate topographic changes (Yamano et al., 2005). For example, to estimate volumetric changes of a sand cay, Green Island in the Great Barrier Reef, topographic information was derived photogrammetrically from multi-temporal aerial photographs. Vertical accuracies of around ±100 mm in height were achieved using this method (Beach Protection Authority Queensland, 1989). This was acceptable for the purpose of monitoring broad-scale fluxes in sediment volume. Photogrammetric measurements can be augmented by beach profile surveys, with interpolation of point measurements using triangulated irregular network techniques to derive continuous vector digital models of terrain and elevation (Biribo, 2012). Alternatively, geostatistical methods such as inverse distance weighting or kriging can provide a raster-based digital elevation model as a continuous representation of the three-dimensional form of a coastal feature from a series of point measurements (see Chapter 6).

Most geographic information system (GIS) software incorporates functions that allow the calculation of the area, surface area and volume of a surface relative to a datum plane. These tools may use either raster or vector datasets as an input. In the case of an input triangulated irregular network (TIN) dataset, each triangle is examined to determine its contribution to the area and volume. The overall area or volume is calculated as the summation of each of the individual triangles. The Topographical Compartment Analysis Tool (TopCAT) provides an extension to these routines for coastal applications, including analysis of changes on sea cliffs and beaches (Olsen et al., 2012). This has been used to compare multi-temporal digital elevation models derived from LiDAR datasets along a retreating coastline in Del Mar, California. In a beach environment, it has also been used to quantify along-shore volumetric sand movements resulting from hurricane Bonnie (1998) in Cape Hatteras, North Carolina (Olsen et al., 2012).

5.8.2 Assessing the Volumetric Changes of a Dredged Lagoon, Diego Garcia, Chagos Islands

Figure 5.5 provides an example of a three-dimensional analysis of volumetric change in the lagoon of Diego Garcia, Chagos archipelago. Diego Garcia is a small, horseshoe-shaped atoll in the middle of the Indian Ocean. An open channel to the north allows mixing of water between the open ocean and a central dish-like lagoon of almost 30 m depth in the centre. The atoll rim is made up of low, flat sandy islands.

The analysis is underpinned by point datasets depicting over 2000 soundings from bathymetric surveys undertaken in 1967 and 1998, before and after the development of a military airbase along the northwestern atoll rim. Survey points were interpolated via kriging to a continuous raster grid for the analysis. A reference surface of mean sea level was employed as a plane against which overall estimates of lagoon volume were estimated. Change statistics were derived by calculating the difference between the two surfaces as a simple map algebra subtraction exercise. Pixel dimensions were calculated for the resulting ‘change raster’. Overall values of volumetric change were calculated from the associated raster attribute tables. A similar exercise conducted using a vector TIN surface derived from the initial bathymetric surveys arrived at comparable estimates of lagoon volume and change (within ±10% for sediment addition and removal).

5.9 Some Pitfalls in the Spatial Analysis of Change

At the heart of any attempt to monitor coastal change over time lies the assumption that the magnitude of change observed represents real change to the features being observed as opposed to an artefact of the process used to detect the change itself. Important analytical considerations include the choice of an appropriate practical method, how change is quantified, the spatial and temporal scales at which this quantification takes place and the estimation of the magnitude of error and uncertainty in the results. Several additional influences could serve to magnify or reduce apparent changes in coastal features, which include the following:

- Implications arising from the use of different technologies to map the location of the coastal features, and inconsistencies that may arise in the representation.
- Methods used to bring two datasets into a common frame of reference (e.g. reprojecting, georeferencing).

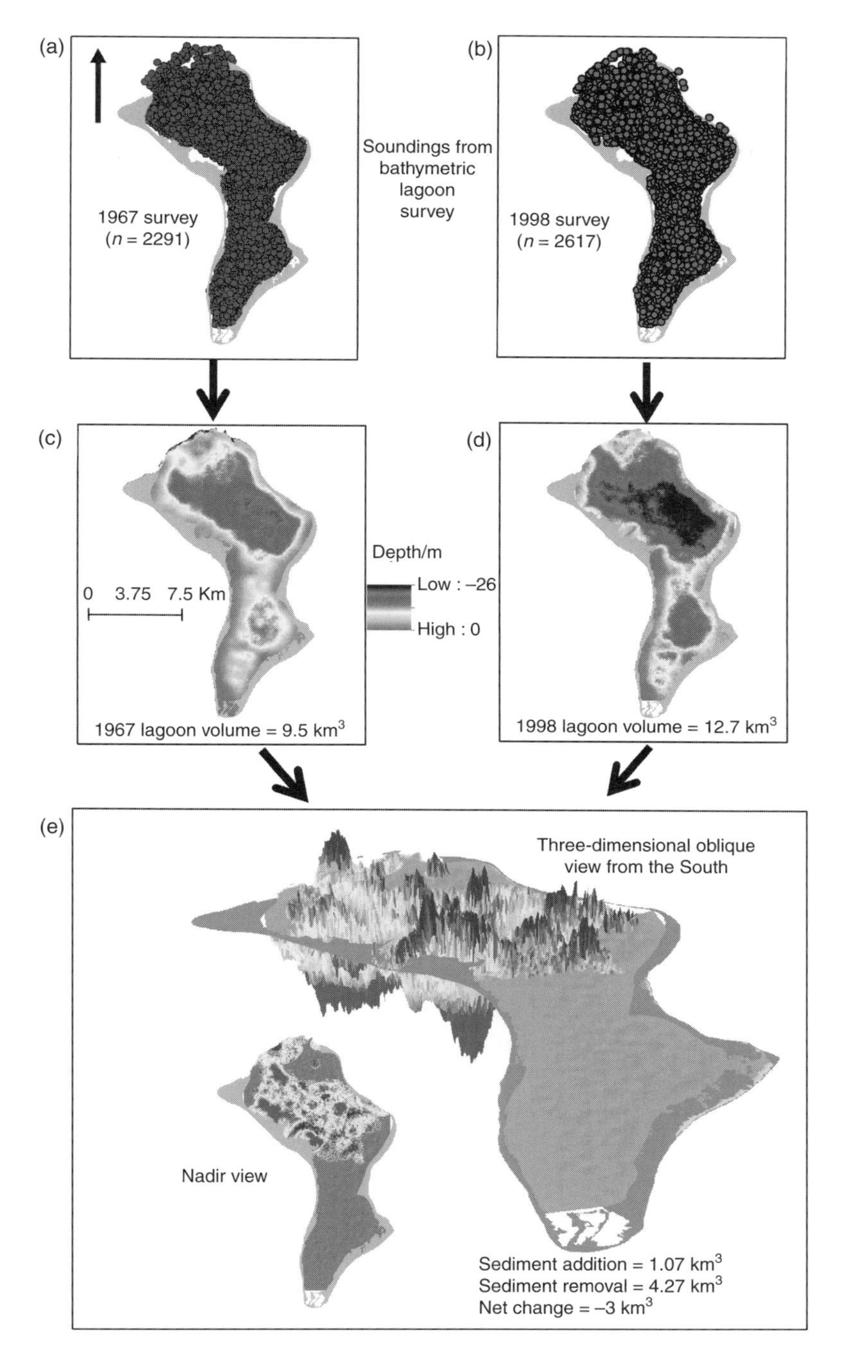

Figure 5.5 An analysis of lagoon volume change at Diego Garcia Atoll as a result of dredging and dumping of sediments during development of a military airbase in the 1970s. (a) Individual echosounder points from the 1967 survey. (b) Individual echosounder points from the 1998 survey. (c) DEM derived in 1967 by interpolating echosounder points. (d) DEM derived in 1998 by interpolating echosounder points. (e) Volume change map derived by subtracting the 1998 DEM from the 1967 DEM. Red (darker) indicates areas of sediment loss from dredging of boat channels; green (lighter) indicates localised areas of sediment gain due to the dumping of dredge spoil. (A black and white version of this figure will appear in some formats. For the colour version, please refer to the plate section.)

- Differences in radiance values of remote sensing datasets caused by factors other than real changes, such as variations in atmospheric conditions, Sun angle or land soil moisture.

Of key importance is the *relative magnitude* of the *total change detected*, and the *total error associated with the procedure* used to detect change real change can only be inferred in a situation where the error associated with the technique does not exceed that measured. For example, in a study comparing aerial photos and satellite imagery to ascertain shoreline change at Majuro Atoll in the Marshall Islands (Ford, 2011), the following five sources of uncertainty were taken into account when extracting shoreline positions using aerial photographs: seasonal variation in climatic conditions, the state of the tide at the time of aerial photo acquisition, digitising error, spatial referencing error associated with the size of the pixels and georectification (Fletcher et al., 2003). The total uncertainty arising from each of these potential sources of error was estimated to fall between ±2.4 and ±2.7 m. It was only once this had been taken into account that any change lying outside this uncertainty band was attributed to net shoreline movement.

A study of shoreline movements around Diego Garcia Atoll in the central Indian Ocean established 69 transects for the measurement of shoreline changes. Comparisons were made between an early vegetation map published of the atoll (Stoddart and Taylor, 1971) and the shorelines mapped from the vegetation edge of an Ikonos image (2005). An initial analysis revealed variable change dynamics around different sections of the atoll rim. Of the 69 rim width transects assessed, 26 were estimated to have eroded and 43 were estimated to have accreted over the time period assessed (Hamylton and East, 2012). Of these 69 change estimates, seven estimates fell within the margin of error calculated for this analysis of ±4.5 m. This accounted for the error associated with digitising the original map and the positioning error associated with the satellite image. This study was later revised because of the small scale (1 : 125 000) at which the original map was published. The margin of error for measurement was adjusted to a wider ±24.7 m. The number of estimates that fell outside this margin was further reduced from 43 to 38. The exercise therefore remained a useful indicator of accretionary dynamics (Figure 5.6). It was only once all sources of error had been taken into account that any estimated change could be attributed to real-world changes in the shorelines.

As with the example of the east Australian coastline (Figure 5.2), if two maps are to be compared to ascertain whether the features they depict have changed, it is important to be aware of the differing ways in which those maps have been constructed. Standardisation of the mapping methods helps to ensure consistency across multiple time steps. This is why the Landsat programme, which has been collecting imagery consistently since 1972, provides an ideal suite of continuous sensors that acquire images between which comparisons can be drawn. Analysis of aerial photography is one technique that has been used to detect decadal changes in coastal environments dating back to the beginning of the twentieth century, when aerial photography began to be used extensively.

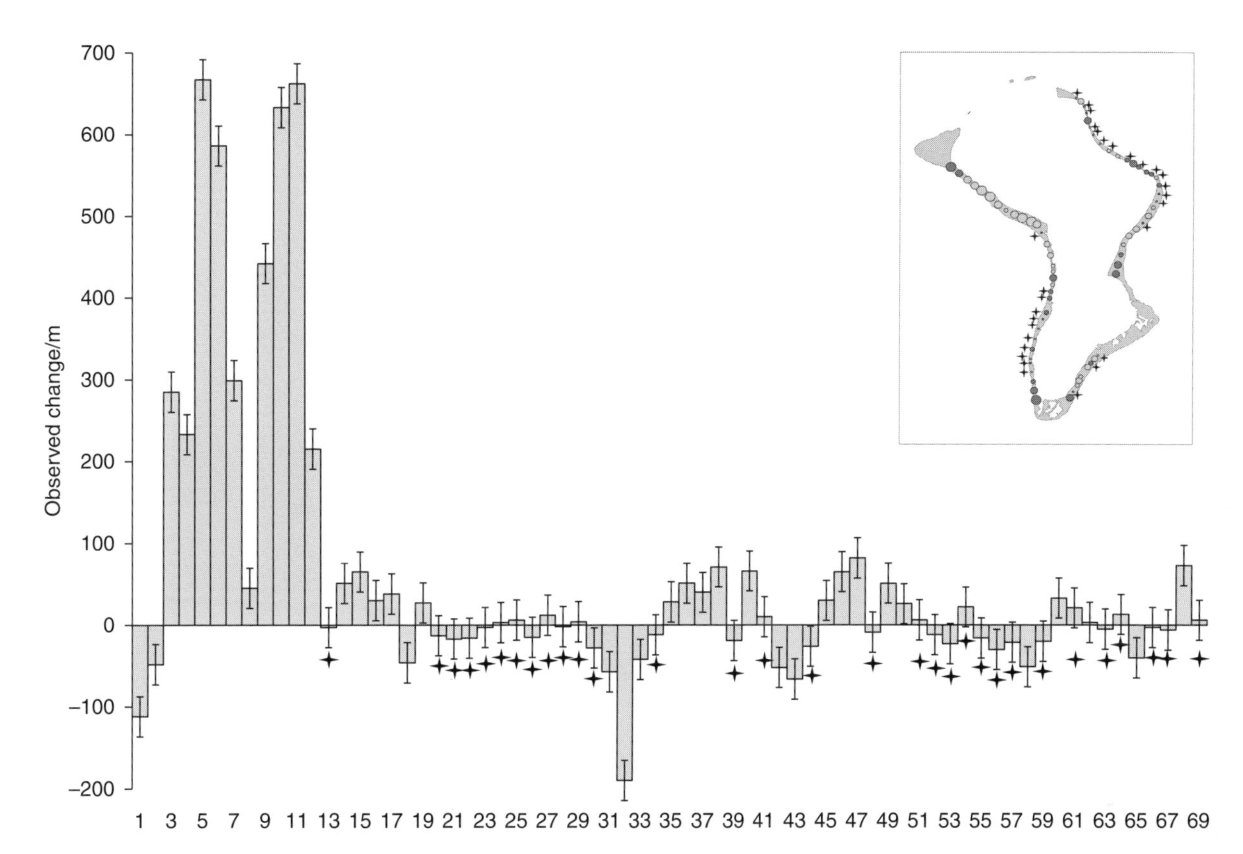

Figure 5.6 Estimates of change in width (metres) of the sandy rim around Diego Garcia Atoll between Stoddart's 1967 survey (Stoddart and Taylor, 1971) and the 2005 Ikonos satellite. Black crosses denote points where the estimated change fell within the methodological error range. On the inset diagram, dark circles denote erosion, while light circles denote accretion. Circle size is indicative of the magnitude of change observed. Transects are numbered sequentially from the northwest, down to the southern tip and then back up to the northeastern point of the atoll (see Figure 5.5 for an indication of scale).

Since aerial photography has the longest history as a coastal remote sensing technology, it represents a longstanding source of information that enables change to be documented through time. In order to compare film-based prints with modern digital data, it is first necessary to scan the photographs at an appropriate resolution (for example, 300 dots per inch), resample, georectify and orthorectify each image. Only once adequate pre-processing has been carried out will it be possible to compare images on a pixel-by-pixel basis. To minimise locational error, care should be taken to ensure that such spatial referencing is carried out in relation to *independent* ground control points. These are easily recognisable (and immobile) features in each photograph. Where two images are to be compared to each other, they must be georeferenced separately, with respect to an independent dataset, prior to comparison. It does not make sense to georeference one to the other before comparison. This will transfer across any spatial referencing errors from one dataset to the other, as well as align features in the images, prior to comparison. Such a step will therefore bias the results of the change detection exercise by minimising any changes observed.

Prior to the comparison of any two maps, it is necessary for them to be brought into the same frame of reference. This applies to the spatial referencing of the image, as well as the features themselves, particularly where they are subject to cyclic or seasonal variation. For example, if the objective of the change detection exercise is to measure changes in algal populations over time, then it follows that the time at which these populations are mapped will have to be carefully selected in relation to any cyclic algal blooms that may take place. In relation to image spatial referencing, there may be several ways in which the images need to be manipulated before we can be sure (or as sure as possible) that the areas or pixels being compared relate to corresponding features. Are the two maps produced at the same scale and in the same projection, or does one need to be reprojected?

The radiance values of remote sensing datasets in an image scene may be influenced by factors other than characteristics of the features depicted themselves. These include atmospheric or water column conditions (see Section 4.10.1), scene illumination due to the time of day of image acquisition and related shadows and weather influences such as the presence of clouds or the amount of moisture due to rainfall. Where change detection proceeds by directly comparing radiance values of a scene, these influences on the data need to be removed to bring the scenes into the same spectral frame of reference.

5.9.1 Sources of Error in Change Detection Studies

Sources of error that may arise when maps are compared over time can be grouped under five broad headings:

- inaccuracies arising from the selected data format (raster or vector),
- positional inaccuracies during georeferencing and rectification,
- comparison of information to which different projections or datums have been applied,

- vector-based inaccuracies during transcription and digitising, and
- inaccuracy in the attribution of labels to land-cover classes (photointerpretation, misclassification).

These five sources of error that pertain specifically to change detection are briefly reviewed here, while Chapter 8 provides a broad overview of sources of error and uncertainty in spatial analysis.

The choice of whether to use a raster- or vector-based approach to change detection will depend on the nature of the change itself. Where a continuous, gradual change within a single feature across a large geographical area is present, it may be preferable to use a raster-based approach. In contrast, an abrupt boundary within a specific feature of interest (e.g. a single rubble or sand spit) might be more precisely captured using a vector-based approach that focuses more explicitly on features. This is because the positioning associated with a specific feature will need to be generalised to the dimensions of a pixel in order to store the information in raster format. The individual coordinates that define each point, line or polygon will be retained in a vector dataset.

It is often the case that a historical record of a given feature takes the form of a hard-copy aerial photograph, or a map pulled from a library or office drawer. Because many maps go back further in time than the advent of aerial photography, they are useful for extending the time range of change detection studies. It may be necessary to undertake a series of processing steps to get the information into a format that is fit for comparison (see Figure 2.6). After scanning a photograph to create a digital file, it will be necessary to geometrically correct an image, that is, to add geographical coordinates to the image. The accuracy of the georeferencing procedure is determined by the number and spread of the ground control points (GCPs), the accuracy with which their image and map positions can be determined and the nature of the transformation equation employed (e.g. polynomial transformations of various orders, localised rubber sheeting, etc.). Problems can arise in the selection of GCPs, particularly where features such as seagrass beds, river deltas, marshes and rubble spits are highly dynamic. While the changeable nature of coastal environments is implicit in their selection as the subject for a change detection study, GCPs need to be stable features that provide a constant location to act as a reliable link between multi-temporal images.

Given that different maps and aerial photographs may have originated from different archival sources, it is likely that they have been positioned with respect to different reference systems. They may need to be brought into the same frame of reference by reprojecting one (or more) of these maps or images prior to comparison. Where elevation information is compared, elevation values must be supplied with respect to a common horizontal and vertical datum (e.g. mean sea level, Australian Height Datum or lowest astronomical tide).

Images must be co-registered accurately because any spatial offsets between the maps may introduce erroneous changes into the assessment. The distinction between global and local positional errors is important because positional errors arising from poor georeferencing may be clustered in particular regions due to the practical constraints of field survey. For example, the collection of GCPs may be restricted to beach and fore dune areas not submerged by a high tide. Local errors will have important implications when overlaying two maps. If georeferencing errors are systematically biased in a given area, then the overlay error will deviate substantially from that predicted by the combination of the independent georeferencing errors (Green and Hartley, 2000). A goodness-of-fit measure between the two sets of coordinates, such as the root mean square value, might therefore imply a level of reliability that is relevant for some areas of the map but not others.

When assessing change by overlaying vector data, the process of digitising and georeferencing mapped lines can add positional errors to boundary locations. Common causes of digitising error include unsteadiness of hand, parallax between the cursor and map, the thickness of the line as depicted on the base map, generalisation of curved lines into a series of short straight line segments, accidental production of vector 'slivers' and the warping, shrinking or distortion of base maps through photocopying (Bolstad et al., 1990). It is possible to establish the accuracy of digitised features such as shorelines by digitising a set of reference points repeatedly and calculating the displacement between these repeated digitisations. This can subsequently be expressed as the standard deviation of point locations along both the x and y axes.

Manual or automated interpretation of photographs underlies many change detection exercises. It is important to standardise the conditions under which photographs were taken to minimise variations that may influence how these images are interpreted. These might include:

- seasonality (particularly in relation to vegetation dynamics),
- lighting conditions,
- atmospheric effects (e.g. clouds),
- scale,
- resolution, and
- processing (development of negatives, orthocorrection, contrast adjustment) (Green and Hartley, 2000).

Variation in such conditions may erroneously be attributed to changes in features, characteristics or boundary locations. Where boundaries are to be defined, the placement of a dividing line along a continuous gradient and decisions of boundary position, shape and precision will inevitably depend upon interpreter subjectivity. Bie and Becket (1973) note that, although most digitisation exercises ostensibly involve drawing boundaries around apparent areas of uniformity, the conscious or unconscious objectives of the interpreter will determine which boundaries are sought and delineated. Where boundaries are manually digitised, it is therefore useful to ensure that the observer is consistent in his or her interpretations. This will mean that the same abstractions are applied to each dataset.

Total error for a change detection exercise should be calculated by taking into account sources of error relating to both datasets at every stage of the exercise (see Figure 5.1). The law of propagation of errors, which assumes that sources of error are random and independent, can be used to combine basic root mean square (RMS) errors from each stage of the change detection flow diagram. This produces an overall standard deviation, or RMS error (Chrisman, 1982; Crowell et al., 1991), as follows:

$$E = \sqrt{e_{\text{dig}}^2 + e_{\text{geo-ref}}^2 + e_{\text{clas}}^2} \tag{5.1}$$

where

e_{dig} = digitising error
$e_{\text{geo-ref}}$ = georeferencing error
e_{clas} = classification error.

5.10 Summary of Key Points: Chapter 5

- Monitoring identifies changes in an object or phenomenon by observing it at different times (also referred to as time-series analysis, change detection and multi-temporal image analysis).
- To assess change in coastal environments, it is necessary to have a baseline dataset recording the historic configuration of coastal features against which change can be measured. Baseline datasets should be reliable and in a comparable frame of reference, and may include maps, aerial photographs of satellite images.
- It is only possible to attribute the difference between a contemporary map and a historical one to coastal change that has occurred if the magnitude of the difference exceeds the uncertainty associated with both the mapping and comparison methods.
- Coastal change can be measured using raster and vector datasets. The success of a change detection exercise depends largely on the nature of the change being observed and the pairing of this with a suitable technique for its detection.
- Sources of error in change detection include the data used, their format and georeferencing, comparison of information in different frames of reference (projections, datums) and inaccuracies in digitising (vector) and classification (raster).

6 Geostatistical Analysis of Coastal Environments

For the purpose of this chapter, the term 'geostatistics' is adopted broadly to mean statistics used to analyse or predict values associated with spatial phenomena from both point and raster datasets. Geostatistics serves several practical purposes, including:

- quantifying and characterising the spatial structure in a dataset (presence, absence and dimensions), and
- interpolating data from discrete point values to a continuous surface, providing estimates of data values at locations where these values are unknown.

6.1 Global and Local Geostatistics

The way in which the geographical referencing associated with data cases is utilised for spatial analysis can vary. Global and local spatial statistical procedures differ in the spatial extents at which data cases are analysed. A local analysis subdivides an overall population of data cases into local samples according to their position within the study area prior to analysis. A global analysis does not subdivide data cases on the basis of their location; rather, it simultaneously analyses them all together to produce summary statistics that apply across the entire sample. Such an approach assumes that the data adhere to the principle of stationarity. This means that their statistical descriptive properties (e.g. mean, variance, autocorrelation structure) do not change across space or time. In such a situation, geographical patterns and processes remain constant across a landscape or map of a landscape. Because many coastal phenomena exhibit some form of spatial structure, the assumption of stationarity rarely holds. A local statistical analysis may therefore be more appropriate, which draws on a subset of data cases.

Figure 6.1 illustrates the difference between a global and local approach to analysing a photograph of Weirs Cove, Kangaroo Island, South Australia. If it were necessary to summarise some property of the image (e.g. image brightness) using a statistical descriptor (e.g. the mean brightness across a smaller grid of points overlaid onto the image), different values would be obtained for the grid point configurations using a global approach and a local approach (Figure 6.1b and c). The global grid would generate an overall summary statistic that averaged out the areas of the image, which have a range of different brightness characteristics. For example, the vegetation in the foreground of the image appears to be dark and would therefore be associated with correspondingly low brightness values, whereas the beach

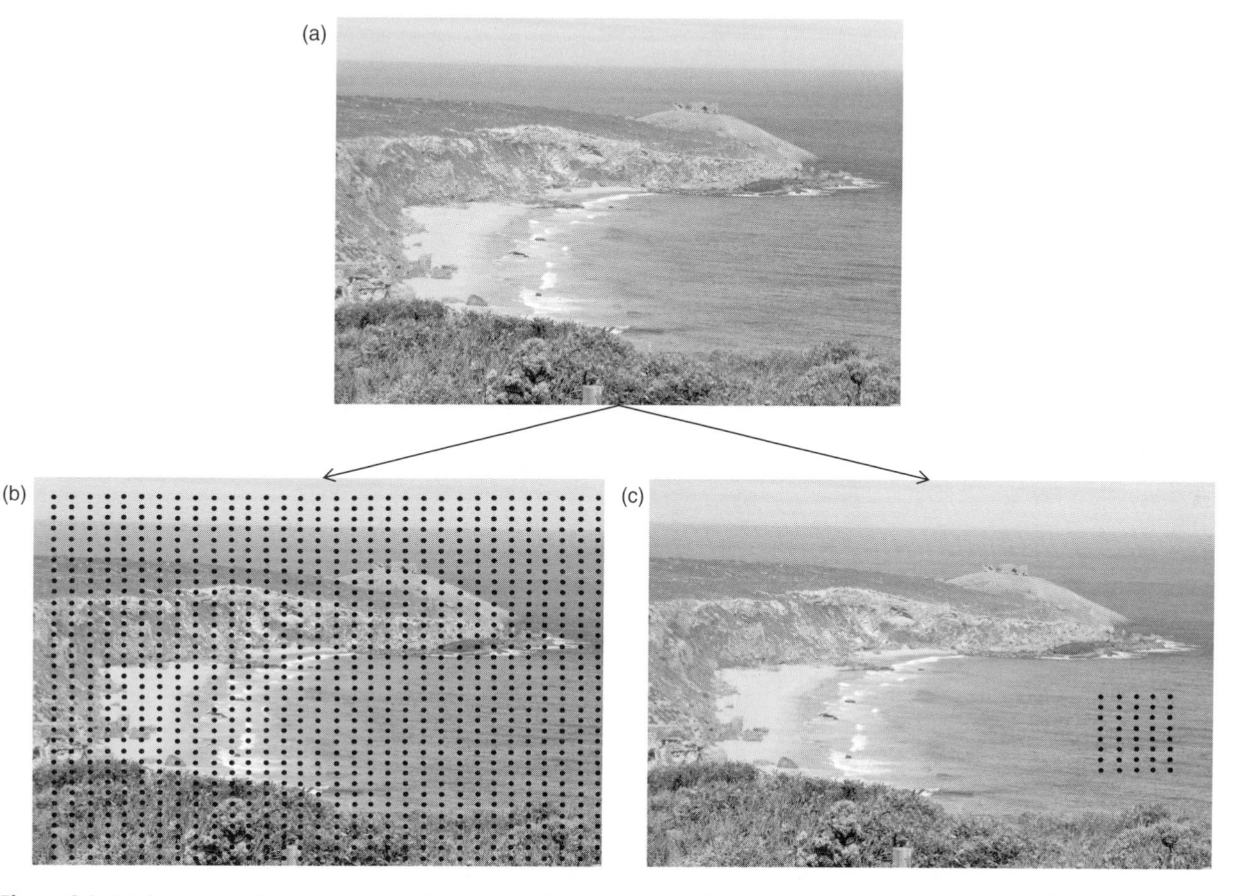

Figure 6.1 (a) A panoramic view across Weirs Cove to Remarkable Rocks, Kangaroo Island (South Australia). (b) Grid points covering the entire image to produce a global analysis of image properties. (c) Grid points covering a subsection of the image to produce a local analysis of image properties; see text for further detail. (A black and white version of this figure will appear in some formats. For the colour version, please refer to the plate section.)

sand, breaking waves and sky areas would have comparably light (or high) brightness values. It may therefore make sense to use a local analysis to capture these differences across the image, as illustrated in Figure 6.1(c). The way in which the grid is defined (e.g. the size and shape) will have important implications for how the differences in brightness are captured by the statistical descriptor. A local neighbourhood may be defined in different ways, depending on the purpose of the analysis and the spatial structure of the phenomenon of interest. For example, the local mean value of a point dataset representing brightness across this image may be expressed by iteratively calculating a value at each location that includes n nearest neighbours, or all points falling within a circle of a nominal radius, or distance band, of that point (see Figure 6.5). By subdividing datasets in this way, it is possible to summarise and express the localised geographical patterning associated with the data values using statistical descriptors.

6.2 Spatial Autocorrelation: a Fundamental Property of Spatial Datasets

Spatial dependence is a fundamental property of most spatially referenced datasets. It arises because of Tobler's first law of geography: *'everything is related to everything else, but near things are more related than distant things'* (Tobler, 1970). Positive spatial dependence means that geographically nearby values of a variable tend to be similar on a map: high values tend to be located near high values, medium values near medium values, and low values near low values. Most coastal phenomena tend to be moderately positively spatially dependent because nearby things are more alike than things in general. For example, patches of mangrove spread across a sandy beach represent interspersed clusters of suitable habitat for many invertebrate species. A measure such as 'organism abundance' might therefore show spatial dependence across such a landscape.

Negative spatial dependence arises when nearby things are less alike than things are in general. This gives rise to an opposing, or chequerboard effect (see Figure 6.2). There are fewer examples of negative spatial dependence in nature, although one example would be the pattern that oil makes as it spreads across the seawater surface. A droplet pattern is formed because the oil is hydrophobic and actively repels surrounding water molecules to form individual droplets that promote the 'water–droplet contact line' (Casagrande et al., 1989). It can also be seen where competition occurs for resources, such as trees competing for light. Negative spatial dependence is less common because of the way phenomena are geographically organised in accordance with Tobler's law. Indeed, negative spatial autocorrelation is impossible to conceive on a continuous surface; rather, it is a scale-dependent phenomenon that can only be expressed on a surface partitioned into spatial units.

Spatial autocorrelation provides a statistical measurement of spatial dependence. The idea of spatial dependence and the application of techniques to quantify spatial autocorrelation underpin numerous spatial analysis techniques. These include geographically weighted regression and interpolation. Spatial autocorrelation is the correlation among values of a single variable directly related to their locational positions on a two-dimensional surface. Autocorrelation is a statistically quantifiable property that expresses the idea of near things being more alike than far things with respect to one single variable (i.e. *auto* refers to 'self', that is, variation within one single variable). It can be loosely defined as the coincidence of value similarity with locational similarity (Cliff and Ord, 1981). Like spatial dependence, spatial autocorrelation arises because of a pervasive *continuity* and *structure* to the real world, in which things rarely change dramatically over short distances (for example, a sand dune is structured by wind, which does not vary randomly across space).

A key objective of geostatistical analysis is the quantification of structure using correlation to demonstrate that attribute values close together in geographical space tend to be more alike than attribute values that are widely separated in geographical space. This is usually achieved by plotting a function of the difference (or similarity) in the value of a single variable measured at two locations against the distance between measurement locations. By comparing combinations of data case pairs across a given area at a range of distances apart, this plot provides insight into the emergent spatial scales of variation across the wider population of data samples.

6.3 Why Measure and Account for Spatial Autocorrelation?

There are two primary reasons to measure spatial autocorrelation. First, spatial dependence introduces a departure from the independent observations assumption of classical statistics. Accordingly, the measurement of autocorrelation indexes the nature and degree to which a statistical assumption is violated. In turn, this indicates the extent to which conventional statistical inferences are compromised when spatial autocorrelation is overlooked. Autocorrelation complicates statistical analysis by altering the statistical properties of variables, changing the probabilities that statisticians commonly attach to their decision-making. In turn, this may lead to incorrect statistical inference (e.g. positive spatial autocorrelation results in an increased tendency to reject the null hypothesis when it is true). It signifies the presence of, and quantifies the extent of, redundant information in georeferenced data. This, in turn, affects the information contribution made by each georeferenced observation to statistics calculated with a dataset. It can be seen from looking at Figure 6.1 that points closer together in the grids that are overlain onto the image will detect image properties that are more similar than those that are further away. Sampling image properties from points that are very close together is likely to result in very low variation in sampled properties, and the incorporation of statistically redundant information.

A second reason to measure spatial autocorrelation is so that it can be incorporated into an explanatory or predictive model. This often leads to an improvement in model performance. It would be shortsighted to simply view spatial autocorrelation as a statistical inconvenience. Cressie (1993) notes that explicitly accounting for spatial autocorrelation tends to increase the percentage of observed variation in a dependent variable explained by a model. This is because it can compensate for unknown variables missing from a model specification. Griffith and Layne (1999) report that building autocorrelation as a spatially explicit model parameter into a spatially lagged regression model may increase the proportion of observed variation explained (R^2), provided that it makes sense to include a lagged form of the dependent variable as one of the independent variables. The resulting improvement in explanatory power of the model will depend on the original model itself. Obtaining additional explanatory power in this way is easier and more reliable than specifying new variables for which additional data must be collected, or from using different statistical methods altogether. Because of this reliability, spatial autocorrelation is also widely quantified to provide an empirical basis for kriging. This is a method of interpolation that requires the spatially autocorrelated variation within a point surface to be modelled in order to estimate the value of a parameter at an unsampled location based on known values of its neighbours (see Section 6.12).

Spatial analysis frequently employs model-based inference based on statistical evidence and reasoning. The dependability of this inference relies upon a range of model assumptions being upheld. These largely relate to the dataset that is being analysed. For example, Table 7.3 lists the assumptions associated with classical regression. One assumption states that model error terms are independent and identically distributed. The presence of spatial autocorrelation in the residuals of a regression model is an indication that this assumption has been violated.

6.4 Measuring Spatial Autocorrelation

Tests for spatial autocorrelation can be broadly classified based on the measurement level of the data to which they are applied (nominal, ordinal, interval or ratio, see Section 2.6). The join-count test provides one of the simplest measures of spatial autocorrelation for application to binary nominal (i.e. presence/absence) data in raster or polygon vector format. This test focuses on the different possible combinations of joins between either polygon or raster boundaries for a mapped distribution. For example, Figure 6.2 illustrates some hypothetical mapped distributions of seagrass patches on sand, represented as raster grids. In the case of mapped seagrass (G, green) and sand (Y, yellow) classes, possible combinations between adjacent grid cells are green–green (GG), yellow–yellow (YY) and green–yellow (GY). A count is made of the number of joins of each type and the result is compared to a frequency distribution of the joins that would be expected under a random pattern (i.e. in the absence of spatial autocorrelation). In the case of positive spatial autocorrelation, the number of joins of the same category will be large in comparison to the number expected in a random map pattern. Conversely, the joins between different categories will be small in comparison to the number expected in a random map pattern (Cliff and Ord, 1981;

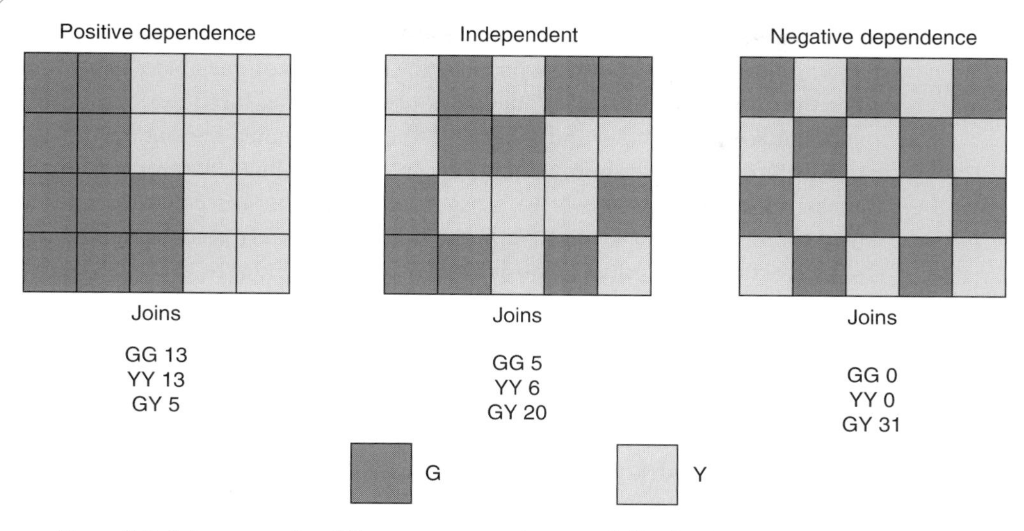

Figure 6.2 Join-counts for different patterns of mapped distributions of seagrass (depicted in green, G) on sand (depicted in yellow, Y). Adapted from Haining (2003). (A black and white version of this figure will appear in some formats. For the colour version, please refer to the plate section.)

Haining, 2003). In the case of negative autocorrelation, the number of joins between different categories will be large in comparison to that expected.

Geary's c statistic and Moran's I statistic are measures of spatial autocorrelation on interval and ratio data. These statistics measure spatial dependence by systematically comparing one point against all others within a defined neighbourhood. Because spatial autocorrelation is the investigation of variation in a single variable over space, these measures emphasise the *difference* in values separated by a specified *distance*. Of the two measures described here, the part of the equation that calculates the *difference* in values of a variable is the *squared difference* between points (Geary's c) or the *cross-product* (Moran's I), which is a function of the difference between the values at locations i and j and the overall sample mean.

Thus, the Geary statistic, c, is based on:

$$\sum_{(i)} \sum_{j \in N_{(i)}} \left(z_{(i)} - z_{(j)}\right)^2 \tag{6.1}$$

where $z_{(i)}$ and $z_{(j)}$ are the values of variable z at locations i and j, respectively, and $N_{(i)}$ denotes the neighbours of case i, for cases where $i \neq j$. The calculated value of the Geary statistic is small and positive if the map pattern has positive spatial dependence, is large and positive if the map pattern has negative spatial dependence, and is intermediate between these extremes if the map pattern is random.

The Moran's I statistic is based on the cross-product, as follows:

$$\sum_{(i)} \sum_{j \in N_{(i)}} (z_{(i)} - \tilde{z})(z_{(j)} - \tilde{z}) \tag{6.2}$$

where $\tilde{z}$ is the mean value of variable z. The cross-product relates to the part of the equation that measures the covariance of the values of variable z at locations i and j. For the purpose of Moran's I, the cross-product is measured by computing the difference between each value and the overall mean value, then finding the product of these two differences and summing across point pairs within a dataset (or neighbourhood subset of a dataset). Thus, the cross-product statistic (or covariance term) denoted by $(z_{(i)}-\tilde{z})(z_{(j)}-\tilde{z})$ is positive if the map pattern has positive spatial dependence, negative if the map pattern has negative spatial dependence, and is close to zero if the map pattern is random.

Figure 6.3 illustrates how this statistic works to discriminate between random and non-random map patterns on the basis of four types of association that correspond to the four quadrants of a Moran scatterplot. Values are plotted with the variable z normalised, such that its mean is 0 and standard deviation is 1 (see p. 42 in Fischer and Getis (1997) for further detail). In Figure 6.3(a), the cross-product formulation produces values as positive with spatial dependence because calculated values are similar, falling into the upper right and lower left quartiles of the plot (i.e. both negative or both positive, meaning that their product will be a positive value). In Figure 6.3(b), the cross-product formulation produces values that are negative with spatial independence because calculated values are dissimilar, i.e. they have a tendency to fall on either side of the mean; therefore, one value is positive and the other negative, meaning that their product will be a negative value and the points will fall into the upper left and lower right quartiles of the plot.

Moran's I statistic is often calculated for a given distance from each point in a dataset. This means that the calculation will only include points that fall within a particular distance or distance band (e.g. within 10 and 20 m). The full formulation of the Moran's I statistic includes a weighting factor to specify whether any pair of values belongs within a given distance class and a normalisation by data variance. The overall value of the corresponding statistic ranges from −1 (tending towards negative autocorrelation) to +1 (tending towards positive autocorrelation) and can be interpreted in the context of its sampling distribution under the null hypothesis of no spatial autocorrelation.

$$I = \frac{\sum_{(i)} \sum_{j \in N_{(i)}} w_{ij}(z_{(i)}-\tilde{z})(z_{(j)}-\tilde{z})}{(1/n)\sum_{j \in N_{(i)}} (z_{(i)}-\tilde{z})^2} \tag{6.3}$$

where $N_{(i)}$ denotes the neighbours of case i, for cases where $i \neq j$ and w_{ij} is an element of a spatial weights matrix (see Section 6.6).

6.5 Functions That Describe Spatial Structure

To quantify spatial structure of all sampled points within a dataset, it is necessary to go beyond the immediate neighbours. The semivariogram and spatial autocovariance functions offer a way to do this for the Geary's c and Moran's I statistics, respectively.

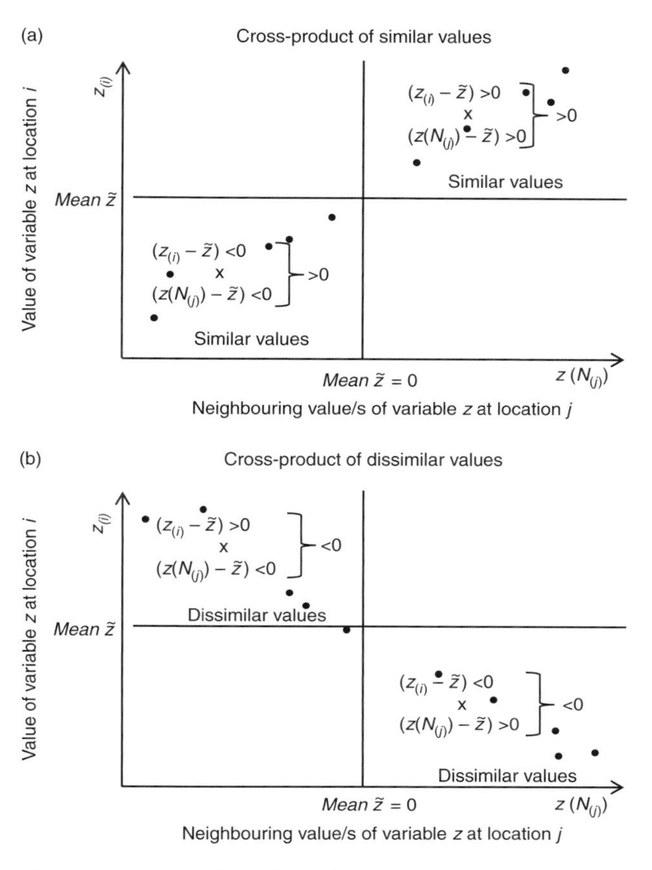

Figure 6.3 A plot of $z_{(i)}$, the value of a variable at location i, against $z(N_{(j)})$, the value of the same variable at neighbouring location j. Note that either a single or multiple values of the variable can be incorporated at location j, depending on how the neighbourhood is defined. The distributions in (a) and (b) show how the cross-product portion of the Moran's I formulation produces values as (a) positive with spatial dependence because calculated values are similar (i.e. both negative or both positive, meaning that their product will be a positive value), falling into the upper right and lower left quartiles of the plot, and (b) negative with spatial independence because calculated values are dissimilar (i.e. they have a tendency to fall on either side of the mean, and therefore one value is positive and the other negative, meaning that their product will be a negative value) and the points will fall into the upper left and lower right quartiles of the plot.

These functions provide a plot of autocorrelation for the entire dataset. They are computed by comparing every location to its neighbourhood. The semivariogram is defined as follows:

$$\gamma(h) = \frac{1}{2}\left(\frac{1}{N(h)}\right)\sum_{(i)}\sum_{(j|d(i,j)=h)}(z_{(i)} - z_{(j)})^2, \qquad h = 0, 1, 2, \ldots \tag{6.4}$$

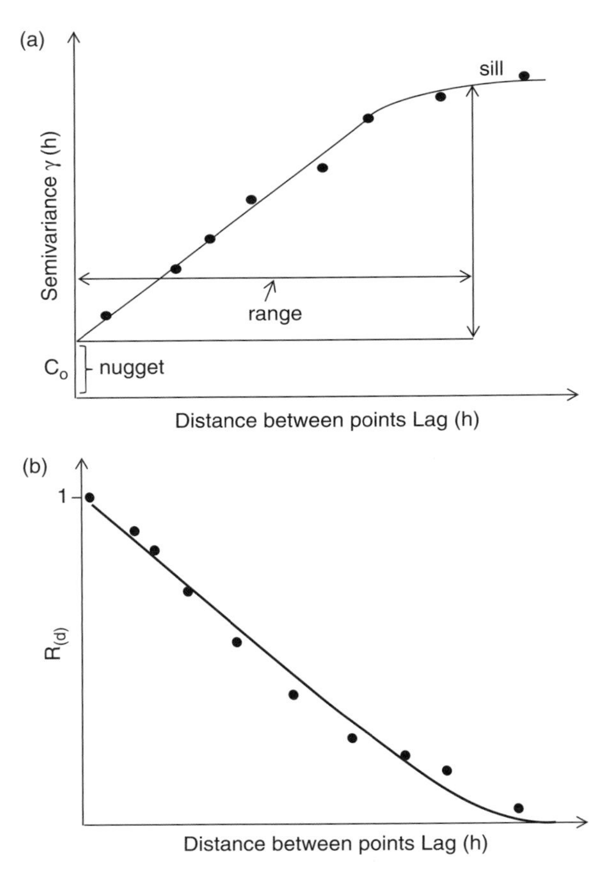

Figure 6.4 (a) The key features of a semivariogram plot of the Geary statistic against distance between paired points: nugget, range and sill. (b) The correlogram plot of $R(h)$, a normalised, global estimate of Moran's I, against distance between paired points.

This computes half the average squared difference in values of a variable, z, for a given number of sites (N) separated by a distance h. As h is varied, a set of values can be obtained and are plotted against h to produce the semivariogram (Figure 6.4a). In practice, for a series of irregularly spaced sample points, there will rarely be many pairs that are separated exactly by the specified distance, h. Instead, a distance band is used and the estimator is calculated on the basis of pairs that are separated by distances lying within this distance band (see Figure 6.5). The point density and geographical area of the dataset being explored will determine both N and h.

It can be useful to fit a function of $\gamma(h)$ to the sample data. This allows the semivariogram to be used in a practical sense (e.g. to derive sample sizes for guiding field survey). Several functions are permissible, although not all are suitable for continuous variables. Functions that are commonly fitted to the points of a semivariogram plot include the spherical, exponential, Gaussian and linear functions, or some combination of these (McBratney and Webster, 1986).

The sill, range and nugget are all features of a semivariogram that reveal information about the spatial structure in the data samples plotted (see Figure 6.4a). The sill represents the stage at which there is a discernible levelling out of the semivariance values. They no longer continue to increase as the distance between points increases. This can be interpreted as the distance beyond which the values sampled at different locations no longer influence each other. This corresponds with a distance 'range' at which points are deemed far enough apart that they no longer influence each other, or adhere to Tobler's first law of geography. The nugget point is the semivariance value for a pair of points separated by distance zero, i.e. at the same location.

The equivalent global estimate for the Moran's I statistic is provided by the spatial autocovariance function. This is computed as follows:

$$C(h) = \left(\frac{1}{N(h)}\right) \sum_{(i)} \sum_{(j|d(i,j)=h)} (z_{(i)} - \bar{z})(z_{(j)} - \bar{z}), \qquad h = 0,\ 1,\ 2, \ldots \tag{6.5}$$

Spatial autocovariance represents a measure of the average cross-product, calculated for a given distance band h in which several point pairs may fall. It is large when values are similar because values of z at both locations i and j will tend to be positive or negative in the cross-product term (the negative values will become positive when multiplied together). Conversely, it is small when values are dissimilar because positive and negative values will offset one another. It is also worth noting that the semivariogram and the spatial autocovariance functions are inversely related (compare Figure 6.4a and b). The spatial autocovariance function forms the basis of the spatial autocorrelation estimate, which can be standardised using the variance of $z_{(i)}$ for a distance of $h = 0$ as follows:

$$R(h) = C(h) / C(0) \tag{6.6}$$

where $C(0)$ is the variance of $z_{(i)}$, to produce an estimate of the autocorrelation at distance h. A correlogram can be constructed by plotting this spatial autocorrelation estimate $R(h)$ against distance (Figure 6.4b). Unlike the semivariogram, this tends to decrease as h increases. The plot of the autocorrelation function (or correlogram (6.6)) has the same behaviour as the autocovariance function (6.5). The difference between these plots is that the autocorrelation function is standardised in the sense that $R(h) = 1$ when $h = 0$.

A phenomenon is said to be 'anisotropic' if it exhibits different values when measured or calculated along axes in different directions. This may be the case when measures of spatial autocorrelation are calculated, if they are not equal in all directions. Thus, a north–south oriented calculation would return a different assessment of spatial structure to an east–west oriented calculation. Results are also dependent on the scale of sampling, with calculated values being a function of the size of the neighbourhood, as well as the density or resolution of sampling points.

6.6 The Spatial Weights Matrix

As we have seen, it is useful to calculate statistics that summarise the spatial patterning in a dataset. A summary statistic can be computed for a given location based on all data cases simultaneously, or within a 'moving analysis window'. Equal importance may be attached to all data cases inside the window at a given location. An alternative approach may apply a weight to the importance of data cases based on their position relative to the location for which the statistic is to be calculated.

A spatial weights matrix is a practical method for applying geographical weights across a spatially referenced dataset. The matrix is a table with one row and one column for every feature in the dataset. The cell value for any given row–column combination is the weight applied to the spatial relationship between those row and column features. A kernel or moving window is placed at each location of interest and data cases within that kernel are weighted. A common objective is to give more weight to data cases that are nearby than those that are further away because of spatial dependence (Section 6.1).

A wide range of 'neighbourhood' functions and configurations can be used to determine the weighting associated with the data cases surrounding a given location. At the simplest level, a spatial matrix may define the neighbourhood of a given location as a basis for including data cases that fall within the defined neighbourhood. Conventionally, these are applied a weighting of 1. Cases falling outside the neighbourhood will be excluded from the analysis by applying a weight of 0. More sophisticated methods might express a decay function associated with a phenomenon of interest. For example, a linear decay function might afford half the weight to a data case that is twice as far as another. Inverse distance or negative power functions address the fact that, as one measure goes up (the distance between a location and the data case), the other measure goes down (the weight allocated to the data case). For example, the square of the inverse distance between the data case and the location of interest is a common function for estimating the weight attributed to a data case. Other functions might include known decay modes such as exponential distance or power distance functions. Where adjacent area data are concerned, a measure of the length of a common border may be used. This could be expressed as a proportion of the perimeter of the areal unit under consideration, a combination of distance and border metrics, or some form of interaction weight (Bavaud, 1998).

There is also a wide variety of neighbourhood configurations that may determine the subset of data to which weights are applied. Within a grid configuration, these include various definitions of contiguity relating to whether grid squares share edges (as in a rook configuration) or corners (as in a queen configuration). The neighbourhood definition may incorporate first- or higher-order relationships that go beyond the first level of neighbours. Distance bands may be specified to include all data points that fall within a range of distances from the location under consideration (i.e. between i and h, and i and $h + \Delta$; see Figure 6.5). For point datasets, nearest-neighbour functions assess each location within the context of a specified number of its closest neighbours. For example, if the specified number of nearest neighbours is

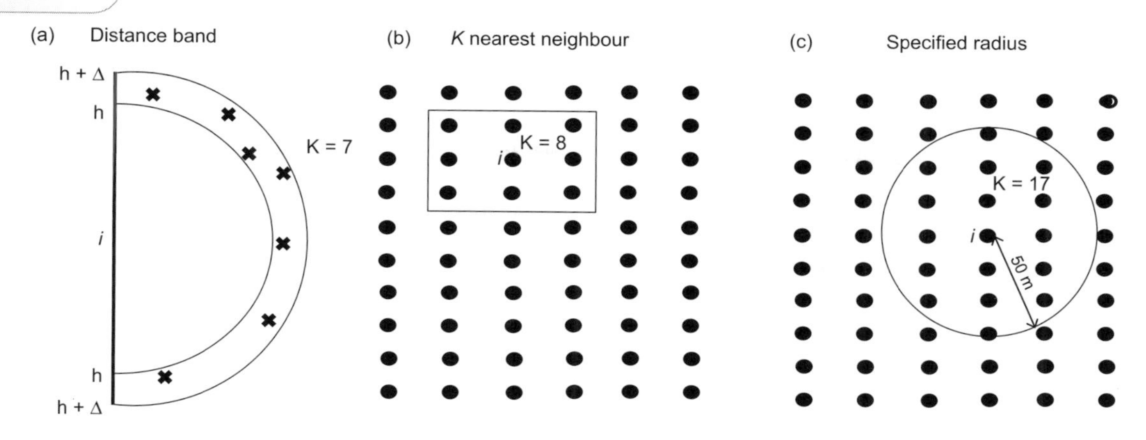

Figure 6.5 Different neighbourhood definitions for the calculation of geostatistics: (a) distance band definition, whereby data cases are selected for inclusion in the neighbourhood providing that they are within a distance band ($h \pm \Delta$) from location *i*; (b) *K* nearest-neighbour definition, whereby a specified number of nearest neighbours are selected; and (c) specified radius, where all points falling inside a given radius are selected for inclusion in the neighbourhood.

$K = 8$, then the eight closest neighbours or data cases to the location of interest will be included in computations for that location. Similarly, a radius can be specified to define a circle inside which all data cases that fall are considered to be neighbours.

6.7 What Does Spatial Autocorrelation Mean for the Analysis of Coastal Environments?

There are many reasons why it may be useful to measure spatial autocorrelation. These include:

- to assess whether (or not) autocorrelation can be detected in a dataset,
- to measure the dimensions of any autocorrelation detected (e.g. the scales of variation in patches observed),
- to include an autocorrelated variable into a model that seeks to explain or predict coastal features as a function of related processes,
- to encourage statistical rigour by including spatial autocorrelation in a model specification,
- to specify a geographically structured process,
- to assess the nature (large vs small scale, internal vs external) of autocorrelated processes underlying an observable feature, and
- to incorporate the spatial structure of a dataset into the design of a sample scheme (e.g. the spatial dimensions of a point grid).

Tables 6.1 and 6.2 summarise a range of studies that have sought to quantify spatial autocorrelation of the biological and geological characteristics of coastal

Table 6.1 Examples of biological coastal studies that have measured autocorrelation.

Study	Reference
Measuring clustering of sedentary invertebrates (abalone, sea cucumbers and urchins) as an indication of the precision level necessary for diver surveys	McGarvey et al. (2010)
Using spatial models to assess the role of environmental variables (median grain size and inundation time per 12.25 hour tidal cycle) in structuring patchiness of intertidal macrobenthic invertebrates in sand flats in the Wadden Sea, The Netherlands	Kraan et al. (2009, 2010)
Characterising meiofaunal and microalgal populations on an intertidal sandflat. Total meiofauna, nematodes, copepods, ostracods, Chl *a* and pheopigments exhibit significant spatial autocorrelation. Correlograms revealed nearly identical patterns, suggesting linked grazer and microalgal resources	Pinckney and Sandulli (1990)
Determining the patch sizes of meiobenthic copepods in Barnstable Harbour, Massachusetts (from Moran's *I*, Geary's *c* and variance-to-mean ratio)	Sandulli and Pinckney (1999)
Using a variant of Moran's *I*, the Mohan's *I* statistic, to define patch scales of meiobenthic harpacticoid copepods on a mudflat in different tidal conditions	Fleecer et al. (1990)
Testing for spatial autocorrelation of 173 benthic species in Puget Sound, Washington	Dethier and Schoch (2005)
Incorporation of autocorrelated error into a spatial model linking benthic community characteristics (abundance, species number, richness, diversity) and environmental variables (Chl *a*, porosity, organic content, depth, redox, sulfide and sediment grain properties) in St Anns Bay, Nova Scotia, Canada	Dowd et al. (2014)
Evaluating the influence of human disturbance from gas extraction on marine invertebrates in the Ionian Sea using Moran's *I*	Fiorentino et al. (2012)
Defining an adequate sampling distance for investigating predator–prey relationships between pelagic seabirds, fur seals and Antarctic krill at Bird Island, South Georgia	Veit et al. (1993)
Assessing the relationship between pelagic fish and ocean temperature across water depth ranges by comparing autocorrelation between fish species and sea surface temperatures	Kleisner (2010)
An exploratory assessment of autocorrelation scale and intensity in seabird counts to gain insight into related drivers such as water temperature or prey abundance	Schneider (1990), O'Driscoll (1998)
Assessing spatio-temporal patterns of connectivity in adult abundance and density of juvenile blue mussels in Québec, Canada	Le Corre et al. (2015)
Examining the relationship between environmental heterotrophic bacteria and phytoplankton in a brackish lagoon (Thau, France)	Legendre and Troussellier (1988)

Table 6.2 Examples of geological coastal studies that have measured autocorrelation.

Study	Reference
Use of autocorrelation metrics to reveal storm-driven sand exchange between along- or cross-shore sediment compartments along the Texas coastline	Morton et al. (1995)
Incorporation of autocorrelation measurements into techniques for determining sediment grain size from digital images taken with a sediment-bed camera	Barnard et al. (2007)
Use of spatial autocorrelation to develop a three-dimensional geological interpretation of palaeo-groundwater rise for the Rhine–Muse delta (The Netherlands) and associated fluctuations in Holocene sea level, subsidence and sediment load	Cohen (2005)
Use of autocorrelation to investigate tidally generated cyclic features in vertical sequences of marginal coastal sandstone and shale deposits (Lower Breathitt Formation, eastern Kentucky)	Martino and Sanderson (1993)

environments. Many biological studies have interrogated the autocorrelation characteristics of benthic invertebrates. This is probably due to their accessible nature, in intertidal sandflats or estuaries, where populations can easily be sampled across environmental gradients and sorted for multivariate analysis. Autocorrelation can subsequently be incorporated into models that explore how these populations' distribution patterns relate to local environmental conditions. Spatial autocorrelation has also been observed in other organisms that inhabit coastal environments, including marine birds (Schneider, 1990; O'Driscoll, 1998), pelagic fish (Kleisner, 2010) and marine bacteria (Legendre and Troussellier, 1988). Geological applications have focused mainly on movements and characteristics of sediment (Morton et al., 1995; Barnard et al., 2007).

An assessment of whether or not spatial autocorrelation is detectable in a dataset goes beyond simply stating whether or not sampled measurements adhere to Tobler's first law of geography. The outcome of such an assessment determines the appropriateness of any statistical techniques to be employed on a dataset. Furthermore, patterning in the structure of any autocorrelation may provide clues as to the identity of the processes that are causing it. For example, in the estuarine fjord environment of Puget Sound, Moran's I detected the presence of spatial autocorrelation in approximately half the benthic population taxa tested. This was attributed to pelagic larval and spore dispersal processes connecting beach segments in Puget Sound via hydrodynamic circulation of this fjordal estuary (Dethier and Schoch, 2005). Where autocorrelation is detected, it can be used to characterise the spatial dimensions of patterning within a dataset. Such an approach employs spatial autocorrelation measurements to visualise trends, gradients or mosaics across a map. Autocorrelation measures have been applied in this way to describe patchiness in macrobenthic invertebrates in the Wadden Sea (Kraan et al., 2009, 2010). Similarly, Pinckney and Sandulli (1990) used semivariograms to reveal the high correspondence in spatial patterning of microfaunal and microalgal populations in the intertidal sandflats of Barnstable Harbour, Massachusetts.

6.8 Measuring the Dimensions of Spatial Autocorrelation

Uncovering the dimensions of spatial autocorrelation provides a clue as to the identity of the underlying processes that structure populations. For example, the spatial scale of patchiness of meiobenthic harpacticoid copepods on a mudflat was measured during different tidal conditions. This used a variation of the Moran's I statistic (the Mohan's I statistic) to reveal that processes of the order of several square centimetres underpin patch dynamics of small metazoans (Fleecer et al., 1990). Similarly, Barnes and Hamylton (2013) sampled sediment cores of benthic invertebrates in grids traversing the boundaries of seagrass meadows for three sites in Moreton Bay, Queensland (Figure 6.6). Semivariograms were constructed for the total number of individuals observed within each lattice of cores. At each of the three sites, these indicated a symmetrical pattern of autocorrelation about the sand/seagrass interface, with spatial lag (or range) distances of 70, 80 and 125 cm at each site (Deanbilla, Polka and Yerrol), respectively. Beyond these distances, no autocorrelation was detectable. This was interpreted as the distance at which the core samples of benthic invertebrates were no longer statistically related to each other.

6.9 Incorporating Autocorrelation Into Coastal Models

Spatial autocorrelation is frequently encountered in ecological data because species abundances are often influenced by multiple external variables (e.g. environmental influences, disturbance) and internal variables (e.g. biological attraction or dispersal). These external and internal variables are themselves autocorrelated (Oden, 1978). Many ecological models therefore implicitly assume an underlying spatial pattern in the distribution of organisms and their environment (Legendre and Fortin, 1989). Such an assumption views spatial autocorrelation as the outcome of some process operating over a geographic landscape, e.g. diffusion, dispersal, marine species interactions and environmental autocorrelation (Section 1.4). Because these processes themselves are autocorrelated, they give rise to a spatial structure that is passed on to the features they influence.

Models provide a way of expressing our understanding of a phenomenon using mathematical terms. They are usually constructed to explain or predict something about the phenomena they focus on. To ignore spatial autocorrelation effects when analysing georeferenced data is ill-advised because the statistical assumption of independent observations is atypical in a geographic world. At the very

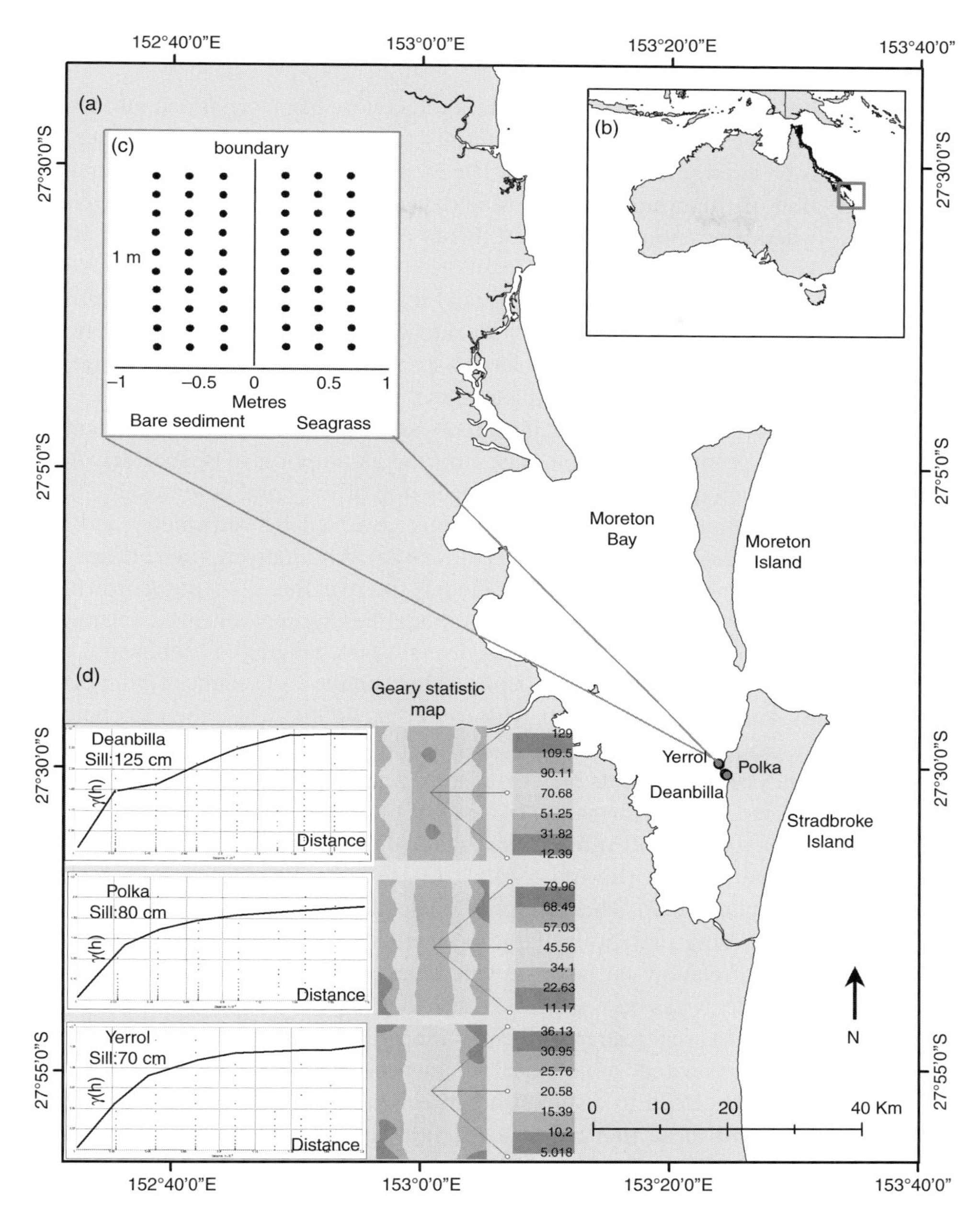

Figure 6.6 (a) Map of Moreton Bay showing the locations from which lattices of core samples of benthic invertebrates were extracted at Polka, Yerrol and Deanbilla. (b) The location of Moreton Bay adjacent to Brisbane along the Queensland coastline, Australia. (c) The configuration of the lattice of sediment cores sampled across seagrass boundaries from Moreton Bay. (d) A series of semivariogram plots for counts of benthic invertebrates sampled from Deanbilla, Polka and Yerrol sites. Coloured squares to the right of semivariogram plots indicate values of the Geary statistic, mapped across the lattice configuration for each sample site. High values of the statistic, shown in red, indicate positive spatial autocorrelation within the bare sediment or seagrass habitat. Low values of the statistic, shown in blue, indicate weaker positive autocorrelation across the sand/seagrass interface. (A black and white version of this figure will appear in some formats. For the colour version, please refer to the plate section.)

least, spatial autocorrelation should therefore be incorporated into a model specification because its presence is necessary for a correct description, rather than because it is of interest in its own right. The focus for a non-spatial analyst may be the estimation of other model terms using data samples that happen to be georeferenced and where conventional statistical theory could be utilised if the value of this term were known or assumed to be zero.

Spatial autocorrelation is usually incorporated into a model by introducing a term specifically to represent it. For example, spatial autocorrelation may be treated as a missing or surrogate variable in a specification that accounts for unexplained variation because of an incomplete model specification. This might take the form of an autoregressive term that introduces some measure of the dependent variable itself into the model equation, based on the assumption of spatial dependence. Such an approach has been used in the prediction of live coral cover around barrier reefs and atolls along environmental gradients governed by bathymetry and hydrodynamics (Hamylton, 2012; Hamylton et al., 2012a). Alternatively, unexplained variation could be 'plugged' with a term that models the structured (i.e. autocorrelated) component of error. This was achieved in a model linking macrobenthic characteristics (abundance, species number, richness, diversity) to changing environmental variables (Chl *a*, porosity, organic content, depth, redox, sulfide and sediment grain properties) across St Anns Bay, Nova Scotia (Fleecer et al., 1990). While the incorporation of spatial autocorrelation into a model specification may require a more complicated statistical analysis, it represents better statistical practice and often improves model explanatory or predictive performance. In terms of practical implementation, standard geographical statistical software packages include GeoDa (Anselin et al., 2006), R-Analysis of Spatial Data (Bivand et al., 2013), PySAL (Rey and Anselin, 2010) and IDRISI (Eastman, 2003). These supply a range of capabilities for estimating autocorrelation.

Assessing the nature (large vs small scale, exogenous vs endogenous) of spatial autocorrelation can also provide clues as to the underlying causal processes. For example, it has been suggested that that large-scale spatial patterns (of the order of 1–10 km) are a response to environmental variables, whereas small-scale spatial patterns are related to biological interactions (<1 km) (Thrush, 1991; Legendre, 1993). It can be useful to distinguish between variables that reflect interaction between the sites and those that reflect a reaction to some other external variable. Exogenous effects refer to those caused by a reaction to external forces. Endogenous effects are a reaction to neighbouring individuals. An assessment of the relative contribution of endogenous and exogenous effects helps to determine the appropriateness of the model specified. When reaction is the major influence, a classic (i.e. non-spatial) process–response model is appropriate. Interactive effects suggest a need for a model with a spatially dependent covariance structure (Cliff and Ord, 1981). Invoking a non-spatial model (e.g. classical regression) as an initial point of reference enables the degree of influence of the external variables to be ascertained. Where the transition to a model that incorporates spatial dependence is accompanied by a marked growth in explanatory power, the theoretical implication that follows is that neighbourhood interactions play an important role. This invites greater consideration of independent variables that reflect interaction between sites.

6.10 Guiding the Design (e.g. Spatial Dimensions) of Sampling Schemes

Spatial autocorrelation plays a crucial role in model-based inference. It can be used to ensure that models are built upon valid statistical assumptions based upon the outcome of a valid sampling design. Sometimes spatial autocorrelation is used to diagnose model misspecification because of a poor sampling design. The presence of spatial autocorrelation indicates that duplicate information is contained in georeferenced data. This reduces the effective sample size and degrees of freedom for a dataset because of the redundancy associated with data samples that are near to one another. In a review of the consequences of spatial structure for the design and analysis of ecological field surveys, Legendre et al. (2002) found that spatial autocorrelation in both the response and environmental variables affects the classical inferential tests of significance. It inflates the goodness of fit between two datasets and potentially leads to higher rates of type I error (incorrect rejection of a true null hypothesis). This can often be addressed by subsetting a large volume of field samples, such that those selected for analysis are adequately separated by a spatial lag. The dimensions of the lag should be large enough to be free from autocorrelation. The lag distance could be determined by the length of the range on a semivariogram calculated on a pilot dataset (Figure 6.4a). This approach was used in an examination of coincident observations of predator and prey populations around Bird Island, South Georgia. Records were collected from a ship every 10 m along 30 survey transects. These were reduced to every kth data point, where $k - 1$ represents the largest spatial lag at which substantial autocorrelation was present (Veit et al., 1993). Similarly, semivariograms have revealed fishing pressure around the UK coastline to be patchily distributed at the regional scale due to the spatially dependent occurrence of targeted species. The dimensions of patches informed practical aspects of fisheries management, such as monitoring fishing activity (Stelzenmüller et al., 2008).

One common feature of spatially referenced information that has been manually collected from coastal environments is that it is often clustered in a manner that reflects different sub-areas of the landscape that were visited. This can create a problem where intensive, localised sampling approaches give rise to geographically clustered samples. These may reflect the practical difficulties associated with visiting many environments. Such difficulties perhaps arise due to the fact that features to be sampled are underwater or in highly energetic areas (e.g. inside surf zones). A SCUBA diver-led survey of sessile benthic invertebrates noted that autocorrelated survey counts of abalone, sea cucumbers and urchins reduced the precision of quadrat surveys (McGarvey et al., 2010). Spatial autocorrelation was quantified by varying inter-quadrat distances between 1 and 100 m across four separate surveys to establish the distance at which autocorrelation declined to non-significant levels. This was identified when samples were a distance of 20 m away or greater. In practice, imposing a spatial lag of 20 m between sample points creates a significant limitation on the sampling density. This limit would apply to an already restricted area visited by a diver whose time and geographical range underwater is limited by the air in their SCUBA cylinder. For example, if safety considerations and the volume of air in a diver's cylinder dictate that a diver can swim only within the boundary of a 100 m × 100 m square, then a spatial lag of 20 m will mean that they are only able to collect 36 samples within that area that will be 'truly' independent.

6.11 Four Practical Reasons to Measure Spatial Autocorrelation Along the Rocky Shores of the Northwestern Mediterranean

Rocky shorelines provide stable, relatively permanent surfaces located in conditions that are favourable for habitation. Tide- and wave-driven water keeps rocks and cliffs well oxygenated and bring nutrients that enhance plant growth. Plants such as seaweeds, lichen and algae provide a primary food source for grazers such as barnacles, limpets and periwinkles. These in turn provide food for crabs, dogwhelks and birds higher up in the food chain (Tait and Dipper, 1998). The littoral zone extends from the high water mark, which is rarely inundated, to lower shoreline areas that are permanently submerged (Viles and Spencer, 2014).

The distribution of organisms across cliffed and rocky coastlines is typically structured by a range of environmental factors, including desiccation, wave action, light availability, temperature, aspect, slope, turbidity, substrate, salinity and a range of biotic interactions (grazing, competition, dispersal strategies) (Barnes and Hughes, 2009). These environmental factors often act in synergy along shoreline cliffs to provide distinct zonation patterns in community distributions (Figure 6.7). This case study illustrates four different reasons for measuring the spatial autocorrelation of benthic communities along the rocky and cliffed shorelines of the northwestern Mediterranean:

1. Quantifying the extent and pattern of regional-scale autocorrelation in benthic communities sampled at 143 stations along the coastline.
2. Quantifying the extent and pattern of local, transect-scale vertical autocorrelation across the littoral zone.
3. Indicating the dimensions of spatial autocorrelation observed in population datasets and thereby signifying the presence of redundant information.
4. Indicating the nature (spatial vs non-spatial) of relationships between observable spatial patterns and the processes that drive them.

6.11.1 Quantifying the Extent and Pattern of Regional Autocorrelation Along the Coastline

Sampling stations were established at 143 sites along the Catalonian coastline to cover a wide range of regional physical conditions (Figure 6.8). Global spatial autocorrelation was calculated as an overall summary metric of the number of species sampled

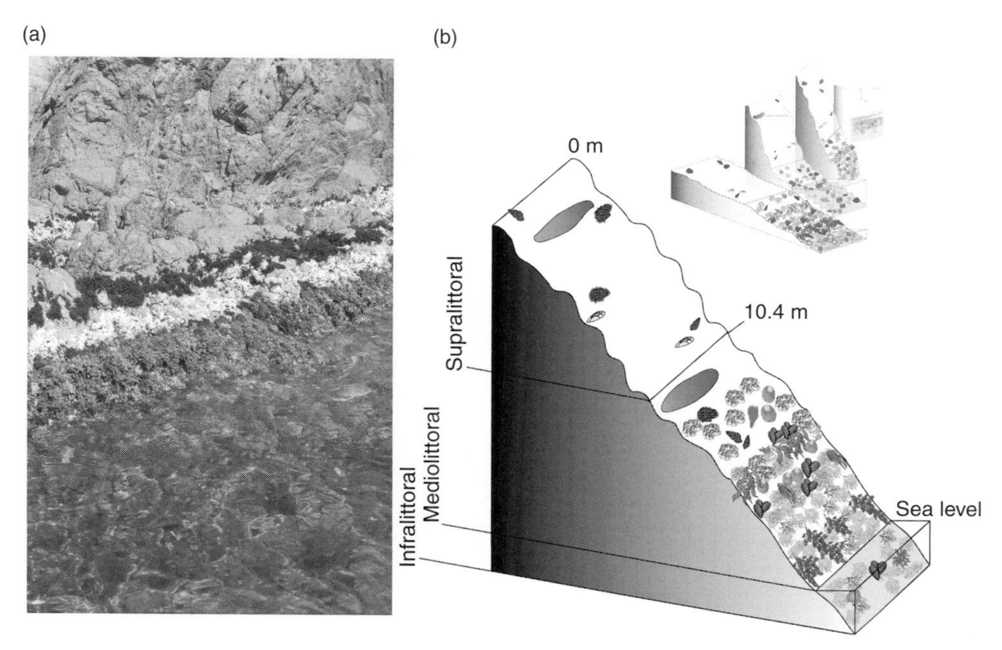

Figure 6.7 (a) Photograph illustrating the spatially structured zonation occurring in species inhabiting the littoral zone along a rocky shore. (b) Patterns of benthic community distribution vary along both horizontal and vertical gradients, depending on factors such as cliff slope (upper right). Reproduced with permission from Chappuis et al. (2014). (A black and white version of this figure will appear in some formats. For the colour version, please refer to the plate section.)

at each transect station using the Moran's I formulation outlined in (6.3). In the first instance, a single summary measure of the overall number of species at each sample station was assessed independently of the vertical sampling across the littoral zone (Figure 6.7b). This compared regional horizontal patterns along the coastline in the distribution of benthic invertebrates (Chappuis et al., 2014). At the global (i.e. all-stations) scale, positive spatial autocorrelation was apparent (Moran's $I = 0.37, p = 0.001$). Given that this statistic ranges from -1 to $+1$, this indicated the moderate extent to which Tobler's first law of geography was being played out across the rocky shore.

6.11.2 Quantifying Local Autocorrelation Across the Littoral Zone

Localised vertical autocorrelation across the littoral zone was quantified for 17 of the 143 sampling stations along the Catalonian coastline. A local measure of spatial autocorrelation was calculated on this subset of transects traversing the vertical gradient along the littoral zone (Figure 6.9). At each site, the vertical transect was defined by the highest point reached by any marine organism, with the lowest point extending as far down the shore profile as practicable, typically to 1 m below sea level. A series of vertically graduated sampling horizons was established in between these extremes.

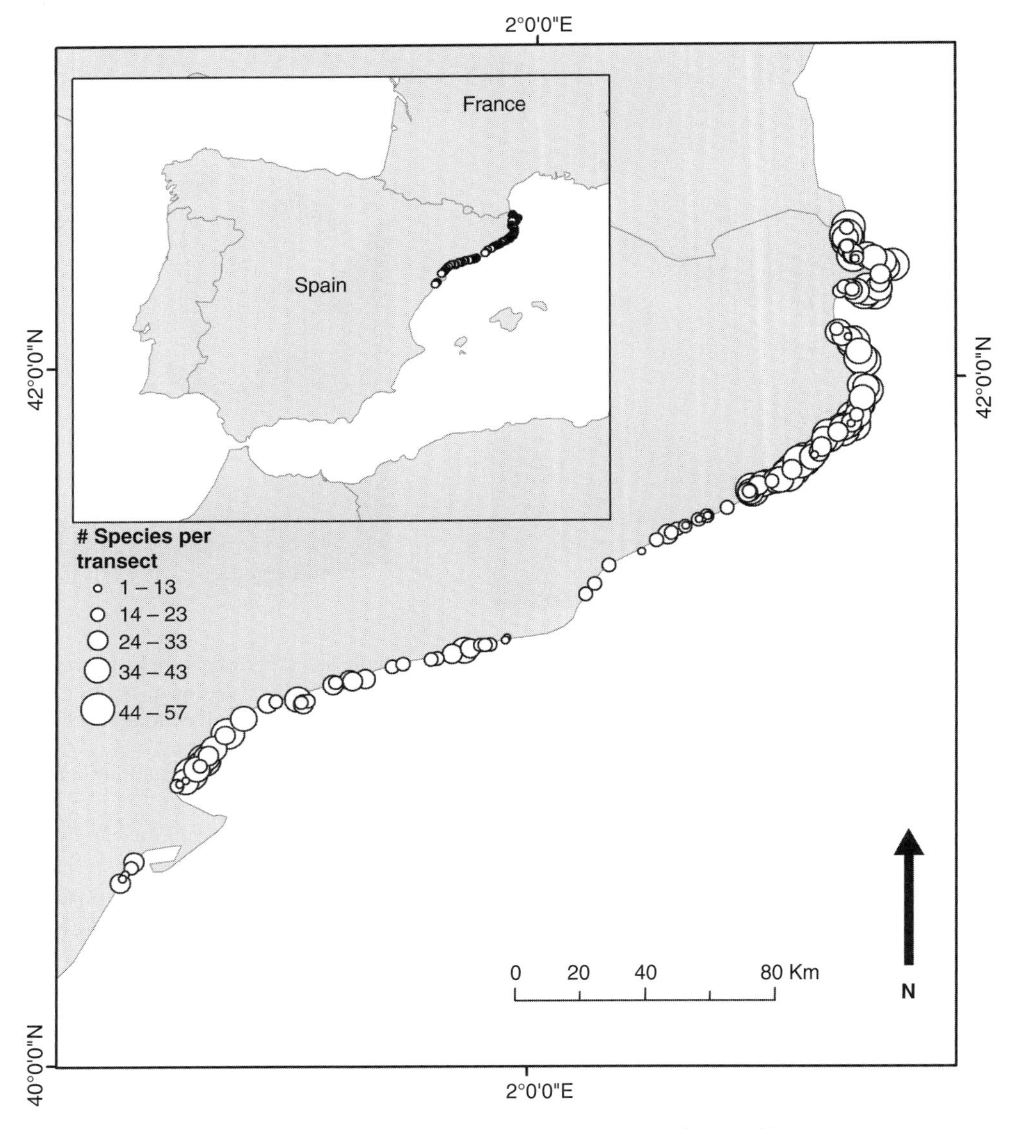

Figure 6.8 (a) Location of sampling stations along the Catalonia coastline, northwestern Mediterranean. (b) Size of red dots indicates the number of species observed at each transect. (c) A univariate Moran's *I* scatterplot of the spatially lagged number of species observed at each station against the number of species observed at each station.

The 17 transects investigated showed wide-ranging spatial structures, with a Moran's *I* that ranged from −0.25 to +0.43, suggesting the presence of both positive and negative spatial dependence. While positive autocorrelation might occur across this zone because of autocorrelated environmental processes shaping the benthic communities, negative spatial autocorrelation might occur because of competition between benthic species attempting to occupy similar niches within the littoral zone.

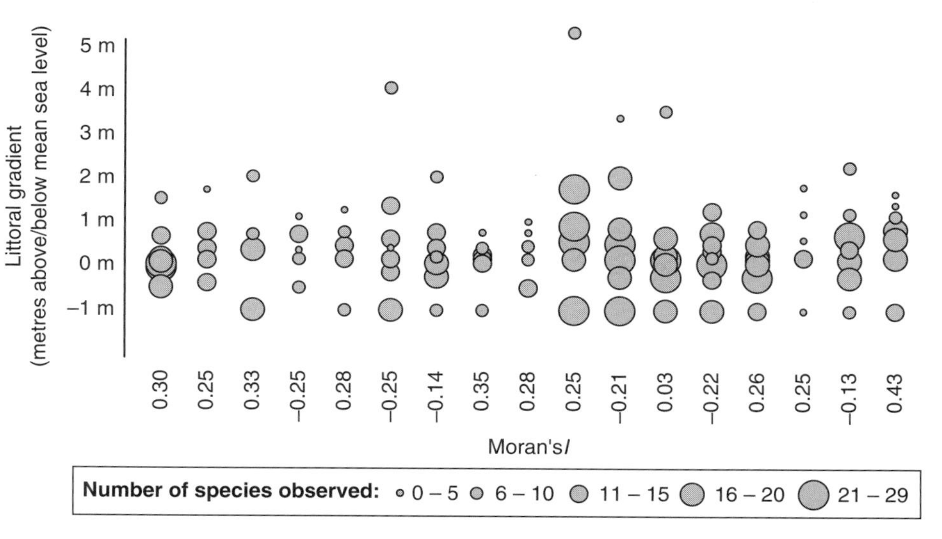

Figure 6.9 A plot of the number of species observed at different sampling horizons across the vertical gradient of the littoral zone. Circle size indicates the number of species observed at each horizon. Measures of spatial autocorrelation (Moran's *I*) are provided beneath each transect.

6.11.3 Indicating the Dimensions of Spatial Autocorrelation in Population Data

A semivariogram based on the Geary's c statistic was generated using the species number data for the whole set of 143 sampling stations. This measured the correspondence between the number of species present for a given distance class, on the basis of the squared difference of the number of species observed for sample sites at point locations i and j (6.4). The semivariogram went beyond a localised measure of spatial autocorrelation to describe spatial structure across a series of spatial lag distances, i.e. through first-, second- and third-order neighbours (Figure 6.10).

The semivariogram was best described by an exponential function with a discernible sill point at $\gamma = 1.7 \times 10^{-2}$, which corresponded with a range of approximately 1.76 km. The sill point represented the point at which the difference between calculated values of Geary's c statistic between the sampling stations levelled off. No further increase in the statistic value was observed as the distance between sampling stations increased. The range (1.76 km) can be interpreted as the distance beyond which the benthic communities sampled along the sections of coastline no longer influenced each other, or were no longer influenced by environmental processes that were autocorrelated at this scale.

In accordance with the dimensions of the sill point and range observed on the semivariogram, a sampling scheme designed to optimise information capture and reduce redundancy along this rocky shoreline would place stations at least 1.76 km apart. This was identified as the distance between two field stations at which sampled littoral communities could be considered fully independent of each other. Defining this range has two practical implications. Firstly, it enables the analyst to identify data cases that

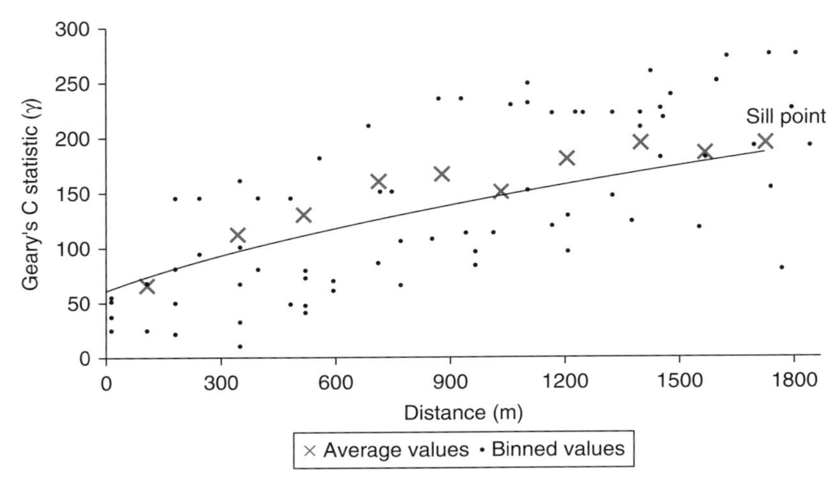

Figure 6.10 A semivariogram constructed on the sampled benthic communities along the rocky shores on the northwestern Mediterranean.

fall inside this range and are thereby autocorrelated. Should the intention be to utilise the dataset further in a statistical modelling exercise, then a spatial lag of 1.76 km between data cases should be implemented to ensure their independence. Secondly, this provides a basis for optimising future field effort given the relative remoteness and difficulty associated with accessing rocky and cliffed shorelines. Needless investment of time, resources and effort in a sampling scheme that is denser than this can be avoided.

6.11.4 Determining Whether Relationships Between Patterns and Processes Are Spatially Structured

Two different forms of regression model were developed to investigate the relationship between environmental factors and benthic community populations sampled. These were an ordinary least squares regression and a spatially lagged autoregressive model. For 17 sampling situations, the number of species observed was treated as a dependent variable and regressed against the independent variables of exposure, substrate, orientation and transect slope. For the spatial model, an autoregressive term was added to incorporate the neighbourhood context effect. This operated through a spatially lagged expression of the dependent variable itself (see Section 7.6.1). This introduced a parameter associated with the interaction effect (i.e. spatial autocorrelation) to the regression equation.

A nearest-neighbour spatial weights matrix was constructed to implement the spatially lagged autoregressive term of the regression equation applied. For each sampling station, this identified those stations that belonged to their neighbourhoods as the four nearest sampling station locations. Values for the numbers of species observed at these stations were then included as a term in the regression model. Values outside the nearest neighbours were excluded from the model. This explicitly built Tobler's

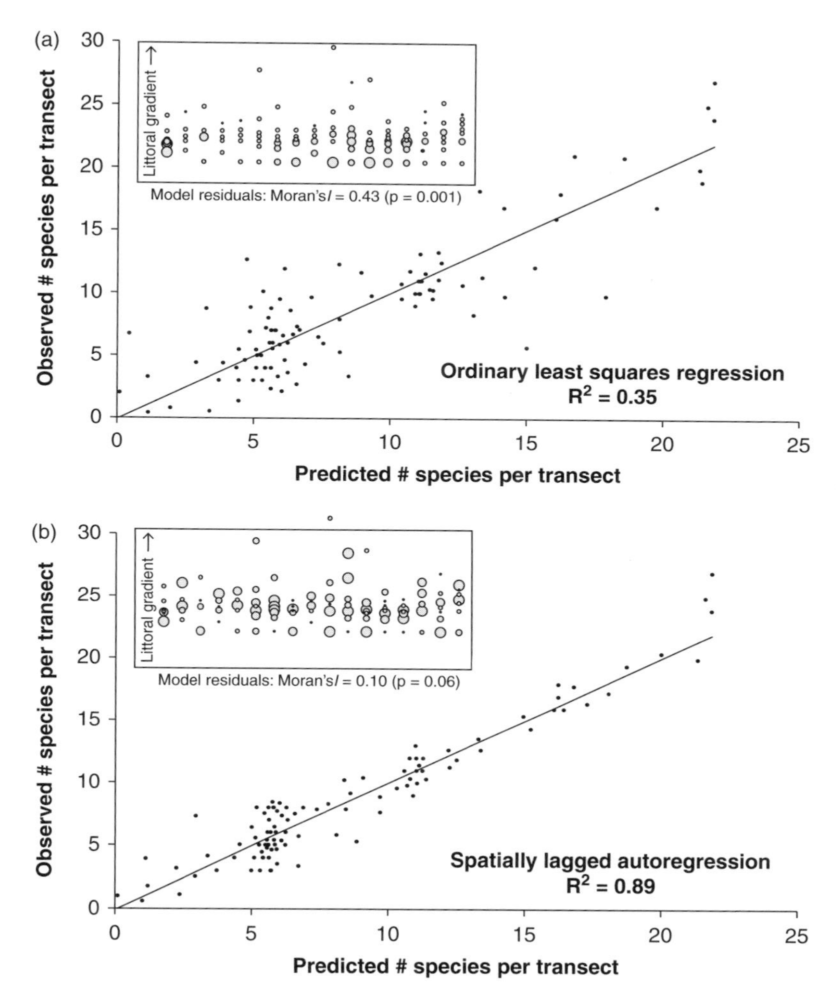

Figure 6.11 Results of two multivariate regression models run to investigate the relationship between the number of species observed along transects spanning the littoral gradient (up to 10 observations along a vertical gradient per transect) at 17 sampling stations against local environmental conditions (exposure, substrate, orientation and transect slope): (a) ordinary least squares regression, and (b) spatially lagged autoregression. Inset boxes show the size and spatial autocorrelation of the model residuals at different sampling horizons along the littoral gradient. Note the higher proportion of variation explained by the spatial model and the lack of autocorrelation in the model residuals (the difference between the observed and modelled value from the regression).

first law of geography into the model because it was akin to drawing on values of the number of species observed at neighbouring locations in order to explain the number of species observed for a given sampling station location.

Once the regression models had been run, residuals were mapped and their spatial structure was measured using Moran's *I*. This assessed whether or not the sample observations were statistically independent of each other, which is an assumption of regression (see insets to Figure 6.11). In the case of the ordinary least squares model, residuals were spatially structured (Moran's $I = 0.43$, $p = 0.001$). This indicated a violation of the independent observations assumption because sampling stations that were close together could not be truly independent of each other. Performing a regression with data from the autocorrelated sampling stations using an ordinary least squares model would probably allocate some of the effect due to interaction to the independent variables. In relation to the model diagnostics, this would have the effect of increasing the likelihood of a type I error, artificially inflating the goodness-of-fit measure (R^2) and underestimating the standard error (Cliff and Ord, 1981). In the case of a regression model, the null hypothesis would be that the datasets being regressed against each other are not related. Residuals of the spatially lagged autoregressive model appeared to be independent of each other (Moran's $I = -0.1060$, $p = 0.06$). This model also explained a greater proportion of the observed variation in species number than the ordinary least squares model (R^2 was 0.89 in the former, compared to 0.35 in the latter). This suggests that the geographical location of the sampling stations was an important factor in determining how many species were present at each sampling station.

6.12 Interpolation to Estimate Unknown Values From Sampled Point Locations

Spatial interpolation is the process of estimating unknown values of a variable on the basis of known sampled point values from a series of nearby locations (Burrough and McDonnell, 2011). The procedure of interpolation results in a transition from a dataset composed of a collection of disparate point samples to a continuous field dataset (often referred to as a 'surface'). Some common applications of spatial interpolation include:

- estimation of values at places where they are not directly measured (e.g. estimating rainfall or temperature across an area where values have been collected from disparate weather stations),
- constructing a continuous digital elevation model from *x*, *y*, *z* measurements collected directly in the field, and
- constructing contours from a digital elevation model.

Some important distinctions relating to interpolations include the difference between global and local approaches, exact and approximate methods, and deterministic and stochastic methods. Global approaches to interpolation make simultaneous use of all sample data in the prediction process. This contrasts with local approaches that utilise only a subset of data, typically identifying points that fall within a moving window across the study area. Exact methods honour the locations and values of the original dataset in the sense that observed values are not replaced. The predicted value at a location where there is a sample is therefore the same as the sampled value. This is not the case with approximate methods, where the value at a location that coincides with a known sample point may be replaced by a predicted value.

Deterministic methods utilise the geometric configuration of surrounding measurements to generate mathematical functions from which a continuous surface of interpolated values can be derived. These functions are derived from measured values of a parameter near the unmeasured location for which a prediction is made. Examples of deterministic interpolators include the calculation of sample means, triangulation and inverse distance weighting. A simple mathematical function is derived based on the distribution of the known, sampled values themselves. This may incorporate the principle of closer measurements being more similar than those that are further apart. But it does not empirically or statistically model how this principle is expressed within the sample data. Thus, not *all* of the spatial information available is exploited. Stochastic (also known as geostatistical) techniques use both mathematical and statistical functions for prediction. The first step for stochastic methods is to statistically determine the spatial structure of the relationship between sampled points. This is subsequently used to predict parameter values for unmeasured points. Stochastic techniques, including ordinary and universal kriging, are often seen as advantageous because their statistical basis allows for quantification of error.

Haining (2003) identifies some ideal requirements of a spatial interpolation method, as follows:

- Exploit the spatial structure in the surface.
- Give the most weight to data at locations close to the site where prediction is needed.
- Incorporate some form of weighting for members of a cluster to avoid any estimate being unduly influenced by clustered measurements from one part of the map.
- Provide some indication of the likely error.
- Honour the known properties of the sample and, in any particular method, when measured values are used to estimate a surface, the estimated surface value should return that measured value (the closer the better).

6.12.1 Three Common Methods for Spatial Interpolation: Inverse Distance Weighting, Splines and Kriging

There are three commonly used methods for carrying out a spatial interpolation: inverse distance weighting, spline and kriging. Inverse distance weighting (IDW) methods are deterministic interpolation methods. They are based on the principle that, if an unknown value is to be predicted for a particular location, the influence of points in a sampled area diminishes with distance from that area. The predicted value is therefore based to some extent on the inverse of the distance between the location where the value is to be estimated and the location of the known, sampled value(s). This principle applies best to phenomena or variables that are spatially autocorrelated, using sampled point values that are evenly spaced. The deterministic function will not return any estimated values above the maximum or below the minimum values of the sampled point values.

There are several variations of IDW that draw on the principle of a value at a given location being more strongly influenced by values at nearby locations, but work differently to express the relative influence or power of samples on the location where the value is to be estimated. For example, the unknown field value at a point can be estimated by taking an average over the known neighbouring point values from the sampled area. Alternatively, a weighting scheme can be devised by introducing a coefficient whereby each known value is raised to the power of a weight that is determined by its distance from the location of the value to be estimated. The rate of decay of the function with distance from a given sampling point will depend on the coefficient employed. This is selected to best represent the surface to be produced (larger functions produce surfaces with more dramatic variation in their values over small distances). Values are predicted for a given location based on a linearly weighted combination of a set of sample points (Table 6.3). One problem with IDW methods of interpolation is that, in a situation where known observation values are clustered (as is often the case with geospatial datasets), the interpolated value will be disproportionately influenced by the clustered values. It is possible to address this through cell declustering. This procedure divides an area into cells and then extends the weight criterion such that it is inversely proportional to the number of samples inside that cell. It may be wise to consider an alternative interpolation approach, such as spline or kriging.

Spline methods of interpolation seek to construct a mathematical function that captures the characteristics of a sampled pattern of point values in a manner that is similar to trend surface analysis (Section 3.6). One unique property of a spline interpolator is that it devises an exact model that passes through all the known sampled point values with the minimum curvature possible. It works well with smoothly varying surfaces and is calculated such that each part of the resulting raster curve is fitted exactly to a small number of data points. Estimated values are constrained so that they fall within smoothing limits. If a smoothing parameter is zero, then the spline passes through the data – it is an exact interpolator. If the value is greater than zero, then the spline is not forced to fit the known data observations – it is an approximate interpolator. All parts of the curve are joined such that the connections between one part of the curve and another are continuous. Spline algorithms can create very smooth surfaces from moderately detailed data and can provide exact or approximate interpolation. Predicted values may fall outside the range of the known sample values. They may therefore produce a surface that contains extreme outlier values. This occurs most commonly in areas where sampled values are sparse. Spline techniques can reveal trends within the entire dataset (global trends) and within a small neighbourhood of sample points around the predicted value (local trends).

In geostatistical terms, kriging is the best of the three interpolation techniques described here in terms of minimising the mean squared prediction error (Burrough and McDonnell, 2011). Kriging can be subdivided into ordinary and universal kriging. Ordinary kriging assumes that there is no underlying trend in the sample data point values. These sample values can be adequately captured by an overall (global) mean. In contrast, universal kriging assumes that there is an unknown trend in the dataset values across the sampled space that must be accounted for.

Table 6.3 A summary of the inverse distance weighting, spline and kriging interpolation approaches to interpolation, including their methodological advantages and disadvantages.

Method	Description	Advantages	Disadvantages
IDW	$Z = \dfrac{\sum_{i=1}^{i=N} Z_i d_i^P}{\sum_{i=1}^{i=N} d_i^P}$ Z = estimated value d = distance between known values and location of unknown values P = weight attributed to known values where Z_i to N are values falling inside the neighbourhood of the location to be estimated	• Insensitive to outliers, as the range is specified by the input sample data • Generates surfaces that are smooth in appearance • Rapid and easy to implement	• Not as accurate as more sophisticated approaches • Not possible to estimate error • Choice of weights does not reflect the statistical properties of the surface
Spline	$Z = \sum_{i=1}^{i=N} (z(x_i) - g(x_i))^2 + \rho J_m(g)$ Z = estimated value $J_m(g)$ = spline function roughness m = degree of polynomial model ρ = smoothing parameter $z(x_i)$ = sample observations smoothing function, g, is dependent on the distance between sample points	• Calculates the smoothing function based on a local set of data for each point	• Is mostly used as an approximate interpolator, i.e. the spline surface does not always fit the data observations
Kriging	$Z(x) = m(x) + \varepsilon'(x) + \varepsilon''$ Z = estimated value $m(x)$ = deterministic variation (constant mean/trend) $\varepsilon'(x)$ = spatially autocorrelated variation (variogram) ε'' = uncorrelated noise	• Subdivides spatial variation into deterministic and stochastic components • Adjusts for clustering of samples • Measures prediction uncertainty	• Requires a normal distribution of values • Can result in extreme estimated values • Ordinary kriging assumes a constant mean and semivariance across all locations

Kriging is an interpolation method that seeks to quantify three components of surface variation:

- deterministic variation with a constant mean or trend,
- spatially autocorrelated variation (autocovariances or variogram), and
- uncorrelated noise.

The deterministic component of the variation for a given spatial phenomenon is modelled from a trend surface that captures trends in a constant model (e.g. linear or quadratic). Universal kriging attempts to model an overriding trend in the data

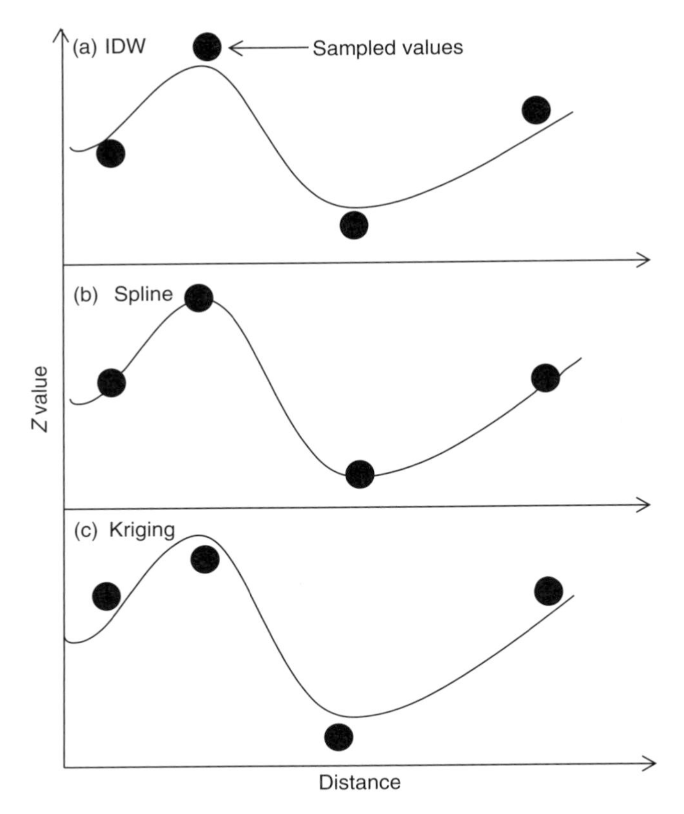

Figure 6.12 A comparison of the surfaces estimated by the different types of interpolation methods: (a) inverse distance weighting, (b) spline and (c) kriging. It can be seen that the IDW method cannot interpolate peak data values; while the spline method creates a smooth surface that passes through known, sampled data values; and the Kriging method produces estimates that are not exact, but provide a close statistical estimate of the sampled values.

using a deterministic function (e.g. a polynomial). This function is subtracted from the original measured points. Spatial autocorrelation is then modelled from the random errors. Universal kriging should only be used when an underlying trend has been observed in the sample data. This provides justification for the need to model this component of variation. The spatially autocorrelated variation is modelled by estimating the autocovariance at different distances apart, fitting a model to describe the empirical autocovariance function (EAF). Values are then obtained from the EAF and substituted into an estimated surface based on the distances between the location to be estimated and the sampled point. Using this approach, weights can be constructed in terms of estimated values in the surface. As with IDW, measured values are weighted to predict values at unmeasured locations. Unlike IDW, weightings are derived from a geostatistical model of the sampled surface and are based on spatial autocorrelation between sample points. A statistical relationship between values at sampled points is determined and this relationship is

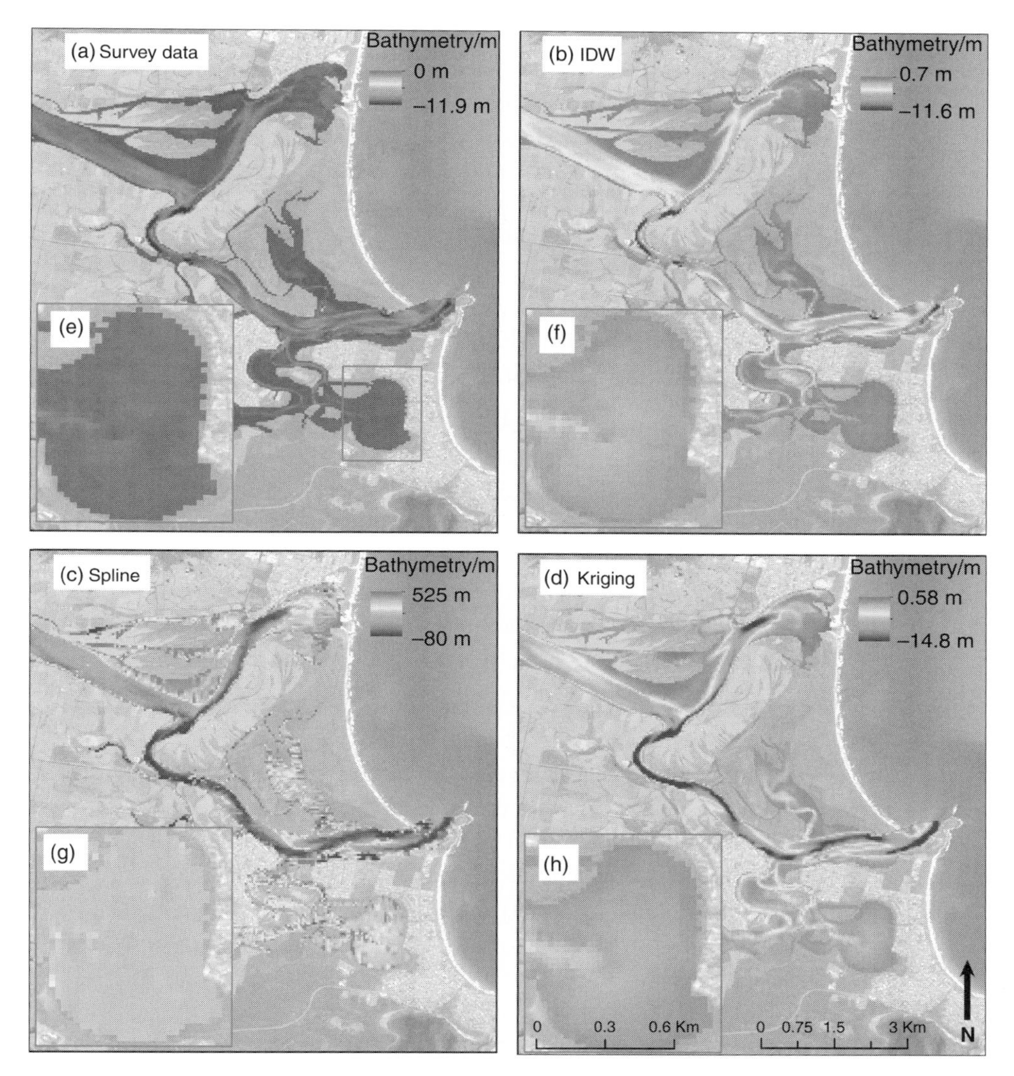

Figure 6.13 A series of interpolations on bathymetric data collected from the Shoalhaven Estuary, New South Wales: (a) raw bathymetric surveyed point data; (b) inverse distance weighting interpolation; (c) spline interpolation; (d) ordinary kriging interpolation; (e)–(h) larger-scale representations of the Curley's Bay area to the south (indicated by the red box in panel (a)). To facilitate comparison, the same colour scheme has been applied to panels (e)–(h). (A black and white version of this figure will appear in some formats. For the colour version, please refer to the plate section.)

used to make predictions about unmeasured points. Associated errors can be calculated and assessed, and adjustments can be made for the clustered sampled values. Figure 6.12 provides a graphic illustration to compare some of the characteristics of surfaces of estimated values derived from inverse distance weighting, spline and Kriging interpolators.

There is no perfect interpolation method. Table 6.3 outlines some of the advantages and disadvantages of IDW, spline and kriging interpolators. It is important to remember that the choice of interpolator must be made with due regard to the dataset and purpose for which it is to be used. Methods should be applied critically, with an awareness of their properties and a clear justification that links choice of method to valid assumptions about the nature of the data and the underlying spatial variation in the specific dataset.

6.12.2 Comparing Interpolation Methods of Bathymetry in the Shoalhaven Estuary, New South Wales

Figure 6.13 is an illustrative example of the three types of interpolation (IDW, spline and kriging) applied to a bathymetric dataset comprised of 52 550 point measurements of water depth surveyed by boat across the Shoalhaven Estuary, New South Wales, Australia. The varying estimations derived from the different interpolation methods can be seen from the mapped results. These are presented as raster grids overlain onto an image of the estuary. The legends show the range of interpolated data values for each of the methods. The following observations reveal the nature of the different interpolation methods:

- The IDW interpolator does not generate values outside the range of the input dataset of known values.
- The spline interpolator results in a surface that passes through the minimum and maximum data points with the minimum curvature. Values can fall outside the range of the input dataset, as can be seen from the values in the legend of Figure 6.13(c), which ranges from −80 m to 525 m across the study area.
- Values estimated by the kriging interpolator are not constrained by the known input values; however, they remain close to the original values.

6.13 Summary of Key Points: Chapter 6

- Geostatistical techniques quantify and characterise the spatial structure of a dataset. They can operate at the global (entire data sample) or local (neighbourhood) scale.
- Spatial dependence is a fundamental property of most datasets. It arises because of Tobler's first law of geography: *'everything is related to everything else, but near things are more related than distant things'* (Tobler, 1970). Positive spatial dependence means that geographically nearby values of a variable tend to be similar.
- Spatial autocorrelation is the correlation between paired values of a single variable related to their relative locational positions on a two-dimensional surface. It arises because of a pervasive *continuity* and *structure* to the real world.
- There are several practical reasons for measuring spatial autocorrelation, including evaluating the statistical validity of analysis undertaken on datasets, incorporating

autocorrelation into explanatory or predictive models, and guiding the design of sampling schemes for data collection.

- Tests for spatial autocorrelation include the join-count test, calculation of the Geary's c and Moran's I statistics and their corresponding semivariogram and spatial autocovariance functions. Plots of these latter functions reveal the nature and dimensions of spatial structure in a dataset (e.g. through features such as the range, sill and nugget).
- Spatial interpolation is a common geostatistical technique that estimates unknown values of a variable on the basis of known point values. This results in a transition from a dataset of disparate point samples to a continuous field dataset (or raster 'surface').
- Three common methods for spatial interpolation are inverse distance weighting (IDW), spline and kriging. These methods have advantages and disadvantages. They should be applied critically, with an awareness of their properties and valid assumptions about the nature of the data and the underlying spatial variation of the phenomena represented.

7 Modelling Coastal Environments

7.1 What Is a Model and Why Model Coastal Environments?

Modelling is a widely used phrase that may mean different things within different settings. Broadly, models are developed to break the world down in a meaningful manner, usually for a predictive or explanatory purpose. They provide a concise, quantitative account of observable patterns and relationships between different components of a natural 'system'. A system can be thought of as a set of objects, together with their attributes and the relationships between them (Von Bertalanffy, 1956). Models generally isolate objects within a given system to investigate how they operate as a whole. In this way, relationships can be derived mathematically through abstract representation and used to concisely express the collective behaviour of the objects within a system (Melton, 1957).

In the context of coastal spatial analysis, the objective of building a model is to find a mathematical expression that optimally characterises some spatial process. This makes it possible to explain or replicate observed variation in a given feature (e.g. the relative patchiness of seagrass beds). Models may help to generate new insight into the underlying processes influencing an observed pattern. A modeller may wish to construct a scenario in which they can experiment on realistic datasets, and simulate 'real-world' actions. Viewing the natural world as a modelled system in this way allows the analyst to examine a range of dynamic outcomes that may arise as the system evolves and responds to inputs. Table 7.1 summarises some different types of models that are commonly applied to coastal environments.

Many coastal modelling exercises are motivated by an increasing desire to bridge the gap between traditional small-scale field experiments, which are necessarily of a limited geographical and temporal scope, and the larger-scale patterns and processes that can be observed in coastal landscapes. To this end, digital maps are commonly used as input for models that seek to explain or predict observable patterns in coastal environments. Both explanatory and predictive models aim to link different components of a coastal system. They infer a causal relationship between them, such that one component (or variable) can be thought of as a dependent variable that responds to either a single or multiple influential drivers (independent variables). For example, a beach modeller may be interested in explaining the relationship between sand deposition (dependent variable) and sediment supply and transport, driven by waves, currents and rock weathering (independent variables). The two variables can be compared across different environmental settings by identifying and interrogating meaningful gradients across landscapes. This can be achieved in a spatially explicit manner using digital maps. Varying distances, gradients and neighbourhood contexts through

Table 7.1 Different types of model that are applicable to coastal environments.

Model	Description	Reference
Conceptual, descriptive	A framework, often expressed in qualitative terms, with respect to which further research can be designed and executed	(Ford, 1999)
Physical	A physical reconstruction of an object or landscape, produced to scale	(Summerfield, 2014)
Empirical	A model based on investigation, observation, experimentation or experience	(Hamylton et al., 2016)
Heuristic	Trial-and-error approaches to experimentation and evaluation to find a solution using loosely defined rules	(Valdes, 1981)
Bayesian	Employs statistical evidence to express a representation of the true state of the world as a Bayesian probability	(Bolstad, 2013)
Numerical	Employs a series of time steps to model the behaviour of a system over time	(Angeloudis et al., 2016)
Probability	For a range of potential outcomes, a probability model computes the likelihood of this outcome occurring	(Harris and Jarvis, 2014)
Cell-based simulation	Employs a grid-based representation of the world to express observed and simulated states of the cell contents; otherwise known as grid cell modelling, cellular automata	(Wolfram, 1994)

a 'space' treatment of the landscape (see Section 1.3) permits exploration of how an independent variable influences a response (dependent) variable.

Maps in raster format represent a useful dataset that provides a foundation for cell-based simulations that utilise geographic information systems (GIS) or modelling software to emulate coastal processes. For example, the type of wetland along a coastline can be modelled on the basis of internal sedimentary processes and external transport of water, sediment and nutrients (Sklar et al., 1985). Similarly, the sediment dynamics of a fluvially derived delta deposit can be modelled by combining a raster grid of a digital elevation model (DEM) depicting height for individual pixels with programs designed to simulate wave and sediment transport behaviour (WAVE and SEDSIM). In this way the nearshore erosion, transport and deposition of sediment can be captured (Martinez and Harbaugh, 1989). At the centre of such a model lies a series of inbuilt rules governing the transport of sediment between adjacent cells. These determine whether grid cells are filled or emptied, and characterise associated topographic fluctuations. This framework can be used to model accretion rates and behavioural thresholds associated with different shoreline features. For example, it has been used to predict the response of both mangrove-dominated estuaries and coral reef platforms to sea-level rise (Oliver et al., 2012; Hamylton et al., 2014). These examples have a common starting point that begins with a map. Algebraic operations are then carried out on this digital map using specialised software. Often these operations draw explicitly on the location of each data value within the dataset, such that individual pixels are manipulated within the geographical context in which they fall within the broader raster grid.

7.2 Four Stages of Spatial Model Development: Observation, Measurement, Experimentation and Theory Development

Figure 7.1 outlines the four stages of model development as observation, measurement, experimentation and theory development. These provide a useful general framework for considering the different activities involved in constructing a spatial model of any coastal environment.

7.2.1 Observation

Observational sciences operate with the world as it is found and are characterised by the absence of experimental control and replication (Section 1.3). Many models begin with the simple act of visually examining features or patterns that are evident

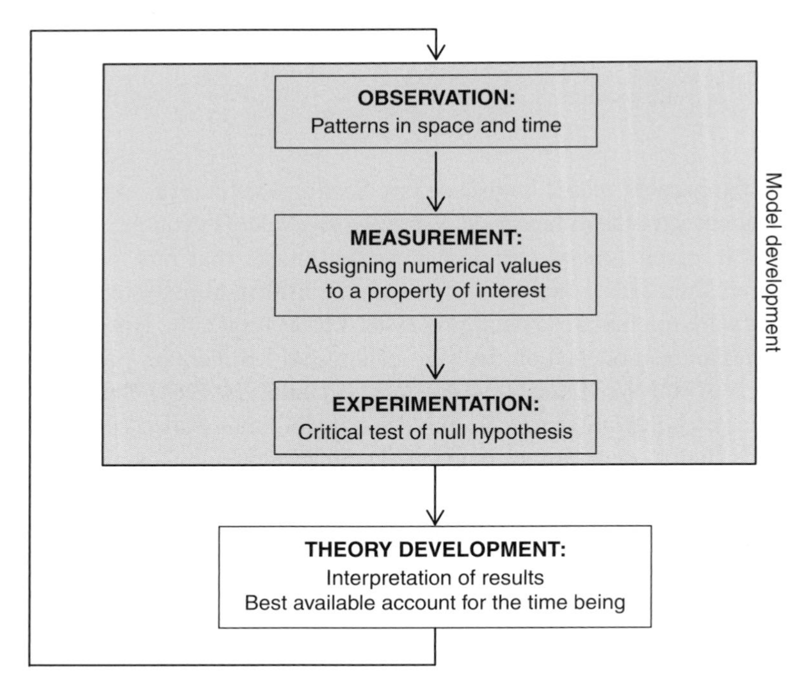

Figure 7.1 The four stages of model development: observation, measurement, experimentation and theory development.

in landscapes. These may be derived from assemblages of multiple features that take on a different form or configuration in different subregions of the landscape. We cannot instantaneously see a coastal environment and resolve constituent features and processes within it. Nevertheless, a map derived from a remotely sensed image provides a digital representation of its structure. Such a map is an important dataset in its own right. It provides an opportunity for a model to be constructed.

The aerial photographs in Figure 7.2 illustrate a range of landscapes along the coastline of France. By visually examining these pictures, it is possible to observe

Figure 7.2 Some examples of French coastal landscapes within which patterns can be observed: (a) Archipel des Lavezzi; (b) Delta de la Leyre, Gironde; (c) Baie des Veys, Manche; (d) Baie de Somme; (e) Grand Cul-de-Sac, Guadaloupe; (f) Marais D'Olonne, Vendée. (Photo credit: Frédéric Larrey.) (A black and white version of this figure will appear in some formats. For the colour version, please refer to the plate section.)

spatial patterns in features such as sand dunes, ridges of sediment across channels and patches of seagrass, saltmarsh and mangroves. Observations may prompt an analyst to pose a theory about relationships between the different landscape elements, and what formed them. For example, an analyst may postulate that the degree of fragmentation in the seagrass beds around the island in Figure 7.2(e) is related to the movement of water currents.

7.2.2 Measurement

Measurement is the assignment of a numerical value to some quality of an object or feature, so as to represent some fact or convention about it in accordance with definite rules (Haggett and Chorley, 1967). This often involves classifying natural phenomena to enumerate their characteristics. In turn, this permits quantification and empirical analysis. This transition relies on the translation of concepts to entities. In the context of spatial data modelling, the process of measurement populates the spatial data matrix. Measurement can be subdivided into two key steps of relevance to the modelling process: conceptualisation and representation (see Section 2.3). These are therefore two key steps of the measurement process. Information can be collected at different measurement levels (nominal, ordinal, interval or ratio) (see Section 2.6). These govern how the information can subsequently be used in a modelling exercise.

7.2.3 Experimentation

Experimentation refers to the control, simplification and manipulation of the world so that causal patterns can be seen more clearly (Richards, 2009). A suitable experimental design can employ a range of observational data from a coastal landscape (e.g. rivers, estuaries and deeper marine environments) to reveal meaningful relationships between them. Broadly, an experiment consists of asking a question about a natural phenomenon, making observations of that phenomenon, hypothesising an explanation for the phenomenon, predicting a logical consequence of the hypothesis, testing the prediction and drawing a conclusion with data gathered in the experiment. A statistical test provides grounds for either accepting or rejecting a hypothesis on the basis of the data collected. Outcomes are interpreted with reference to the existing body of knowledge on the processes investigated and may reveal additional insight that has come to light by empirically analysing the data collected.

Most models follow a path of logical reasoning in which a hypothesis is proposed and tested. A model cannot be intended to verify the hypothesis that is under scrutiny. This is because the possibility remains that further observations will be made that are contrary to the model. In the context of modelling drainage basin geomorphology, Melton (1957) outlines a series of generalisations that provide a useful framework for interpreting the findings of coastal models. It is possible to say that: (i) when various properties of coastal environments are analysed, occurrences are found that would not reasonably be expected to result from operation of chance or sampling error alone; (ii) these non-chance occurrences may represent in some manner the true laws of nature, and hence formal model specifications can be posed; and (iii) some of the non-random occurrences revealed in the case studies may reflect the agencies of cause and effect, but interconnections of causative factors and the presence of random error make drawing simple conclusions of cause and effect inappropriate.

Although a model may explain a substantial proportion of the variation that occurs in a coastal landscape, this does not necessarily imply predictive success. It is possible, for example, that a model may reveal a relationship that is actually determined by a covariate of the variable examined. A covariate is a variable that displays a relationship to a variable of interest, which may or may not be of direct interest to the study. It is also impossible to assure that the model has specified all relevant variables. Chorley (1964) suggests that the points at which our mathematical analogues differ from their real-world prototypes are of particular interest, perhaps more so than those with which they agree. These deviations have greater potential to help us adjust our understanding of the landforms and processes in which we are interested, such that this can be expressed with greater mathematical precision.

7.2.4 Theory Development

Having revealed statistically significant relationships in a given modelling exercise, the analyst is prompted to look for an explanation that goes beyond an empirical account in the form of an equation. A theory must be constructed to explain experimental results. In this way, models are continually revised in a circular dependence of theory and observation. Rather than defining strict physical laws, this calls for a redefinition of modelling as an activity that incrementally advances scientific knowledge. This knowledge, or theory, may represent the best available account for the time being, paving the way for new hypotheses and experiments that provide even more stringent tests of the model. As philosopher Karl Popper observed, the essence of science is a growing body of theory that has resisted attempts at falsification (Popper, 2014).

7.3 Structure vs Function: a Framework for Modelling Coastal Environments

Coastal models are often described as empirical because they are constructed by building up a dataset of many observations. These may be scattered across a digital map in locations where coincident information on dependent and independent variables can be collected. An empirical exercise can therefore be undertaken to uncover a statistical relationship between variables that represent some measure of system response, usually expressed in the *structure* of the coastal landscape, and those that summarise aspects of *function*, or underlying processes driving these structures (Mather, 1979).

Structure commonly refers to the observable characteristics or interconnected features of landscapes, i.e. the sizes, shapes, numbers and configurations of components. Function refers to the interactions between these elements, i.e. the flow of energy, materials or organisms among the components. This framework corresponds well to a broader model specification in which functions can be treated as independent variables and structures are treated as dependent variables. Such a framework can be implemented by quantifying the result (variously referred to as response, structure, form, landform or habitat) and driving variable (function or process) multiple times, using location as a basis for linking the datasets. In this way, a statistically powerful collection of data samples can be built up through which the relationships between the two can be empirically reconstructed. This provides a framework for building models that enable us to generalise about the collective behaviour of multiple interconnected features within coastal environments. Such a framework for studying the mutual co-adjustment of structure and function has been adopted by numerous disciplines, including landscape ecology (Forman and Godron, 1986; Farina, 2008), marine metapopulation patch dynamics (Kritzer and Sale, 2010), landform geomorphology (Verstappen, 1977; Mather, 1979) and coastal morphodynamics (Wright and Thom, 1977; Woodroffe, 2002). The wide applicability of this approach to geographical phenomena is demonstrated by its correspondence to general systems theory. For example, process–response systems demonstrate how morphological structure is related to processes because of throughputs of energy or mass (Chorley and Kennedy, 1971).

Techniques like regression and spatial regression can be employed as an intuitively appealing statistical technique for modelling. These techniques work best where a relationship can be expressed as a single, linear function. They can be difficult to apply when feedbacks are present. Feedbacks occur when a coastal environment responds to a perturbation, either by exacerbating the effects of that perturbation (positive feedback) or minimising it (negative feedback). These responses arise when structural properties exert an influence, in turn, over the functions that have controlled their form. Feedbacks give rise to what are termed 'nonlinear' responses, because the relationship between the dependent variable (x) and the independent variable (y) can no longer be expressed as a simple linear function. Thus, if a graph is plotted to characterise the structural response of a coastal environment (y axis) over a range of values for some specific functional driver (x axis), the pattern that emerges may not be a simple straight line of direct or inverse correspondence between the two (i.e. of the form demonstrated in equations (7.1) and (7.2)). Other functions may be more appropriately fitted to the dataset, such as polynomial (first-, second- or third-order), power, exponential, Gaussian or logarithmic models. For example, second-order polynomial models have been used to model the presence or absence of shallow water seagrass and the production of calcium carbonate by coral on a reef around Lizard Island on the Great Barrier Reef (Hamylton et al., 2013; Saunders et al., 2014). This relationship between structure and function is explored, with seagrass or coral representing the structural component (dependent variable) and wave energy representing the functional component (independent variable). In both cases, a positive relationship exists in which water circulation associated with waves enhances metabolic processes of seagrasses and coral. This relationship appears to be governed by a threshold point at which the mechanical stress associated with water currants exceeds the metabolic benefits, whereby the association is reversed to an inverse relationship (as captured by the shape of a second-order polynomial model).

7.4 Measuring and Modelling Coastal Processes

All models begin with a theoretical link between some sort of response variable (the dependent variable) and a variable or set of variables that collectively represent a driving force of influence over this variable (the independent variables). A critical part of successful modelling, for either explanatory or predictive purposes, is the conceptualisation and representation of a set of *functionally relevant* independent variables that will do a good job of explaining variation in the dependent variables. Constructing an initial conceptual model can help a modeller to consider the full range of processes that may influence the dependent variable, which may be any aspect of a coastal environment, from a grain of sand to a barnacle (see Figure 1.8).

It is important to match the spatial scale at which information is collected on independent variables with the response of the phenomenon being modelled. For example, an immobile coral that is attached to the seafloor will experience the environment at only one location. Conversely, a mobile pelagic fish or a grain of lightweight sand may interact with a range of environments over different spatial and temporal scales. As we have seen in earlier chapters, processes that influence shallow marine environments operate along relatively predictable environmental gradients. These processes can often be captured with spatial technologies such as remote sensing instruments.

Table 7.2 summarises some common environmental processes acting across a range of geographic scales in coastal environments. Sections 7.4.1 and 7.4.2 describe in further detail techniques for generating spatially referenced datasets on coastal processes. These include representations of wave energy and the suspension of sediments in coastal waters across the Al Wajh Bank, a semi-enclosed lagoon in the Saudi Arabian Red Sea.

7.4.1 Modelling Wave Energy From Wind Conditions and Local Fetch Lengths at the Al Wajh Bank

Wave energy influences many characteristics of coastal environments by moving sediment and delivering energy and nutrients to organisms. This example of modelling environmental processes illustrates a wave exposure model that employs linear wave theory to predict wave power inside the Al Wajh Bank, a semi-enclosed lagoon along the Saudi Arabian coastline (see Figure 7.3). Predictions are made on the basis of wind and fetch conditions. Fetch is the distance over which the wind blows uninterrupted from a specific direction along an open-water body (Bishop et al., 1992). It is a key determinant of wave conditions, because the further wind can blow uninterrupted over a water surface, the greater the time window for wave generation. Wave energy can be calculated using linear wave theory by incorporating information about

Table 7.2 Common environmental processes that act at the macro-, meso- and microscale in coastal environments.

Scale	Variable	Example measurements and/or models
Macroscale (e.g. ocean basin)	Movements of the Earth's crust due to tectonic activity	Geodetic measurements of plate motion with the global positioning system (GPS)
		Plate trajectories can be determined using palaeomagnetic, geometric or seismic methods (Vincent, 1997)
	Fluctuations in sea level	Sea-level fluctuations can be measured at local scales using tidal gauges, or at broader global scales using satellite altimeters
		The broader outlook offers a perspective on the varying rates of sea-level rise for different regions (Church et al., 2004; Nerem et al., 2006)
Mesoscale (e.g. a coastal embayment)	Seawater properties, including sea surface temperature, salinity, seawater chemistry	Measurements of seawater properties can be made locally using instruments such as thermometers or hydrometers directly for a specific purpose, or continual measurement can be logged over time at monitoring stations
		Across larger geographic scales, satellites can be used to measure seawater properties, including temperature, salinity, aragonite saturation, chlorophyll and nutrient levels (Woody et al., 2000; Robinson, 2004)
	Climatic conditions	Weather and climate conditions, such as wind direction, speed, frequency and wave characteristics, can be recorded at weather stations based on islands or mounted to oceanic buoys, or measured from satellites (Strong and McClain, 1984; Gilhousen, 1987)
	Wave energy generated by displacements of the ocean surface due to storm swell or local winds	Wave energy can be measured locally from weather buoys, or modelled using satellite data, or a combination of wind data and local fetch conditions (Cruz, 2007; see Section 7.4.1)
	Water depth	Can be measured directly using sonar instruments, or modelled using remote sensing approaches (see Section 4.12) (Calmant and Baudry, 1996; Riegl and Guarin, 2013)
	Water currents	Currents driven by tidal fluctuations and movements of water bodies around topographic features, i.e. over barriers or through channels, can be measured directly with an acoustic Doppler current profiler (Johnson et al., 2002)
Microscale (e.g. within a kelp bed)	Ambient levels of light	Ambient levels of sunlight are known to reduce exponentially as a vertical axis is traversed down the water column
		The amount of light reflected off a reference panel can be measured with a spectrometer at different depths (Mobley, 1994)
	Turbidity, i.e. the presence of suspended sediment in the water column	Turbidity can be measured directly, either by lowering a Secchi disc from a boat down into the water column or by filtering, drying and measuring the quantity of dried sediment from seawater samples
		The amount of suspended sediment in seawater can also be modelled from the longer-wavelength bands of optical satellite images (Glasgow et al., 2004)

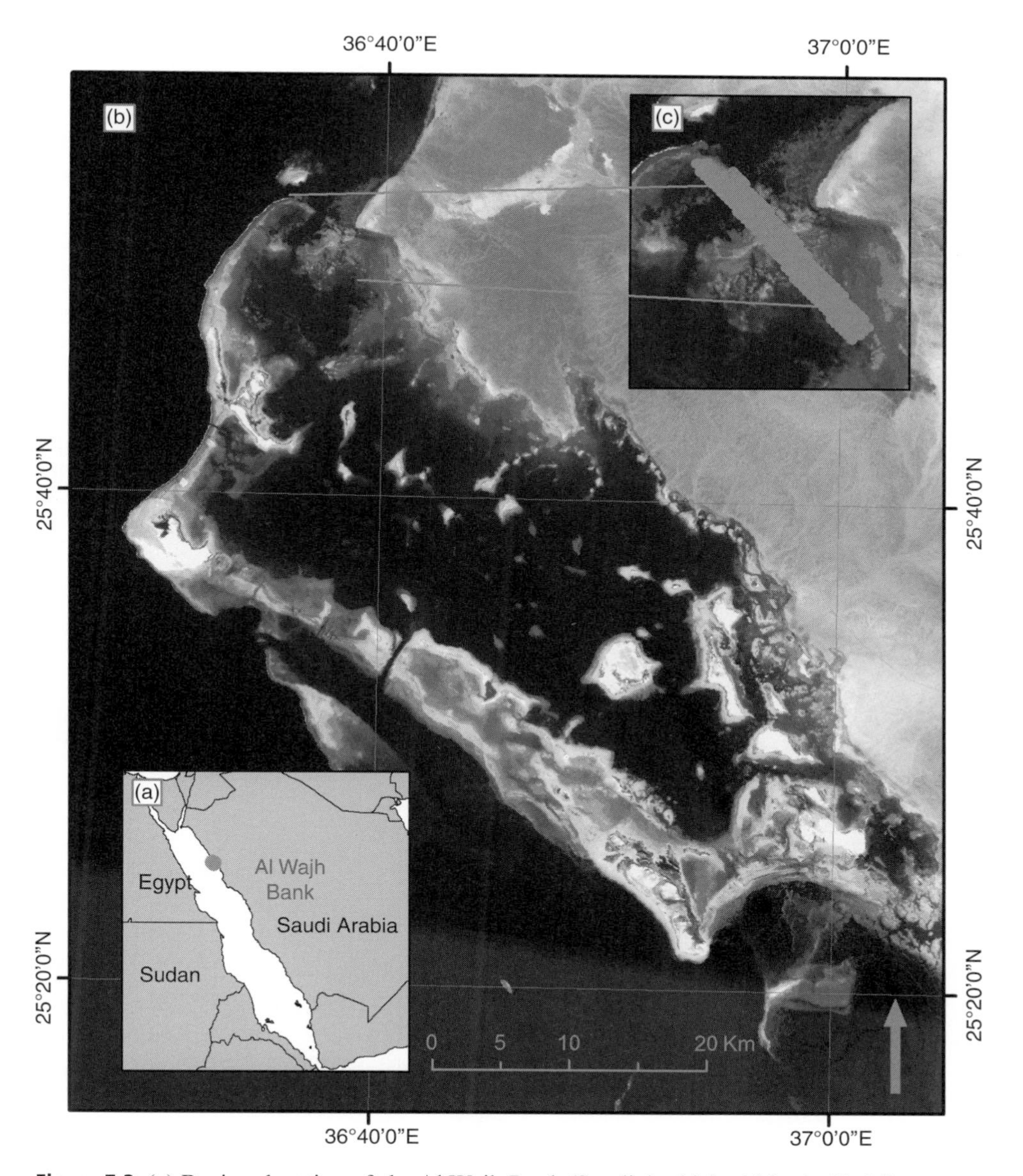

Figure 7.3 (a) Regional setting of the Al Wajh Bank (Saudi Arabia) within the Red Sea. (b) The Al Wajh Bank, a north–south aligned semi-enclosed reef complex. (c) The location of the study area covering a strip of water at the northern end of the Al Wajh lagoon. (A black and white version of this figure will appear in some formats. For the colour version, please refer to the plate section.)

wind conditions (strength and frequency) with information on the fetch setting. An enclosed area that is sheltered from winds will have low fetch values associated with it, whereas an open, exposed area will have high fetch values. For illustrative purposes, the model is applied to a study area defined by a point grid that covers a strip of water inside the Al Wajh Bank. This strip of water lies inside an opening in the Bank that results in a range of exposure conditions. For a central band of the study area, an open channel increases exposure to winds from the north (the prevailing direction of winds in the eastern Red Sea).

To measure fetch direction and length, a map delineating the areas of water and surrounding land was used to characterise the fetch lengths across eight cardinal and subcardinal directions for a regular grid of points plotted across the study area. From each point, eight lines spaced 45° apart radiated out to a distance at which they intersected with the coastline. The lengths of these lines represented the fetch distances from each direction that could be input into the linear wave transform model. Figure 7.4 illustrates the fetch distances calculated for each of the cardinal directions (north, east, south and west) across the study area.

To convert fetch lengths into estimates of wave exposure, it was necessary to combine them with information on the local wind climate. Data on the wind speed and direction for the study area were taken from a nearby weather station at Bahrain Island,

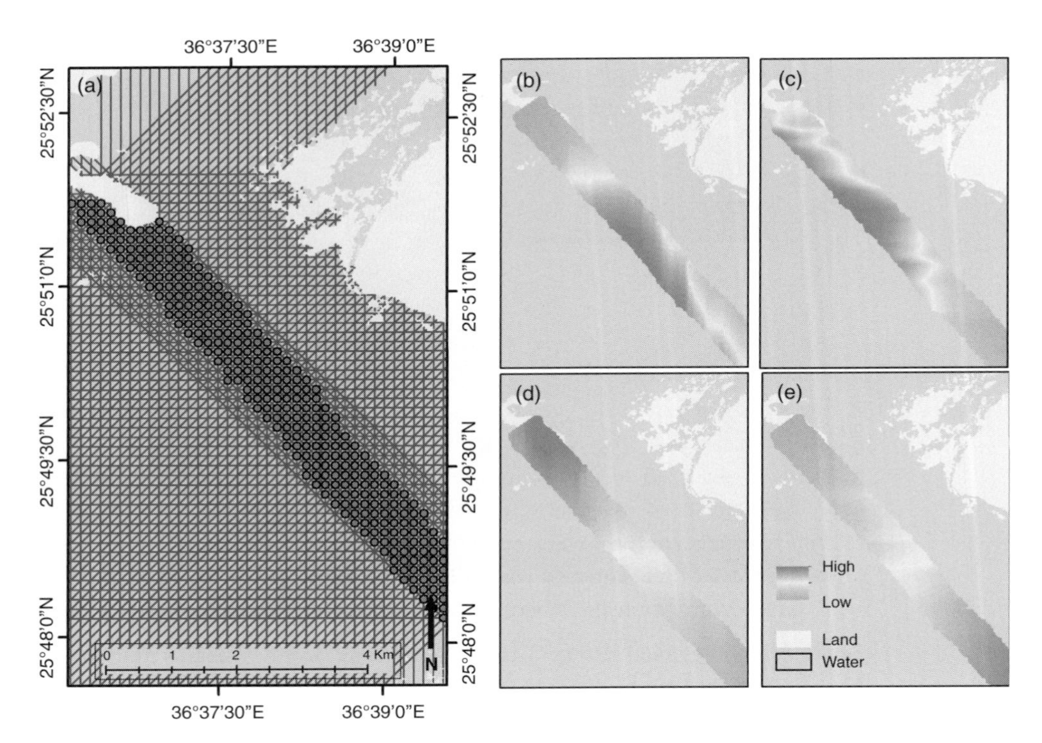

Figure 7.4 A series of maps illustrating high and low fetch distances from each cardinal direction in relation to land surrounding the study area, with emergent land features depicted in orange. (a) Fetch lines in all cardinal and subcardinal directions emanating from a grid of points covering the study area. Relative fetch values within the study area (red indicates high fetch, green indicates low fetch) for fetch directions (b) north, (c) east, (d) south and (e) west. Note the gap in land masses to the north, resulting in a non-fetch-limited area with respect to this direction in the centre of the study area. (A black and white version of this figure will appear in some formats. For the colour version, please refer to the plate section.)

Saudi Arabia (26°16′N, 50°37′E). This station spanned a time series of weather measurements from January 1951 to October 2005. The wave exposure model uses a stepwise procedure to predict wave energy from wind data and fetch characteristics of the area. The model begins by calculating wave period and significant wave height from the wind data and fetch distances across all eight directions for every grid point. From these parameters, linear wave theory can be used to derive horizontal and vertical particle orbital velocity at the seafloor. In a final step, wave energy can be calculated from these velocities. A total output value for wave energy can be calculated by summing the contributions from each directional component, weighted by the probability of the wind blowing from that direction (Ekebom et al., 2003; Harborne et al., 2006).

7.4.2 Modelling Water Turbidity From Optical Satellite Imagery at the Al Wajh Bank

Water turbidity is an important characteristic of many coastal environments. This example illustrates how the concentration of suspended sediment, which drives water turbidity, can be modelled from a combination of water samples and remote sensing data. It is possible to estimate the concentration of sediment suspended in sea surface waters from the relationship between suspended sediment concentration (SSC) and the reflectance values of a remotely sensed satellite image. As the concentration of suspended sediment inside a water body increases, the spectral reflectance properties of a water body change. Overall image brightness increases and the wavelength of peak reflectance shifts from a maximum in the blue towards a maximum in the green. The reflectance peak becomes broader until, at very high levels of SSC, the colour of the water approaches that of the sediment itself (Campbell, 2002). This relationship makes it possible to estimate SSC from remotely sensed imagery by relating measurements of suspended sediments taken directly in the field to the brightness values associated with a corresponding satellite image (Curran and Novo, 1988).

Suspended sediment concentration was estimated for the study area previously defined at the Al Wajh site using a remotely sensed image taken from an AISA-Eagle airborne hyperspectral instrument. The wavelength at 665 nm (in the red section of the visible spectrum) was selected, as this has been found to correlate strongly with suspended sediment elsewhere (Curran and Novo, 1988; Binding et al., 2005). A field dataset of 50 water samples was collected spanning the geographical extent of the study area. Water sample collection was timed to coincide with the airbourne remote sensing survey. To estimate the SSC of these water samples, the dry weight of suspended sediment was then measured in a laboratory. The range of SSC measured from the lagoon water samples extracted was 5–60 mg l^{-1}. The distribution indicated that turbid waters, associated with elevated levels of suspended sediments, were in shallower areas (water depth < 1 m), whereas deeper water tended to be clearer.

The 50 water samples were split in half for calibration and validation purposes. Half of the data values were used to calibrate or establish a relationship between radiance on a remote sensing image and suspended sediment. The remaining half of the water samples were used to validate these estimates. To calibrate the relationship, sample points were overlaid onto the extracted wavelength of the hyperspectral remotely sensed

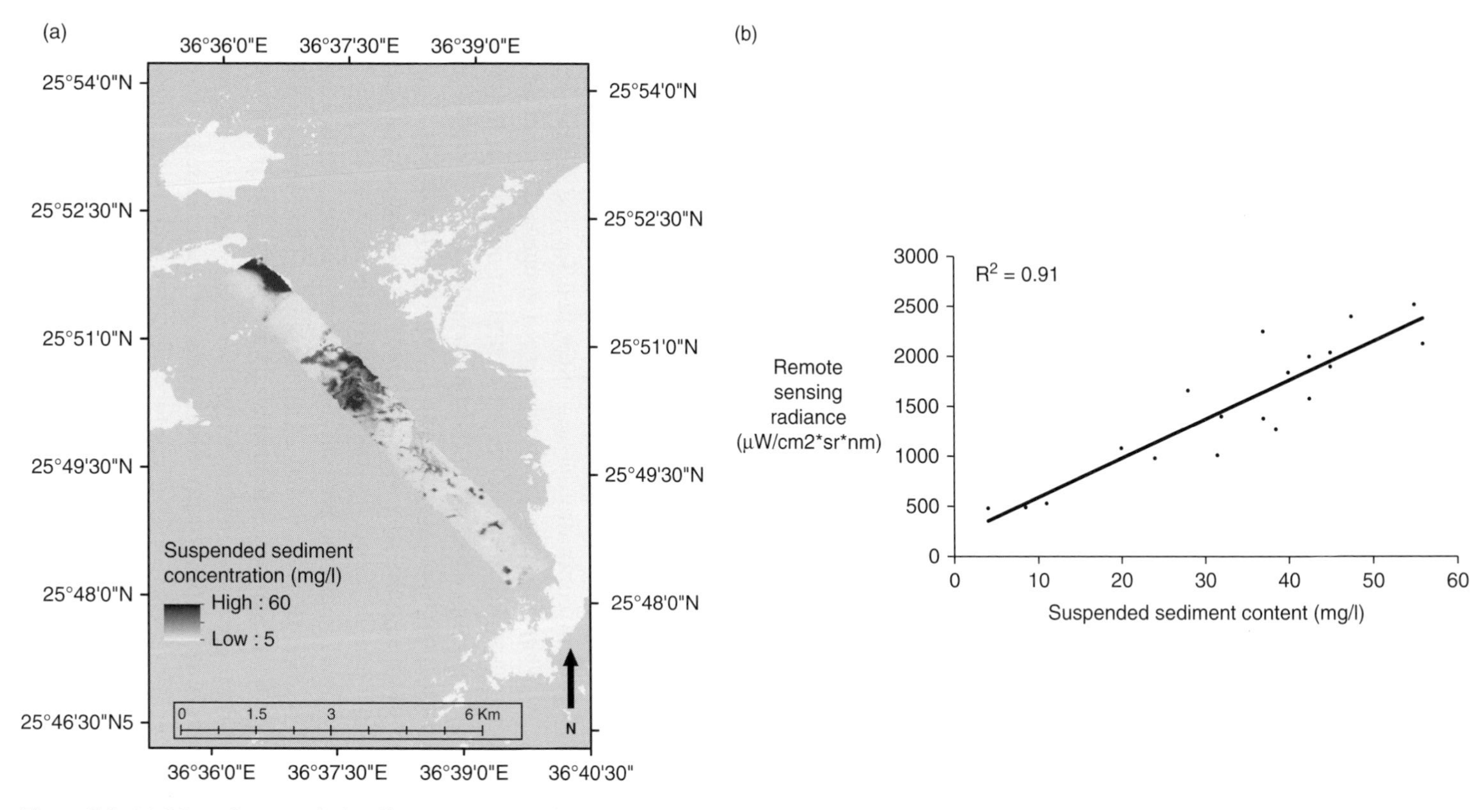

Figure 7.5 (a) Map of suspended sediment concentration across the reef systems inside the Al Wajh Bank, Saudi Arabia, as modelled from remote sensing radiance of the surface waters. (b) Empirical relationship between suspended sediment content of 19 water samples taken from the study area and the associated surface water radiance. (A black and white version of this figure will appear in some formats. For the colour version, please refer to the plate section.)

dataset, which had been corrected for environmental influences. Radiance values were extracted and plotted against the sediment content of each water sample to establish a linear relationship between the two. The association between radiance as measured on the imagery and sediment content of the water was strong ($R^2 = 0.91$, Figure 7.5b). On the basis of this strong relationship, suspended sediment concentration was modelled across the rest of the area covered by the remote sensing image (Figure 7.5a). This was achieved by applying the algebraic expression that was represented by the linear equation to each image pixel. In this way, the relationship between the suspended sediment concentration of the collected samples and the remote sensing image radiance was used to estimate SSC from radiance (Figure 7.5b). This enabled the suspended sediment concentration at all other locations covered by the image to be estimated.

A final step was to validate the estimates of SSC. These estimates corresponded with the locations of the remaining 25 water samples for which suspended sediment concentration had been measured and that were reserved for validation. Estimated values were compared with measured values of SSC to assess their validity via a linear regression. Comparisons indicated a close relationship between the two ($R^2 = 0.81$).

7.5 Developing Spatial Models

7.5.1 Explanatory Models

In the context of spatial analysis, an explanatory model seeks to construct an explanation for an observed pattern in a natural environment using empirical data. Explanatory modelling begins with a theory that a phenomenon of interest (dependent variable Y) is caused by a different phenomenon (independent variable X). The relationship between X and Y can be expressed as a function F, such that $Y = FX$ (see Figure 7.6). The function F can be represented by a series of qualitative statements, a conceptual model, a plot or an equation. To operationalise F into an explanatory model, a study must be designed to collect data that will enable an estimate of the function, f, to be derived empirically. The degree to which the estimated function, f, resembles the real-life function, F, will depend on a range of factors, including what phenomena are specified, how the phenomena are conceptualised and represented, the quality of the data collected and decisions made when the relationship was specified.

7.5.2 Predictive Models

Predictive modelling is the process of applying a statistical model to data for the purpose of predicting new observations. These predictions can be made across either space or time. Broadly, the goal is to predict the output value for a dependent variable (Y) for new observations, given information on input values or independent variables (X). Such a definition includes both spatial prediction (e.g. interpolation methods) and temporal forecasting.

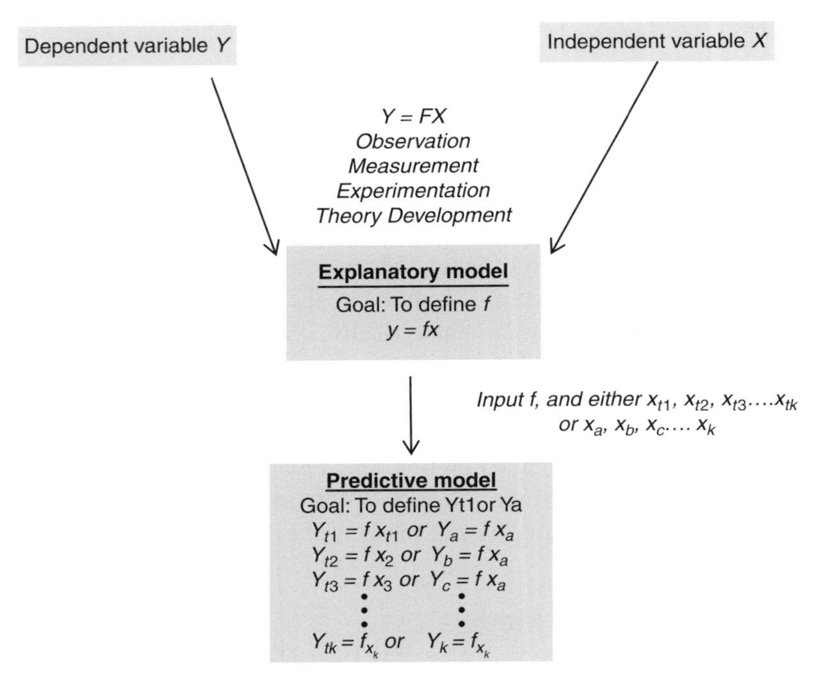

Y = value of dependent variable
X = value of independent variable
y = sampled value of dependent variable
x = sampled value of dependent variable
x_{t1} = anticipated value of independent variable at time = t_1
x_a = anticipated value of independent variable at location a
F = function linking dependent and independent variables
f = estimated function linking dependent and independent variables

Figure 7.6 The key differences between and explanatory and a predictive model.

7.5.3 A Comparison Between Explanatory and Predictive Models

Figure 7.6 illustrates the difference between explanatory and predictive models. One key difference between explanatory and predictive objectives is that the goal in explanatory modelling is to match the 'real' and estimated functions, F and f, as closely as possible. If this is achieved, the statistical inference will support the theoretical hypothesis. Defining a meaningful function with good explanatory power and accompanying statistical significance is therefore the main aim of explanatory modelling. The focus is on uncovering relationships in data that are already accessible to the analyst. Explanatory modelling is therefore commonly used for theory building and testing.

Predictive modelling is rarely used for developing scientific theory. In contrast, predictive modelling places a focus on measurable variables x and y, and f is used as a tool for generating good predictions of Y values (see Figure 7.6). Rather than examining accessible data, our understanding of existing relationships is used to extrapolate into the unknown. In this sense, predictive modelling relies on the prior construction of an explanatory model. Indeed, explanatory and predictive modelling approaches are often employed in a sequential manner. Once statistical inference has been used to construct an explanatory model, the terms depicting the independent variables in the explanatory model can simply be substituted to derive a predictive model. For example, known sampled values can be replaced by future anticipated values, or values can be estimated at field locations where sampling has not been undertaken. In this way, the relationship derived empirically by an explanatory model can be employed for predictive purposes.

An assessment of the response of coral reefs and seagrasses to sea-level rise provides an example of a predictive modelling approach (Saunders et al., 2014). First, a species distribution model was developed to establish a relationship between wave energy (independent variable) and seagrass distribution (dependent variable). This corresponded to an explanatory model. A second stage then examined how sea-level rise would affect the distribution of seagrass through the deepening water depth over the reef and consequent increased levels of wave energy inside the reef. This corresponded to a predictive model. Known contemporary values of wave energy were substituted for anticipated future values, resulting in a transition from an explanatory model to a predictive one. The focus moved on from defining the function relating seagrass to wave power (the objective of the initial *explanatory* species distribution model), to defining potential future distributions of seagrass on the basis of anticipated future levels of wave energy. Similar principles have been applied to predict other characteristics of coastal environments, for example, the prediction of future beach forms on the basis of wave and sand characteristics (Wright et al., 1985).

7.5.4 Specifying Models by Combining Elements of Spatial Analysis

Specification is the process of formulating a model. Modelling the features of coastal environments requires dynamic phenomena such as wave energy or the distribution of seagrass beds to be expressed as if they were static properties. In practice, this means that models can only be developed through selection, simplification and ordering of information at a series of levels. This might include identifying which variables to include and defining the type of algebraic manipulation to be performed. The need to specify a model forces the analyst to conceptualise the system scope and to identify which structural components are to be included. They will also need to identify what functional relationships between structural components are to be included (and excluded). This conceptualisation step is therefore fundamental to the explanatory or predictive performance of a model and, by extension, our understanding of a given coastal phenomenon. The axiom of correct specification states that a set of independent explanatory or predictive variables must be unique, complete, small in number and observable (Leamer, 1978).

The process of developing a model involves running many trial models and examining associated statistical diagnostics for an indication of the performance of individual variables within the model. In a spatial model, features such as residuals can be mapped for clues as to missing variables. Individual model components can then be adjusted before the model is re-run (see Section 8.8 for an example). This trial-and-error method of model building usually follows one of two approaches. In the first approach, the specification originates from a simple, well-defined initial model and progresses towards a more general model by adding variables and autocorrelation parameters. This approach to model building is known as 'simple to general' (Gilbert, 1986). For example, the relationship between the structural form of patchy seagrass beds (e.g. their 'complexity') and wave energy can be modelled, with increased model performance as more variables are added to characterise wave energy, including aspect relative to wave approach, water depth and longshore distance from wave impact (Hamylton and Spencer, 2007). By choosing to amend rather than discard an original specification, the analyst demonstrates faith in the original predictor or explanatory variables. It can be useful to retain minor variables and avoid excessive simplification to preserve a reality that is indeed complex (Chorley, 1964).

The second approach to model building begins with a higher number of variables and discards those that have a low power of explanation to arrive at a simpler specification of high performing variables. The principle of parsimony (sometimes referred to as Occam's razor) prioritises a simpler model that provides an adequate description of the phenomenon under consideration over a more complicated model that may explain more variability in the data, but be impractical to implement. An approach that begins by investigating a broad range of potential influences and reduces these to fewer, high performers is intuitively appealing. Yet the feasibility of this approach depends on access to greater volumes of information and the potentially inefficient investigation of variables that may not be utilised in the final specification.

An analyst who undertakes statistical modelling with a specific explanatory or predictive objective in mind will interpret the performance of their model through diagnostics such as the proportion of variation in the sample population that is explained by the model (R^2). Non-spatial assessments of performance traditionally focus on the relationship between independent and dependent variables. A spatial statistical model must additionally account for the variation between data observations at different locations, represented by terms such as the mean or correlation structure (or both). Assessment of model specification may therefore consider additional criteria such as autocorrelation of model residuals. The analyst must choose between complex models that achieve a high level of explanation and simpler models that achieve lower but still adequate levels of explanation. The art of modelling therefore lies in a combination of knowing when to stop adjusting a specification, and appropriately judging what constitutes an 'adequate' performance.

7.5.5 Employing Statistical Inference on Georeferenced Data in a Spatial Model

Statistical inference enables an analyst to make statements about the properties of a large population of data samples on the basis of a smaller sample. The objectives of statistical inference are to test hypotheses, or derive estimates of a given phenomenon. Empirical reasoning is used to deduce propositions about the phenomena represented by the model. These propositions might take any number of forms, including:

- providing a statistical basis for acceptance or rejection of a stated hypothesis,
- estimating a value of some parameter of interest at an unsampled location using a spatial interpolator,
- deriving confidence levels at which a probability can be stated that a true parameter value may be contained within a given interval, and
- clustering or classification of data points into groups on the basis of probability.

All statistical inference requires assumptions to be made concerning the properties and distributions underlying the data on which they are based. A selected sample is only one of many possible samples that can be drawn from a population. The larger the sample, the higher the probability that the inference made from it will be observed in the wider population.

Classical statistical inference provides a formal framework for both hypothesis testing and parameter estimation. In the case of hypothesis testing, a null hypothesis is assumed to be a state of nature that is either true or false (Fotheringham and Rogerson, 2013). A test statistic is applied to the data to determine the truth or falsehood of the null hypothesis. For a statistical test, the *p*-value is the probability, calculated under the null hypothesis H_0, of obtaining a test result as extreme as that observed in the sample. If this probability is regarded as small, H_0 should be rejected; otherwise, it should not be rejected.

Classical statistical inferential tests typically compare a sample mean to a population mean to determine the probability of the observed sample outcome being applicable to the wider population. The appropriateness of a test used to draw this comparison depends on the number of variables or samples and the measurement level of the data. Tests for comparing the sample mean with a population mean include the *Z*-, *t*-, *F*- or ANOVA tests. Methods for assessing bivariate associations include the chi-squared test (for nominal or ordinal data) or the correlation coefficient (for interval data). In the case of parameter estimation or clustering or classifying a dataset, the spread of data observations can be analysed to define an upper and lower bounding limit for a given parameter, called a confidence interval. The probability that the parameter is contained within the confidence interval can then be estimated (see Section 8.5.1).

The methods of statistical inference described above are classical methods that make a series of assumptions about the data on which the inference is drawn. The use of geographically referenced information introduces some special dataset characteristics that need to be taken into account when considering statistical properties of the sample data. In the presence of spatial dependence, data often fail to satisfy standard assumptions. This may result in a number of effects on the results. For example, where *n* observations are made on a spatially dependent variable, those observations that are closer together may not provide much additional information about the process. Because of Tobler's first law of geography (Section 6.2), data points that are close together in space are likely to be similar. It follows that they will contribute less than the amount of information that would be carried if the observations were independent of each other. This leads to an underestimation of the sampling variance of the statistics. Where the analyst's objective is to place confidence intervals on statistics, these intervals will be misleadingly small, giving an overestimated impression of the precision with which a value can be estimated. If the analyst's objective is to test a hypothesis, there is an increased probability of rejecting the null hypothesis when it is true (i.e. committing a type I error).

Statistical tests can be modified to address the problem of spatial dependence. One simple modification is to reduce the sample size in the presence of spatial dependence so that it measures the equivalent amount of independent information in the sample. This will have the effect of enlarging the estimate of sampling variance of the statistic under the null hypothesis. In practice, the structure of the spatial dependence can be modelled using trend surface methods and semivariograms (Section 6.5). If a reliable model of the spatial dependence can be constructed, the modelled trend can be subtracted from the original values to de-trend the spatially dependent dataset. Conventional methods that assume independence can then applied.

7.6 The Use of Regression in Spatial Analysis: a Brief Background

Section 7.3 outlined that a common objective of coastal models is to derive a statistical relationship between structural features of coastal landscapes (dependent variables) and their functional drivers (independent variables). Correlation analysis assesses whether there is a linear relationship between two variables. A correlation coefficient will characterise the direction (positive or negative), shape (linear or otherwise) and intensity (strength) of a relationship. A correlation coefficient typically indicates a strong relationship if it exceeds 0.5, a moderate relationship if it falls between 0.30 and 0.50, and a weak relationship if it falls below 0.30 (Haining, 1993). The most commonly used measure of correlation is the Pearson correlation coefficient, which measures the covariability of two variables. With random samples, the statistical significance can be tested (e.g. using a T-test) to determine how likely it is that the correlation found in the sample data reflects a correlation that is observed in the wider population. For example, at a probability level of less than 0.05, there is a smaller than 1 in 20 chance that the correlation observed is due to chance alone. It may be the case that two variables are related, but not in a linear fashion (e.g. by a quadratic or cubic association). A Pearson correlation coefficient would not detect a nonlinear relationship.

Spatial analysis often derives a statistical relationship between a set of sampled field measurements and corresponding values or combinations of values in a continuous remotely sensed dataset (Mather and Koch, 2011). This kind of analysis is often performed when the values of a biological or physical parameter have been sampled at a few, localised areas and are to be determined for a larger, more continuous area from remotely sensed imagery. The statistical technique of regression is commonly used to achieve this.

The objective of ordinary least squares regression (also referred to as classical regression, or simply regression) is to explore the relationship between a dependent variable and one or more independent variables. Bivariate regression explores the relationship between two variables (one independent and one dependent variable). Multivariate regression combines several independent variables into a single function to explain variation in a single dependent variable. Regression is used to express a dependent variable (Y) as a function of one or more independent variables (X), plus an error term (e), by inputting data cases into the following equations:

$$Y(i) = \beta_0 + \beta_1 X_1(i) + \beta_2 X_2(i) + \cdots + \beta_k X_k(i) + e(i), \qquad i = 1, \ \ldots, \ n \tag{7.1}$$

where n denotes the number of data cases, values for the independent variables are treated as fixed, $e(i)$ is an independent normally distributed error term, subscripts i and k refer to location i and the total number of independent variables, respectively, and the coefficients β_0 to β_k are estimated empirically from the model.

A linear regression derives the straight line that best fits the relationship between the two variables. As with other statistical inference techniques, the technique of regression makes a number of assumptions about the properties of the sample dataset on which it is based. These are summarised in Table 7.3. Where data are spatially dependent, these assumptions often fail to be met (Section 6.3). This calls for the regression model to be adjusted to account for spatial autocorrelation in the data cases. It may be necessary to specify an alternative geographically weighted regression model.

7.6.1 Geographically Weighted Regression Using Spatially Lagged Dependent Variables

The transition from classical ordinary least squares to a geographically weighted regression involves the introduction of a spatially explicit term into the regression equation (7.2). This may take the form of a spatially lagged dependent variable (sometimes referred to as an 'autoregressive' term) as an explanatory variable. Alternatively, a spatial error model component may be introduced, in which errors are modelled as a simultaneous spatial model parameter (see Section 8.10). A spatially lagged dependent variable essentially means that a term is added to the regression equation that represents the value of the dependent variable itself, at a given distance away (i.e. spatial lag) from the location for which the variable is being modelled. These spatially explicit formulations of the regression equation do not require information on any additional covariates; rather, they utilise the location information of each existing data case to construct an additional, geographically weighted, term for the regression model. This characterises and makes use of the structured component of variation in the population of data samples that is unexplained by the classic regression model.

In geographically weighted regression, data cases refer to geographic locations. The regression equation draws explicitly on the relative location of each case. This is achieved through the construction of a spatial weights matrix. For each case, a spatial weights matrix will express the relationship it has to other cases in the dataset by virtue of its location. This will commonly specify whether other cases fall into the neighbourhood of a data case on the basis of a user-defined neighbourhood. Those data

Table 7.3 The statistical assumptions of classical regression and associated diagnostic tests to determine whether these assumptions have been met.

Assumption	Summary and test
Correct specification of the model	There must be a theoretical justification for the presence of a relationship between the dependent and independent variables
Measurement level of variables	Variables need to be expressed at a measurement level of interval or higher (i.e. ratio)
Adequate number of cases per independent variable	A minimum of five cases are used per independent variable
Absence of outliers	There should be no outliers in the sample dataset that is employed. Tests to identify outliers include a scatterplot, frequency distribution or Mahalanobis distance. For a given number of degrees of freedom, this identifies data cases exceeding a critical Mahalanobis distance value, suggesting that they are outliers
Linearity	The dependent variable needs to be linearly associated with each independent variable. This can be examined visually from a scatterplot
Multi-collinearity	Independent variables cannot be too highly correlated with each other, i.e. $r > 0.7$. Tested with a correlation matrix
Normality and constant variance of residuals	Model residuals need to be normally distributed. Tested with a histogram of residuals or *p*-plot
Homoscedasticity of residuals	Residuals must be constant when plotted against the independent variable(s) (i.e. there must be homogeneity of variance). Tested with a scatterplot of residuals vs independent variable(s), or with a Breusch–Pagan test
Independence of observations	The sequence of observations must not be correlated with the residuals, i.e. there must be no autocorrelation in the residuals. Tested with a Durbin–Watson test

cases that fall within the neighbourhood will receive a greater weighting within the regression model than those that are outside the neighbourhood. This operationalises Tobler's first law of geography (see Section 6.2). A spatially lagged dependent variable can be introduced into the regression model to utilise the spatially dependent nature of the process being modelled, that is, the tendency for similar values to be close together. For example, cases may be included or excluded depending on whether they belong to a data point's neighbourhood, such that $w_{ij} = 1$ when i and j are neighbours and $w_{ij} = 0$ otherwise (Anselin and Bera, 1998). The values of the dependent variable at neighbouring locations are introduced into the equation accordingly:

$$Y(i) = \beta_0 + \beta_1 X_1(i) + \beta_2 X_2(i) + \cdots + \beta_k X_k(i) + \rho \sum_{j \in N(i)} w(i,j) Y(j) + e(i), \qquad i = 1, \ldots, n \tag{7.2}$$

where $w(i, j)$ is the aforementioned weights matrix, $Y(j)$ is the value of the dependent variable at the adjacent location and ρ is a parameter associated with the spatial interaction effect. As with the preliminary model runs, the diagnostics recorded include the explanatory power of the overall model and of each independent variable, and the results of testing that the regression assumptions are met. An alternative approach may be to model the spatially structured component of the error term itself, which can be achieved using a similar specification (see Section 8.10).

7.7 Explanatory and Predictive Modelling of Live Coral Cover Around Lizard Island, Great Barrier Reef, Australia

7.7.1 Case Study Background

This case study outlines the development of both an explanatory and a predictive model of the spatial distribution of live coral, a key component of the marine ecosystem around the Lizard Island Group within the Great Barrier Reef Marine Park. The case study is based on a common data availability scenario in which 365 field observations of live coral are accompanied by some remote sensing datasets that provide continuous coverage of the area around the Lizard Island Group.

There is particular concern about reductions in live coral cover within the Great Barrier Reef Marine Park due to its functional role in providing habitat for fish and invertebrates, producing carbonate sediments for beaches and attracting many tourists. Effective management of the broader marine ecosystems requires the distribution of live coral to be mapped around the Lizard Island Group. The island group comprises three granitic islands that sit approximately 30 km off the northern Queensland coastline. A complex of reefs has developed around these islands. A narrow fringing reef occurs around much of the main island, but a broader barrier reef occurs between Palfrey and South Island, and a similar barrier reef extends north from South Island (Figure 7.7).

7.7.2 Model Overview

It is not possible to undertake detailed *in situ* ecological field sampling or to collect many, continuous observations of the seabed over broad synoptic scales (10–100 km). This would be labour-intensive and expensive over such large geographical areas. The primary objectives of this case study were therefore to construct an explanatory and a predictive model of live coral cover around Lizard Island. In the first instance, the aim of the explanatory model was to use the 365 field measurements to define a relationship between live coral (y) observed in the field and terrain characteristics (bathymetry, rugosity and bathymetric position index) depicted in the remote sensing datasets (x_1, x_2 and x_3). This relationship served as an estimate of the function, f, linking live coral to terrain characteristics (Figures 7.6 and 7.8). This differed from the theoretical 'real' function, F, which links the dependent and independent variables, the value of which remains unknown but can be estimated empirically (Figure 7.6). In the second instance, the aim of

Figure 7.7 Aerial photograph looking north over the three islands that make up the Lizard Island Group, northern Great Barrier Reef. Lizard Island is in the background, with Palfrey Island (left) and South Island (right) in the foreground. (A black and white version of this figure will appear in some formats. For the colour version, please refer to the plate section.)

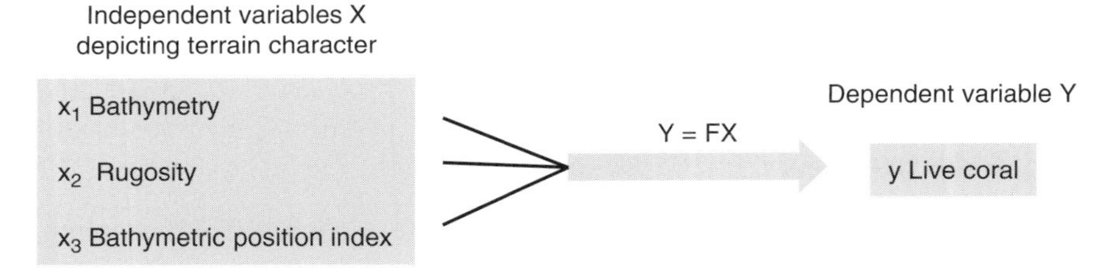

Figure 7.8 Conceptual model of the relationship between the cover of live coral (*Y*) around Lizard Island and three aspects of terrain character (independent variables *X*). The relationship is defined as a theoretical function (*F*). Note the difference in notation between theoretical variables and functions (upper case) and those that are estimated or sampled from the available datasets (lower case).

the predictive model was to apply the function, *f*, derived from the explanatory model to the remote sensing datasets (which provide values for x_1, x_2 and x_3) to predict the cover of live coral at unsampled locations around the reef. This yielded continuous modelled estimates of live coral across the entire reef complex around Lizard Island (Figure 7.9).

Table 7.4 Datasets employed to develop the explanatory and predictive models of live coral cover around Lizard Island.

Data layer	Description
Spatially continuous remotely sensed image layers	
Bathymetry (x_1)	A raster digital elevation model depicting above and underwater height of the terrestrial and marine environments of Lizard Island (datum: mean sea level). Incorporates data from seven different sources (GPS, LiDAR, single-beam bathymetry, remote sensing-derived bathymetry, Laser Airborne Depth Sounder (LADS II), multi-beam bathymetry and nautical charts)
Bathymetric Position Index (BPI) (x_2)	Defines the elevation of each raster cell with reference to the overall landscape (e.g. slopes, depressions, crest lines, flat areas). Negative BPI values for depressions, positive values for crests and zero values for flat areas and constant slopes (see Lundblad et al., 2006; Guinan et al., 2009)
Rugosity (x_3)	The ratio of the reef surface area to the planar surface area (Nellemann and Cameron, 1996). A measure of seafloor complexity.
Discrete point shapefiles from field collected datasets	
Ground referencing video footage (y)	A point shapefile of 365 spatially referenced underwater video snapshots of the seafloor. Percentage cover of live coral estimated from each video

Data include 365 underwater video records that show the composition of the seafloor community. These have been viewed to derive an estimate of the percentage cover of live coral. Alongside these field data, several remote sensing datasets provide continuous, synoptic cover of the entire reef system. These include bathymetry from a DEM (pixel resolution 1 m) and associated terrain variables of rugosity and bathymetric position index (BPI), which can be derived from the DEM (see Table 7.4 for a description). The degree to which the estimated function f corresponds to the theoretical function F will depend on how well a representative range of coincident live coral and terrain characteristics has been captured in the 365 data points. It will also depend on whether terrain character is accurately represented by the DEM and associated derivatives of rugosity and BPI. This can subsequently be evaluated through a validation exercise.

Figure 7.9 illustrates the key components of both the explanatory and predictive modelling exercises. Terrain character was represented by the DEM, rugosity and BPI layers, each of which is a continuous remote sensing dataset. The explanatory model used a spatially lagged autoregressive function to define a relationship between live coral and terrain character ($R^2 = 0.72$, $p < 0.005$). A nearest-neighbourhood weights matrix was used to specify whether or not data points fell within the same neighbourhood. Only those data points falling within the immediate neighbourhood of eight pixels were counted in the autoregressive model (see Figure 6.5b). The predictive model applied this function to the remote sensing datasets in order to predict live coral continuously around Lizard Island. A final validation step then revealed a good correspondence between predicted values of live coral cover and an independent set of field observation of the cover of live coral ($R^2 = 0.78$).

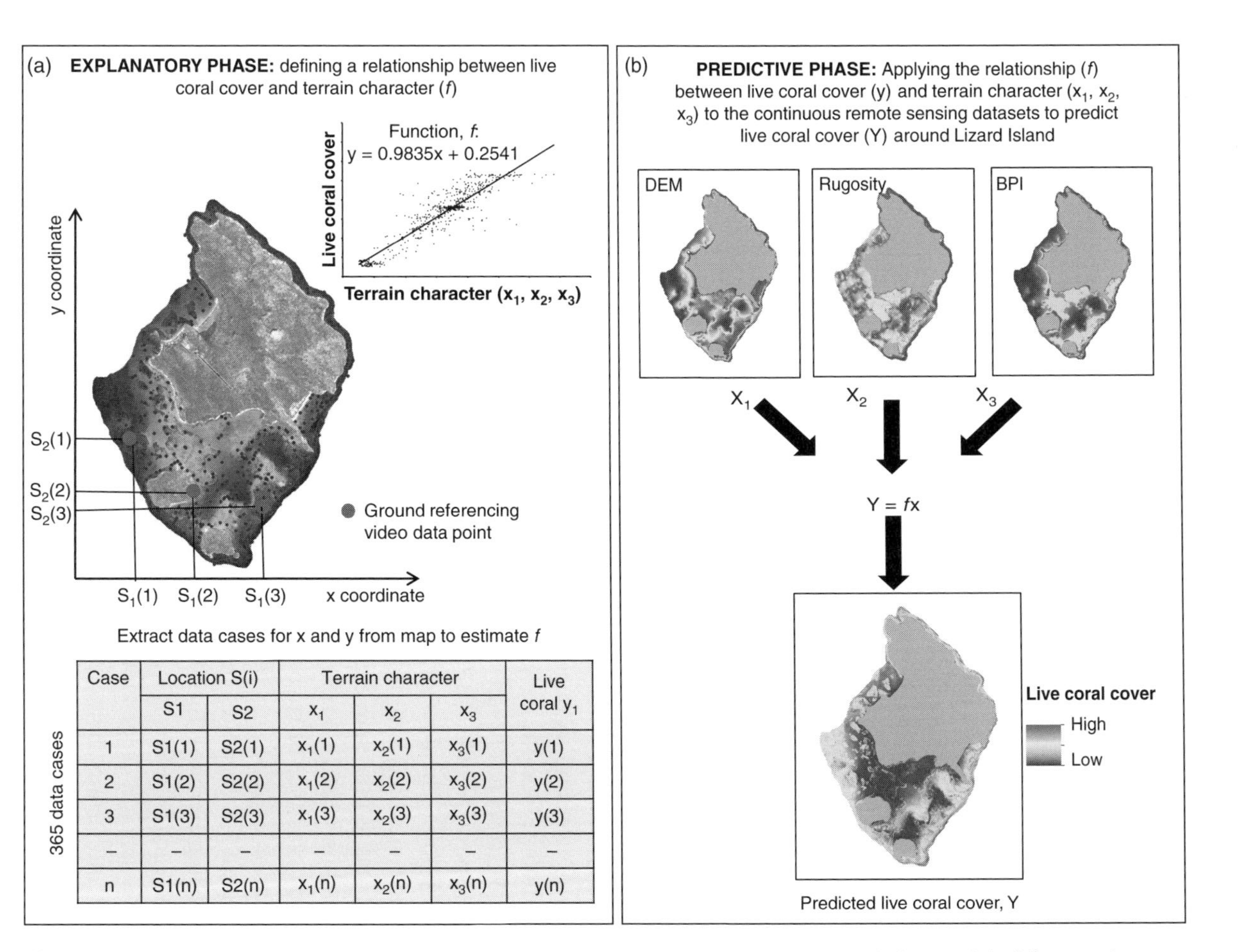

Case	Location S(i)		Terrain character			Live coral y_1
	S1	S2	x_1	x_2	x_3	
1	S1(1)	S2(1)	$x_1(1)$	$x_2(1)$	$x_3(1)$	y(1)
2	S1(2)	S2(2)	$x_1(2)$	$x_2(2)$	$x_3(2)$	y(2)
3	S1(3)	S2(3)	$x_1(3)$	$x_2(3)$	$x_3(3)$	y(3)
–	–	–	–	–	–	–
n	S1(n)	S2(n)	$x_1(n)$	$x_2(n)$	$x_3(n)$	y(n)

Figure 7.9 A schematic overview of the steps taken to develop an explanatory and predictive model of live coral cover around Lizard Island. (A black and white version of this figure will appear in some formats. For the colour version, please refer to the plate section.)

7.8 Summary of Key Points: Chapter 7

- Models provide a concise and meaningful account of observable patterns and relationships between different components of a natural 'system'. The development of empirical spatial models can be broken down into four stages: observation, measurement, experimentation and theory development.
- Coastal models are often cast within a framework that uncovers statistical relationships between coastal *structure* (response) variables and *function* (driving) variables. Structure refers to the observable characteristics or interconnected features of landscapes, i.e. the sizes, shapes, numbers and configurations of components. Function refers to the interactions between these elements, i.e. the flow of energy, materials or organisms among the components.
- Explanatory and predictive models are two common types of spatial model. An explanatory model constructs an explanation for an observed pattern using empirical data (it expresses an independent variable, y, as a function of one or more dependent variables, x). A predictive model generates a new observation by applying a statistical model to data (it predicts a dependent variable from one or more independent variables given a function that relates the two).
- Statistical inference is used to make statements about the properties of a large population of data samples on the basis of a smaller sample.
- Regression is a statistical technique for measuring the relationship between two or more variables. In spatial analysis, each data point is a location, so this can be included in a geographically weighted regression.
- In spatial autoregression, a spatially lagged dependent variable can be introduced into the regression model. This utilises the spatial structure of the process being modelled to improve the regression.

8 Addressing Uncertainty in the Spatial Analysis of Coastal Environments

8.1 What Is Uncertainty in Spatial Analysis?

It is important to recognise that any form of spatial analysis is not complete until an evaluation of uncertainty associated with the final result has been carried out and reported. Uncertainty refers to the inevitable limitations that are associated with spatial analysis. These occur because of the impossibility of exactly expressing the phenomenon that is the subject of analysis. This lack of certainty can be evaluated. Acceptance and recognition of the uncertainty associated with any analysis leads to honest and open communication between the analyst and those interested in using the results of the analysis. Evaluation of uncertainty, for example, by devising a statistical measure of error, provides a useful basis for understanding the limitations that govern how the results can be used. More broadly, a critical evaluation of both the methodology and the results associated with any analytical exercise brings confidence to that exercise.

Because uncertainty is unavoidable, it is the job of the analyst to minimise error as much as is reasonably possible and to generate a reliable estimate of its magnitude. A qualitative evaluation might simply state whether the results agree with existing theory or whether a new phenomenon has been observed. A quantitative approach may attempt to calculate the accuracy or sensitivity associated with a given analysis or model. Alternatively, it may analyse the distribution of a set of measurements to estimate the probability that the 'true' value lies within a specified range.

Accuracy is defined as the difference between an anticipated value (i.e. a measured or modelled quantity) and its 'true', or accepted, value (Mather and Koch, 2011). This definition can also be applied to the term 'error'. An analytical exercise may be underpinned by a range of different error sources (e.g. human mistakes, instrument limitations and measurement error). An estimate of accuracy is therefore an overall expression of the difference between an anticipated value and a true value. It may be impossible to resolve the contribution of errors from multiple sources from this overall difference alone. This highlights an important difference between error and accuracy: *error* relates to a single value and is data-based, whereas *accuracy* is a quantity derived from an ensemble of values and is based on a statistical model (Atkinson and Foody, 2002).

The term 'precision' is often used interchangeably with 'accuracy', although it has a different conceptual meaning. A precise measurement is one where the spread of results is small, either relative to the average result, or in absolute magnitude (Hughes and Hase, 2010). An instrument that generates measurements to a high number of decimal places is said to have a high precision. This is because the absolute magnitude of variation of measurements taken with that instrument may be small. Figure 2.8 illustrates the difference between accuracy and precision for a series of point measurements.

8.2 Sources of Error in Spatial Analysis

The nature of an error will depend upon the underlying source, which in some cases may not be identifiable. Nevertheless, it should be possible to work out whether an error is stochastic or systematic. Stochastic errors cannot be predicted precisely because they have a random probability distribution. A stochastic or random error may arise when repeated measurements (ostensibly of the same phenomenon) are scattered over a range. Statistical techniques exist for quantifying random error. For example, the standard deviation is a commonly used measure of the spread of a given distribution (see Section 8.4). A higher spread may or may not be attributable to an underlying error.

Stochastic scatter in a dataset may be attributable to the design of a given analysis, or to the noise imposed by instruments employed by the analyst. Most instruments will have a fundamental noise limit determined by a combination of the engineering of the instrument and the laws of physics. For example, the ability of a remote sensing instrument to detect light is a function of the charge-coupled detector (CCD) array within the sensor head, and the way that incoming light passes through the diffraction grating and camera lens (see Section 4.8). Most instruments do not approach their fundamental noise limit; rather, they operate with a higher noise level that could be reduced through careful experimental design. For example, it is possible to improve the accuracy of coordinate measurements taken with the global positioning system (GPS) by establishing a base station at which a GPS device is left to take repeated measurements of its position in the same location over time. By taking an average of the location measurements, it is hoped that the random errors associated with GPS technology, such as atmospheric interference (see Table 2.4), are minimised. After error associated with the design of an analysis has been reduced, any error remaining will be due to the fundamental noise level of a given instrument. This may become the limiting feature of an analysis. Where this is the case, it may be necessary to adopt a new instrument (e.g. to make the transition to using differential, as opposed to non-differential, GPS).

A systematic error causes a measured or estimated quantity to shift away from the true or accepted value and is also known as a 'statistical bias'. This shift may have a consistent direction and magnitude in absolute terms, or it may represent a proportion of the value being measured. In the latter case, its magnitude may therefore increase with larger measurements. It may be possible to reduce a systematic error by estimating its possible size, and adjusting measurements accordingly. Figure 8.1 illustrates systematic error in modelled estimates of water depth when compared against independently measured values of depth. Most modelled estimates lie below the red line of correspondence, indicating a consistent tendency for the model to underestimate water depth. This tendency to underestimate also increases at greater water depths.

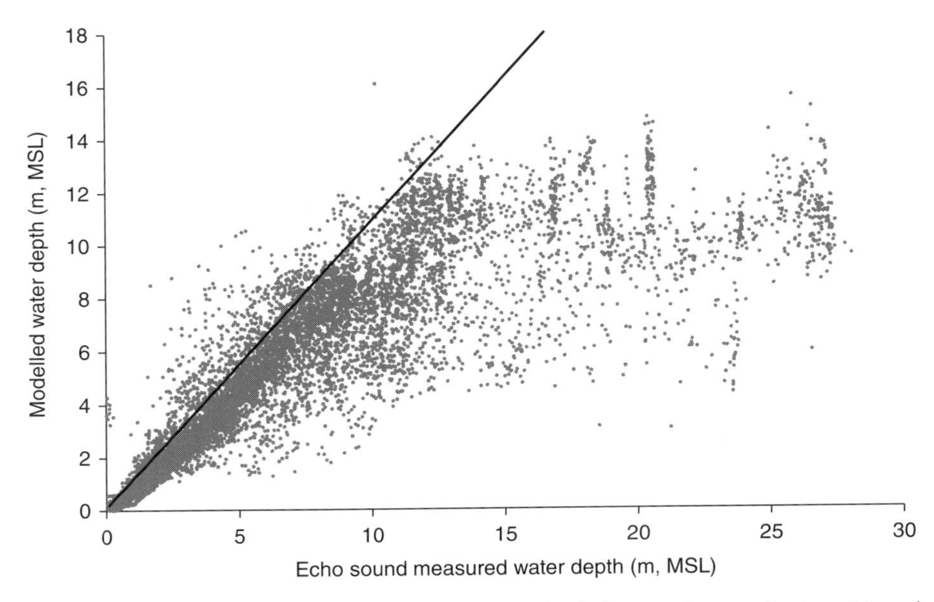

Figure 8.1 A plot of water depth measured with a single-beam echosounder (x axis) against estimates of water depth derived from a model (y axis). Most of the points fall below the red line of correspondence, indicating a systematic error in which the model underestimates water depth, particularly in deeper waters.

The root mean square error (RMSE) is a commonly used metric for quantifying errors. This captures both random and systematic sources of error, as follows:

$$\mathrm{RMSE} = \sqrt{\frac{\sum_{i=1}^{n}(z_i - \tilde{z}_i)^2}{n}} \tag{8.1}$$

where

n = number of data points
z_i = estimated value at location i
$\tilde{z}$ = true or expected value at location i.

Standard statistical techniques do not exist for quantifying systematic components of error. This is because the nature in which they are systematic often varies between applications. For example, the systematic error associated with a GPS-derived estimate of position has little in common with the systematic error associated with a modelled estimate of water depth. It is therefore left to the analyst to devise evaluation procedures that might provide some insight as to the origin of systematic discrepancies for a given analysis. Where data are georeferenced, spatially explicit approaches can be used to characterise and correct for any geographical structure associated with error terms (see Section 8.6). This is because the error measurements themselves can be plotted on a map and evaluated in their geographical position.

Another class of error arises from mistakes made by an analyst. Mistakes can be difficult to detect and, by their very nature, often display no overriding structure. They may arise from incorrect use of equipment, e.g. the failure to properly calibrate an instrument, misreading a scale, or confusion over units. More fundamental mistakes may arise from the failure to correctly specify a variable through inappropriate conceptualisation or representation (see Section 2.3).

8.3 Measuring the Length of Britain: the Influence of Scale on Uncertainty

The uncertainty associated with any measurement is itself a function of the spatial scale at which that measurement was taken. An illustrative example is provided by the collection of length measurements in order to answer the question *'How long is the coast of Britain?'* This was famously posed by Benoit Mandelbrot in 1967 (Mandelbrot, 1967).

This question can be answered by digitising the coastline as a vector dataset composed of many lines. The overall length can then be calculated by adding together the lengths of all the individual lines. If several digitisations are carried out at different scales, it becomes apparent that the greater the scale at which the coastline is viewed during the digitisation procedure (i.e. the more 'zoomed in' the analyst is to the coastline), the longer the overall length that is estimated. This is because the more you zoom in on the coastline, the more detail is apparent. Thus, a line digitised at a larger scale more closely follows the intricate meandering shape of the coastline. As a consequence, the estimated length increased as a function of scale.

Figure 8.2 illustrates the lines digitised to estimate the length of the mainland UK coastline at a scale of 1 : 1 000 000 (the largest scale), 1 : 10 000 000 and 1 : 30 000 000 (the smallest scale). It can be seen that a greater degree of error is associated with the line digitised at the smallest scale. This is because it is not possible to see the detailed meanderings of the coastline. Indeed, the lines digitised at this smaller scale often miss the coastline altogether. Such an error can be quantified through repeat digitisations of the coastline and statistical analysis of the distribution spread associated with the resulting coordinates belonging to the vertices of the digitised lines (see Section 8.5.1). To get a more accurate measure, it is necessary to increase the scale. The estimate of length will continue to increase as the scale is increased. Even if the length is measured of every boulder, rock, pebble or grain of sand, this estimate will continue to increase as the scale or precision increases, approaching infinity. This phenomenon makes it very difficult to provide a definitive answer to the question of how long the coast of Britain is!

8.4 Evaluating Uncertainty in Spatial Analysis

Statistical methods for evaluating uncertainty in spatial analysis can be divided into methods that are geographic (i.e. they draw explicitly on the spatial referencing information associated with the data points to evaluate uncertainty) and those that are

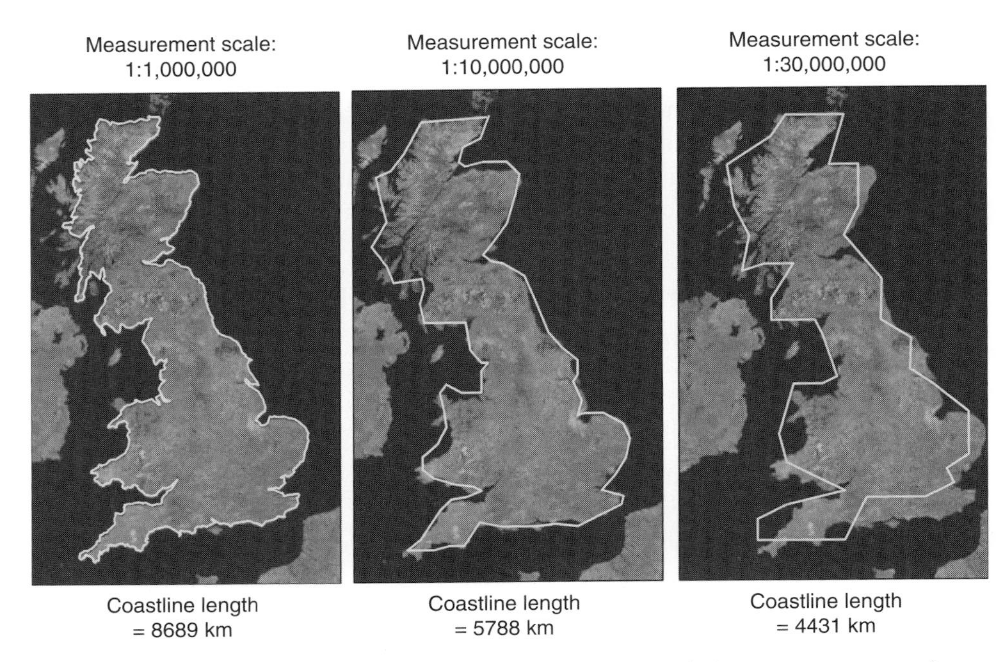

Figure 8.2 Three digitisations of the UK mainland coastline carried out across a range of scales. The coastline was digitised in order to estimate its length from the resulting vectors. The resulting length estimate increases as a function of the scale at which the digitisation was carried out. (A black and white version of this figure will appear in some formats. For the colour version, please refer to the plate section.)

non-geographic and are therefore based on non-spatial statistical techniques. As a point of departure it is useful to consider some simple descriptive statistical tools for quantifying two important characteristics of a spread of data points: (1) our best estimate of the true or accepted quantity being measured (or estimated), and (2) our estimate of the uncertainty associated with this value.

Our best estimate of the true or accepted value from a series of repeated measurements may be the mean (or average) of all the readings. The mean is often thought of as very reliable because it reduces the effect of any random error by dividing it up among the data points; however, excessive outliers may unduly influence the mean value in either a positive or negative direction. A more 'resistant' descriptor of the true or accepted value may therefore be provided by the median. This is the central number that separates the higher half from the lower half of all values when they are arranged in a sequence from lowest to highest.

Statistical descriptors of the uncertainty associated with a set of measurements can be estimated from the distribution or spread of the scattered data. The more similar the data values, the more precise the measurement or estimate is considered to be. Statistical descriptors of uncertainty quantify precision by evaluating the width of the distribution of a spread of data points. For example, the standard deviation uses all of the available data to estimate the width of the distribution of n

data values on the basis of the sum of the squares of deviations (*d*) from the dataset average, as follows:

$$\sigma_n = \sqrt{\frac{d_1^2 + d_2^2 + \cdots + d_n^2}{n-1}} \tag{8.2}$$

Note that the deviations are squared, so that their average is not zero (as would be expected to be the case for a set of random errors). The distribution is normalised to n data cases to represent the independent values of the deviation with which the variance can be found. The variance is the square of the standard deviation. It is therefore calculated as part of the process of calculating the standard deviation.

The standard error expresses the fact that, for a given set of measurements, the distribution is calculated in such a way that it is centred on the mean of the dataset. As the number of data cases increases, the number of measurements involved in calculating the mean is increased and the mean is therefore better defined. In the absence of a systematic error, the standard error, α, should therefore decrease as the number of data cases increases, and it should reduce by a factor of $\sqrt{n}$ with respect to the sample standard deviation. In a situation where a large number of data cases are likely to add reliability to a dataset and the analyst can be reasonably confident that these are free from outliers, it may be preferable to use the standard error instead of the standard deviation:

$$\alpha = \frac{\sigma_n}{\sqrt{n}} \tag{8.3}$$

In the presence of either spatial dependence (Section 6.2) or a systematic error, the standard error does not necessarily decrease as the number of data cases increases (Cressie, 1993). It may therefore be necessary to invoke spatially explicit approaches to uncertainty evaluation that account for the geographical distribution of the data cases.

8.5 Non-Geographic Approaches to the Evaluation of Uncertainty

8.5.1 Evaluating the Probability of Uncertainty From a Statistical Distribution

Probability statements associated with uncertainty rely on the adequate characterisation of a spread of data as a particular ‘type’ of distribution. The normal or Gaussian distribution is most commonly associated with random errors. It describes a distribution of data that is evenly distributed about a mean value. This distribution would be expected when a quantity being measured or estimated is subject to small, independent errors that may be positive or negative and that collectively make an additive contribution to the overall uncertainty. For a repeated set of measurements, or estimations, a histogram plot of the value for a given measurement or estimation, x, against the frequency with which that value occurs (or the probability of that value occurring) can therefore be characterised as a Gaussian density function curve (Figure 8.3).

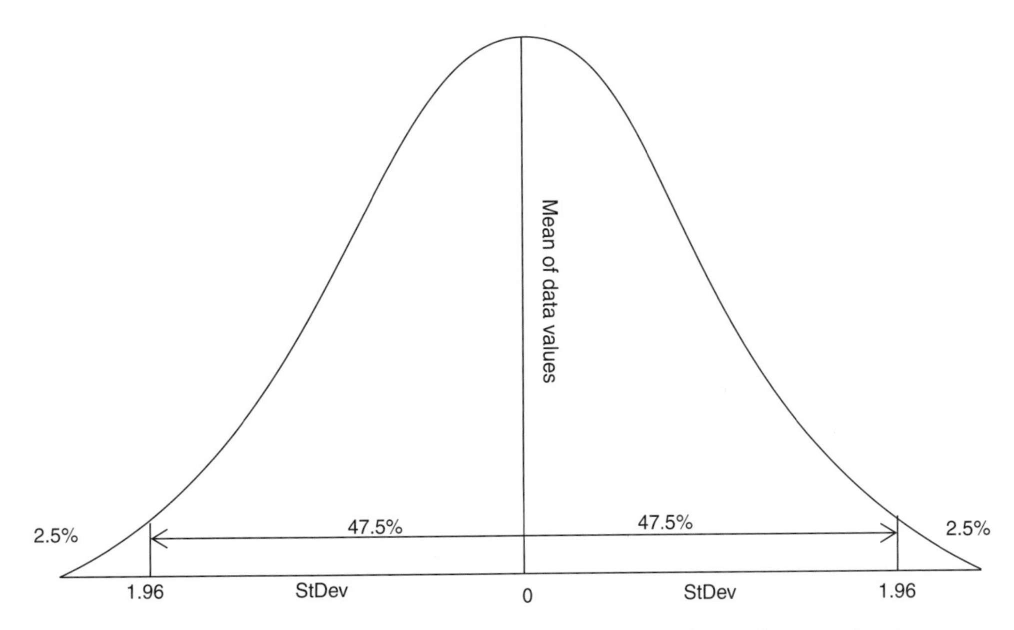

Figure 8.3 A standard normal distribution or Gaussian curve. The total area under the curve is 100% and the central limit theorem states that 95% of this area lies within ±1.96 standard deviations of the mean.

Conventionally, the area under a Gaussian curve sums to one (which is sometimes expressed as 100%). For a specified set of data ranges, the fraction of data which is expected to lie within this range can be determined from the form of the curve. For example, the 95% confidence interval indicates the range of values within which we can be 95% confident that our measurement or estimated value will fall. For a standard normal distribution (or Gaussian distribution), the central limit theorem states that 95% of the area under the curve lies within ±1.96 standard deviations of the mean. For a given dataset, it is therefore possible to calculate the mean, standard deviation and a 95% confidence interval for any value within the dataset (see the case study in Section 8.12). For example, if we take a value of 7 from a dataset with a standard deviation of 2, we can be 95% confident that the true or accepted value lies somewhere within the range of 7 ± 3.92, that is, between the values of 3.08 and 10.92. This is known as the *confidence interval* for this particular spread of data values. As long as the data spread can reasonably be characterised as a Gaussian curve, and the mean and standard deviation are known, this simple procedure can be used to calculate the confidence interval for any data point.

In the case of spatial analysis, data values can be thought of as point locations that can be mapped in geographic space. A stationary spatial process is one in which statistical properties such as the mean, variance and autocorrelation do not change over space or time. Yet most coastal phenomena are non-stationary because their mean and standard deviation values are subject to geographic variation. It follows that the uncertainty associated with analysing them is also non-stationary. It makes sense, therefore, to map associated measures of uncertainty such as confidence intervals, as these can be calculated for data values at different locations. The results can be mapped to illustrate the geography of uncertainty across a dataset (see Section 8.9).

8.5.2 Sensitivity Analysis of Uncertainty

Some types of spatial analysis attempt to model an outcome on the basis of several inputs. For these types of analysis, it is useful to move beyond the objectives of estimating a true or accepted value, or the distribution of a dataset, and instead to test the sensitivity of the analysis itself. Both explanatory and predictive modelling adopt a conceptual framework that attempts to relate variables representing some sort of input to variables representing an output. The specification of these models relies on assumptions about the relative extent to which variation in the values of each input variable can influence the outputs, i.e. the overall outcome of the analysis or model result. Sensitivity analysis is used to test the response of a model to changes in its parameters or variables (Goodchild et al., 2005).

A sensitivity analysis examines each model input variable in turn to work out the extent to which variability in the input values translates to variability in the model outcome or result. The source of variation in the input parameter may be related to an error, or simply due to fluctuations within an expected, natural range of variability. In either case, fluctuations in the value of an input variable or parameter may propagate through a model to influence the result in a small way that is proportionate to the initial level of fluctuation in the input variable. A very small fluctuation in an input may produce a much larger, disproportionate change in the output or result. In this case, the model itself will be deemed to be sensitive. The sensitivity of a model will ultimately depend on what the model does with the data values, i.e. what mathematical functions are applied to manipulate the data values and produce an output. An insensitive model produces an output that varies in proportion to the variation experienced by the input, whereas a sensitive model may be disproportionately affected.

As a general rule, if changing an input parameter or variable by 10% produces a less-than-10% change in results, the model can be said to be relatively insensitive to that parameter (Goodchild et al., 2005). For example, a recent model simulated the response of islands to environmental change on the Great Barrier Reef (GBR) from 2010 to 2100. The levels of tropical cyclone activity, an input variable used in the model, were subjected to a sensitivity analysis (Figure 8.4). One of the reasons for doing this was because of uncertainty associated with future trajectories of cyclone activity. Input values for tropical cyclone activity were adjusted to five different levels that ranged from a 12% decrease to a 60% increase on current levels. The objective of the overall model was to simulate the effect of environmental change on island volumes (output variable). A range of input variables were used to represent environmental conditions. These included cyclone frequency, tidal range, reef platform area, vegetative area and wave energy. The sensitivity analysis revealed a trend whereby islands began eroding as cyclone activity increased, with an overall transition from an average change of +16 km^2 to −7 km^2 in volume of the 22 islands

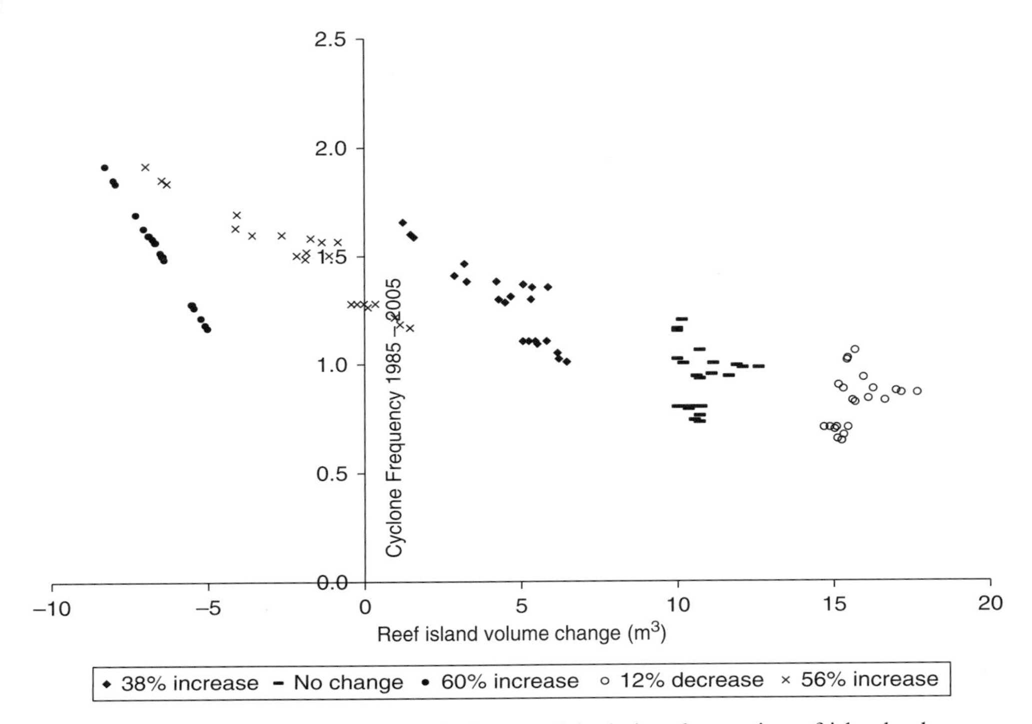

Figure 8.4 Results of a sensitivity analysis for a model relating changes in reef island volume (output variable, x axis) to changes in cyclone activity (y axis) for a group of 22 islands on the Great Barrier Reef. Potential scenarios for future tropical cyclone activity ranged from a 12% reduction to a 60% increase on current levels.

tested over the simulation period. Each 10% change in cyclone activity equated to an average change in simulation model output of 8% island volume. This insensitivity suggests that, despite the uncertainty of projected future cyclone activity, potential future island dynamics on the GBR could be constrained by the relationship between islands and the wider suite of local environmental conditions that were included in the model.

The practical value of a sensitivity analysis is that it focuses the analyst's attention on those input variables or parameters that have the greatest influence on results. Every effort should be made to reduce error and increase precision of the values associated with these input variables. It is also useful as it identifies variables that do not make much difference to the output of a model, on which it makes little sense to place effort refining input parameters.

8.5.3 Error Propagation

The application of uncertainty evaluation to spatial analysis is often motivated by the need to understand the reliability of the results associated with that analysis. Results of spatial analysis often take the form of a map generated by bringing together datasets with different origins. These may have been generated at different scales, often of varying quality, which are then subject to various overlay operations and arithmetic procedures (Unwin, 1995). Propagation refers to the transmission of error through the different stages of a model.

Error propagation can be evaluated either through numerical simulation, or through cumulative assessment. The properties of the error associated with an arithmetical procedure can be derived through repeated numerical simulations of that procedure in which random distortions are added to inputs and the impact on outputs is observed. An empirical evaluation of probability may be repeated hundreds or thousands of times as part of a Monte Carlo simulation to determine the effects of error propagation. In the case of a spatial analysis where the results take the form of a map, an alternative approach may be to perturb either the variables or operations that give rise to the final map to evaluate how much these must be changed before the observed relationships begin to break down (Haining, 2003).

The propagation of error through an analysis can also be quantified independently for the different stages of the analytical process, and then calculated as the cumulative error by summing across all stages of the model. For example, Ford (2011) assessed shoreline changes on Majuro Atoll in the central Pacific Ocean by digitising aerial photographs. Uncertainty was evaluated by summing up error values from the different components of the analysis, including the manual digitising procedure, image pixel size and georectification of the aerial photographs. Such an approach assumes that the cumulative propagated error is adequately represented as a linear additive function of the different components, which is often not the case.

8.6 Geographic Approaches to the Evaluation of Uncertainty

The approaches to error evaluation that have been described up until now are stationary assessments in the sense that a single model is fitted to all data and applied equally over the geographic area of interest. An alternative, non-stationary approach allows the error estimates to vary with geographic space. This is often a more realistic approach because errors are dependent on local circumstances. Descriptive statistics can be calculated on regional subsets of data in order to better describe geographic variation. Passing a moving window or kernel over a grid of pixels or set of mapped points provides a practical way of subsetting data values in order to calculate and interpret statistics such as the mean and standard deviation for localised areas.

A geographically variable evaluation of uncertainty allows the analyst to identify the subregions of their study area for which their results are more reliable or less reliable than others. This helps to identify areas where an analyst can be confident in the results of a model, and areas where caution is advisable. It also provides a visually accessible picture of the spatial distribution of model performance. In turn, this picture provides an opportunity for model improvement because it yields useful clues as to the processes that contribute towards model uncertainty.

8.7 Mapping Residuals to Visualise the Spatial Distribution of Uncertainty

One of the simplest geographically explicit ways to evaluate the performance of a model is to map the residuals of an initial model run (Franklin, 2010). For each data value put into the model, the residual is calculated as the difference between the value of the modelled output or estimate, and the true or 'accepted' value. Because data values represent points in space, the residuals can be plotted on a map. Where a model seeks to predict an outcome, such a step can be useful for further developing and improving models because it renders spatial patterns of their performance observable. In doing so, it invites the analyst to consider postulates or variables that might not otherwise have been included at the model specification stage. If an analyst wishes to build a multivariate model by sequentially testing different combinations of variables, mapping the residuals of a model test run may provide clues as to the identity of missing covariates. This is because geographical areas where the model is under- and overpredicting an outcome are revealed. The case study in Section 8.8 provides an illustrative example of how mapping residuals enables a model to be re-specified, using the spatial properties of the data themselves as a guide to the appropriate specification.

8.8 How Mapping Residuals Can Help Model Specification: an Example of Vegetation

Coastal zones span environmental gradients that are reflected in the communities that inhabit them. For example, zones of coastal vegetation transition towards more terrestrial communities with increasing distances inland. The cover of vegetation may be linked to broad-scale environmental factors such as the underlying geology, local weather, temperatures and elevation. This case study models such gradations of vegetation across the land mass of southern Patagonia. Vegetation cover, the model output, is estimated from a MODIS satellite image using the normalised difference vegetation index (NDVI; see Chapter 5). To construct an explanatory model of the observed distribution of vegetation for the study area, a map of the NDVI was compared to potential environmental controls (elevation, temperature and rainfall) using a spatial regression model. The explanatory model was constructed in a staged manner. Two regression models were trialled, so that the residuals of the first model run could be mapped in order to assess their spatial distribution. This provided clues as to the distribution and identity of missing model covariates that could be included in a second model run.

Data for this exercise were obtained from Google Earth Engine (the NDVI) and WorldClim Global Climate Data, a topographic digital elevation model (DEM) of the study area, average temperature and rainfall from 1950 to 2000 (Hijmans et al., 2005).

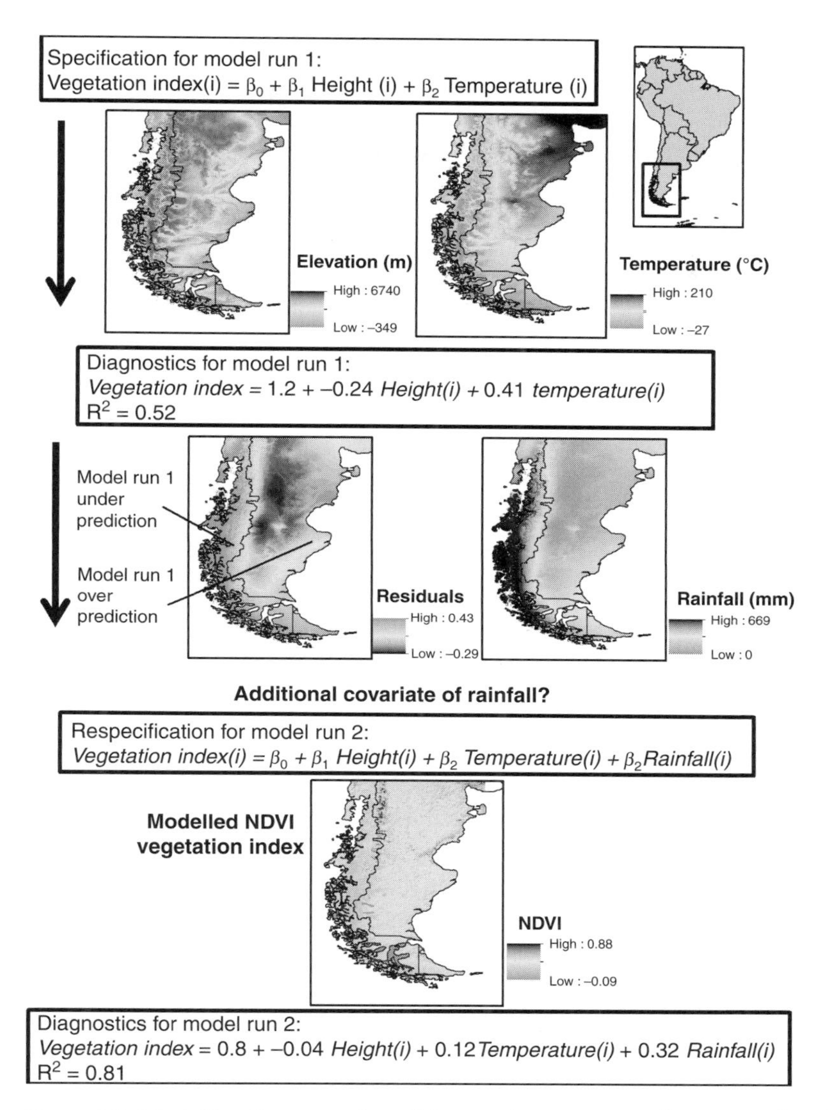

Figure 8.5 Modelling the vegetation of Patagonia across a gradient traversing the coast, a mountain range and the Argentine steppe-like plains. An illustration of model specification and re-specification using mapped residuals to indicate the potential identity of a missing covariate. (A black and white version of this figure will appear in some formats. For the colour version, please refer to the plate section.)

In the first model run, the relationship was assessed between the vegetation index (dependent variable) and the two independent variables of ground height (topographic elevation as measured by the DEM) and average temperature. The assessment was carried out for each pixel in a 5 km grid covering the study area (Figure 8.5). Diagnostics for this model run revealed a relationship of moderate strength between the two ($R^2 = 0.52$). The residuals map indicated that the model was overpredicting the presence of vegetation in the east and underpredicting the presence of vegetation in the west. The eastern side of the study area coincided with dry Argentinian plains, whereas the western side was associated with the higher Chilean Andes coastal mountain range. These environments would have been subject to markedly different local rainfall regimes. A comparison of a rainfall map with the residuals map revealed an inverse relationship between the two, suggesting that rainfall might be a useful additional covariate in the model (see Figure 8.5). This was confirmed by a second model run, which incorporated rainfall and had an increased explanatory performance (R^2 increased to 0.81).

Mapping the residual values from the first model run provided a visual clue as to the structure of information that was missing from the model. Model development was therefore guided by the additional insight gained from interrogating the geographical structure of the model error, as represented by the residuals. A clear east to west structure was apparent in the residuals. This gave rise to a subsequent search for environmental variables with a similar geographical structure to the one suggested by the residual map. Any environmental variables with such a structure (e.g. rainfall) could potentially be a useful covariate. The suitability of rainfall as an additional independent variable was confirmed by the improved diagnostics of the second model, for which the rainfall variable emerged as significant predictor ($p < 0.05$). Residual mapping therefore emerged as a useful activity in the process of model development, nested between two separate model runs. Most models that represent the end point of a given spatial analysis are the product of several model test runs conducted throughout the development phase. Mapping residuals is an example of how the error associated with a model can be put to productive use when taking such an approach.

It is worth noting for this example that, while the statistical diagnostics indicate the importance of rainfall for explaining the distribution of vegetation across Patagonia, additional information is necessary to establish proof of a causal link. This can in part be addressed by drawing on the fact that plants need water to survive as a theoretical reason for performing a regression of the variables against each other in the first place (thereby meeting one of the assumptions of regression as outlined in Table 7.3).

8.9 Mapping Confidence Intervals

An assessment of uncertainty might itself take the form of a map depicting the geographic variation in uncertainty. This geographic variation might arise from the non-uniform distribution of the factors that underpin error. Where regional measures of data distributions such as the standard deviation can be estimated, it follows that associated confidence intervals can be calculated, allowing probability statements to be made about different regions (see Section 8.5.1). A confidence interval map might specify the upper or lower confidence interval or the width of the confidence interval band (that is, the upper minus the lower confidence interval) for a corresponding geographic area. Instead of providing a single statement of uncertainty that is applied uniformly across a geographic area of interest, an uncertainty map will reflect the varying degrees of confidence that can be applied to a model output at different locations.

Where it can be reasonably assumed that the values of a point dataset vary smoothly across geographical space, it is possible to map associated confidence intervals simply as a function of the local mean and standard deviation. For a raster grid of predicted values, local confidence intervals can be calculated by passing a 3 × 3 pixel moving window across the model output values. This window defines a local area for which the mean and standard deviation of the values associated with the nine pixels centred on the pixel can be considered. By moving the window, these calculations can be iteratively made across all pixels in the entire raster grid. Confidence intervals can then be expressed based on the localised mean and standard deviation. Because the predicted variable exhibits variation across the map, the resulting confidence intervals will also exhibit variation depending on the localised spread or distribution of the data points. In accordance with the assumption that model output values only vary smoothly over geographic space, any widening in the spread of model output values within the grid will have the effect of increasing the standard deviation and widening the range of data values within which the confidence limit can be applied, thereby resulting in a lower level of certainty for the model output values.

Figure 8.6 illustrates a confidence interval map generated from a raster grid representing modelled water depths over a reef platform. In this map, the model output values for which confidence intervals are calculated are estimates of bathymetry, or water depth, over the Sykes reef platform (Great Barrier Reef). These confidence intervals were derived from an arithmetical procedure similar to that described in Section 8.5.3 in which numerical simulations were performed, with random distortions added to inputs. These simulations were repeated and the impact on outputs was observed. Introduced distortions were comparable to pixel-to-pixel variations due to the water surface, atmosphere and instrument, and were assessed from a covariance matrix over a deep water area (Hedley et al., 2012). The Sykes reef platform rises to a shallow plateau in the centre (coloured deep orange on the left-hand bathymetry map) and falls to deeper depths around the periphery on the right-hand side (coloured dark blue), with a steeper drop-off to the east and a network of sand shoals at a moderate depth in the west. Because of a smaller signal-to-noise ratio in areas of deeper water, the model that produced the estimates of water depth is more reliable in shallower areas than deeper ones (Goodman et al., 2013). This is reflected in the wider confidence interval width around the edge of the map (coloured red), as compared to the narrower interval above the shallow platform in the centre of the map (coloured green).

8.10 Geographically Weighted Regression of Spatial Error

In the case of a regression model, it has been shown how the transition from a classical ordinary least squares model to a geographically weighted regression model involves the introduction of a spatially explicit term (Section 7.6.1). One alternative choice for

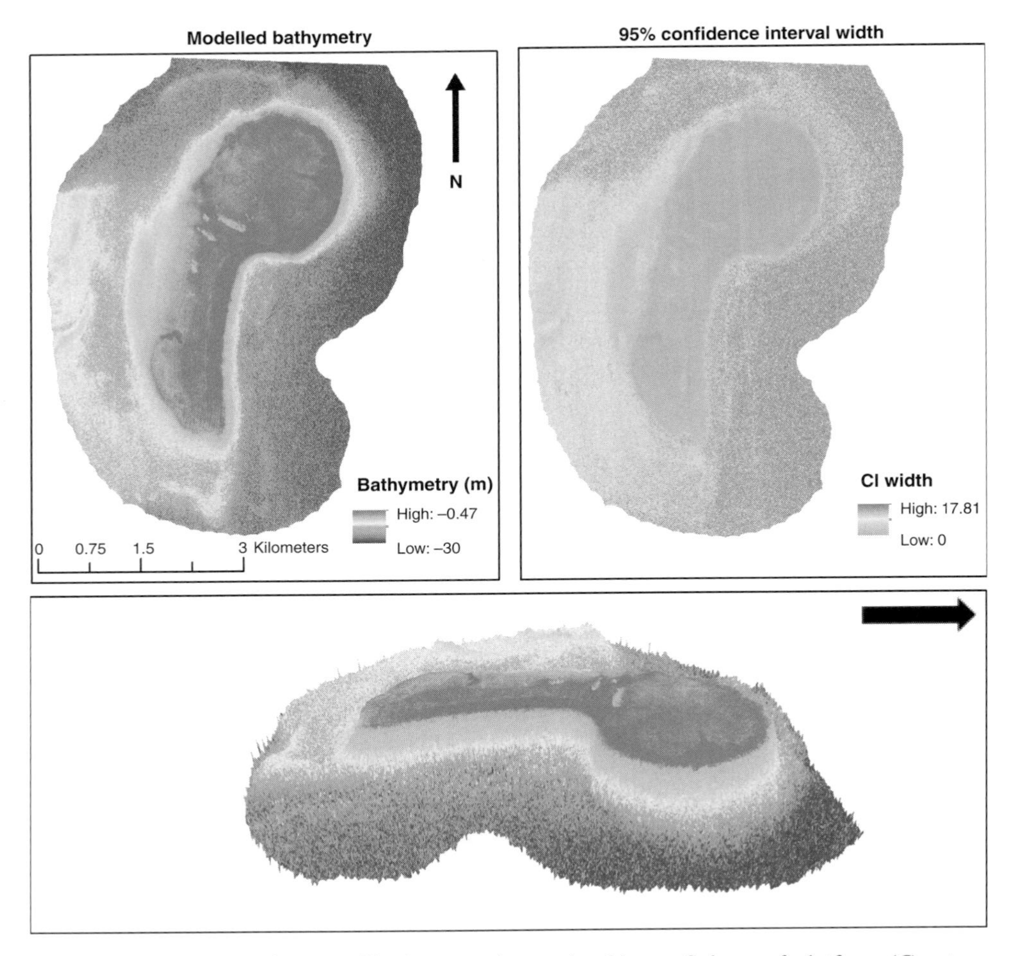

Figure 8.6 Modelled estimates of bathymetry (water depth) over Sykes reef platform (Great Barrier Reef). Sykes reef is a submerged platform extending from a shallow shelf in less than 1 m water depth to a deeper base in approximately 30 m water depth. The 95% confidence interval width illustrates the widening of confidence intervals associated with an increased spread of data points and higher levels of uncertainty in deeper areas. The lower image shows an oblique three-dimensional view of the reef platform bathymetry model, as seen from the eastern aspect. (A black and white version of this figure will appear in some formats. For the colour version, please refer to the plate section.)

introducing a spatially lagged dependent variable for this term may be to develop a spatially correlated error model. This model can be used to split the error term in (depicted as $e(i)$ in equation (7.1)) into two components, one of which can be associated with a spatial pattern and the other of which cannot. These are referred to as the *spatially structured unexplained* and *unexplained* components, respectively. This will enable the spatial structure of error to be built into the model without the cause necessarily being known.

Most errors have a geographical structure associated with them that can be modelled using spatially explicit techniques. Given a set of data values for which dependent and independent variable information is known, multivariate regression is

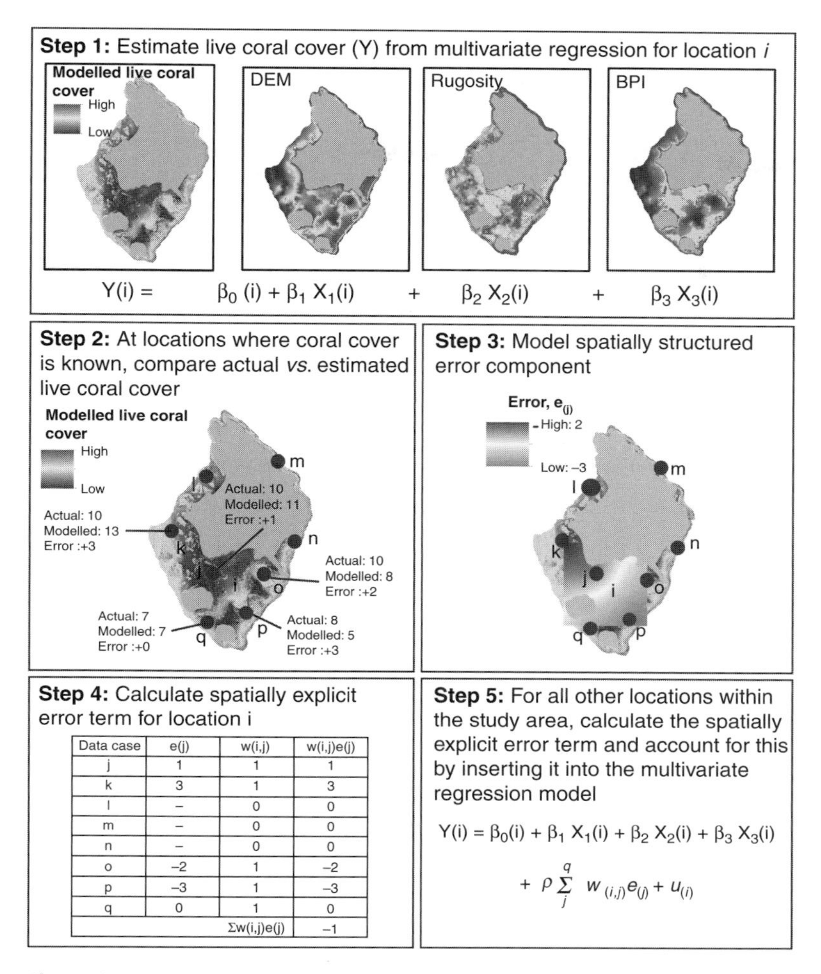

Data case	e(j)	w(i,j)	w(i,j)e(j)
j	1	1	1
k	3	1	3
l	–	0	0
m	–	0	0
n	–	0	0
o	–2	1	–2
p	–3	1	–3
q	0	1	0
		Σw(i,j)e(j)	–1

Figure 8.7 The steps taken to develop a spatial error model of live coral cover around Lizard Island (see text for further information). (A black and white version of this figure will appear in some formats. For the colour version, please refer to the plate section.)

traditionally used to empirically define the relationship between the independent and dependent variables using the form of (7.1). The spatially structured unexplained and unexplained components can be separated out of the error term associated with the standard multivariate regression equation. A spatial error model can then be constructed under the assumption that the unexplained variation is multivariate normal and could be expressed alongside a spatial autoregressive component. This is achieved by incorporating the error at a neighbouring location as a spatially explicit term into the model. This is the form of the spatially correlated error model, as follows:

$$e(i) = \rho \sum_{j \in N(i)} w(i,j)e(j) + u(i) \tag{8.4}$$

where $e(i)$ is the sum error for location i, ρ is a parameter associated with the interaction effect, $e(j)$ is the sum error (i.e. both spatially structured unexplained and unexplained components) of the linear regression model for case j and $u(i)$ is the unexplained error component. The i and j indicate that the calculation is being performed for the data case at location i, drawing on information from the data case at neighbouring location j. The spatial error model allows the spatial structure of error to be built into the model, without the cause necessarily being known. Such a step makes use of information from the error term of the neighbouring cases within the explanatory model. This spatially explicit term is represented by the '$\rho\sum_{i\in N(i)} w(i,j)e(j)$' portion of (8.4). The $w(i, j)$ represents a weights matrix populated with values 1 and 0 that identifies neighbouring points for which the error is taken into account, the ρ term expresses the localised strength of spatial association in the error term between neighbouring locations, the $\sum_{j\in N}$ sums all the neighbouring points (j) that fall within the neighbourhood of i, and $e(j)$ represents the error at the neighbouring location. Collectively, these terms allow the analyst to characterise and express the spatially structured component of the error term within the regression equation itself.

Figure 8.7 illustrates the steps taken to develop a spatial error model of live coral cover around Lizard Island. Following on from the model outlined in Section 7.7, predicted values of live coral cover are generated (step 1), then compared to an independent set of measured values (step 2), so that the spatially structured component of the error can be modelled (step 3). This spatially structured component of error is then estimated at all other locations within the study area and incorporated into predictions of live coral cover (step 5).

This represents an alternative approach to the introduction of a spatially lagged dependent variable (described in Section 6.8). As with the geographically weighted regression that was previously described, the introduction of a spatially explicit error term usually results in an improved fit between modelled and sampled data in comparison to the classic ordinary least squares model.

8.11 Approaches to Digital Map Validation

For digital maps generated from classifying remotely sensed images, a validation exercise assesses the extent to which the thematic content of the map corresponds with an independent dataset. Validation involves comparing remotely derived information against field-collected data, often referred to as a ground reference (or 'ground truthing') dataset. Ground reference datasets can be a georeferenced photograph, or simply a record of the analyst having visited a given location in the field, recorded their position with a GPS and made a note of the ground-cover type at that location. It is important that the field-collected validation dataset remains independent of the digital map. It cannot be used as part of the classification procedure that made the map in the first place (e.g. to guide training of a supervised classification algorithm, or to interpret the output classes of an unsupervised algorithm). The feasibility of applying a given validation technique is dictated by the level of measurement to which the associated phenomenon has been mapped.

Table 8.1. Error or confusion matrix showing the results of producer, user and overall accuracy assessments. The matrix is a tool for comparing classes assigned by a ground reference dataset with those assigned to an image through a digital classification procedure. Row labels are generated by the classification procedure, while column labels are assigned independently to a ground reference dataset. See text for further explanation.

		Validation class							Sum	User accuracy (%)
		1	2	3	4	5	6	7		
Mapped class	1	**252**	67	21	8	0	0	0	348	72
	2	10	**273**	0	12	0	0	0	295	93
	3	5	12	**189**	14	21	17	0	258	73
	4	0	0	0	**399**	42	12	0	453	88
	5	0	21	0	25	**1659**	122	0	1827	91
	6	0	0	0	40	126	**1701**	84	1951	87
	7	0	0	0	32	147	315	**357**	851	42
Sum		267	373	210	530	1995	2167	441	**5983**	
Overall accuracy										**81**
Producer accuracy (%)		94	73	90	75	83	78	81		

To validate a map composed of nominal categories (i.e. 'seaweed' vs 'sand' vs 'rock'), it is necessary to have an independent set of ground reference points that also identifies whether the ground cover for a given point location is seaweed, sand or rock. It is possible to overlay all the different points representing individual ground reference records onto the digital map and extract the corresponding ground-cover class identified by the map for the same location. This provides a useful basis for comparing the classifications assigned to individual pixels by these separate procedures in relation to the classes mapped. Where lots of pixels are compared, several statistical measures of accuracy can be derived. These are quantified by constructing an error (or confusion) matrix. This represents the degree of accuracy of each class mapped in an $n \times n$ matrix (Table 8.1). The elements of row 1 in this matrix provide the number of pixels that have been identified independently from the ground referencing dataset as being members of classes 1 to 7, represented by the columns. The row number identifies the corresponding classes that were assigned to these pixels by the classified digital map. For example, eight pixels were identified by the ground reference dataset as belonging to class 4, while the map identified these same pixels as belonging to class 1. They therefore represent incorrectly classified pixels. The bold numbers along the diagonal axis of the matrix indicate pixels that were assigned to the same class membership by both the ground reference dataset and the classified map. These can be thought of as correctly assigned pixels.

The overall accuracy of the classified map is defined as the proportion of all the points that were classified by both the map and the ground referencing data as the same class (Congalton, 1991). For example, in Table 8.1, 5983 ground referencing points were assessed in total, for which 4830 were classified into the same classes (represented in the diagonal shaded boxes) and 1153 were classified into different classes (represented elsewhere in the matrix). The map can therefore be said to have an overall accuracy of 81%. As the name suggests, this provides an overall measure of accuracy that can be applied to the map as a whole.

Class-specific measures of accuracy can be derived as the producer and user accuracies. The producer accuracy represents, for each class, the proportion of ground reference data points that were correctly classified by the map. The user accuracy represents the proportion of pixels mapped that were correctly classified in relation to the ground reference data points. The values of the producer and user accuracies are represented in the bottom row and right-hand-most columns of the matrix, respectively. Other statistical measures of accuracy exist that can be derived from the error matrix. These include the kappa and tau coefficients (Stehman, 1996). The tau coefficient, τ, estimates the proportion of pixels classified correctly and compares this estimate to that which would be achieved by chance (Ma and Redmond, 1995). By calculating statistical measures based on the error matrix, all of these approaches incorporate both spatial and thematic aspects of the data. Congalton and Green (2008) review a range of statistical measures for estimating the accuracy of classifications of remotely sensed data.

The calculation of accuracy metrics for digital maps introduces a need to establish what level of accuracy is acceptable for a given map product. Such matters should be judged on a case-by-case basis, depending upon the practical objective for which the map was generated in the first place. For example, it is generally accepted that a map for coastal 'management purposes' is useful if it has an overall accuracy of 60% or greater (Green et al., 2005). At a coarser descriptive resolution (i.e. four ground-cover classes), classified coastal maps tend to be more accurate, with a corresponding drop in accuracy as more classes are added (see Figure 4.6).

8.12 Evaluating the Uncertainty of Tsunami Inundation Maps: the Vulnerability of Manhattan, New York

This case study illustrates how the uncertainty associated with a dataset can translate into a substantially different result for any analysis that is carried out on this dataset. In turn, this can have considerable implications for the practical objectives associated with the analysis. An assessment of the land area that would be inundated in the event of a 6 m height tsunami wave is undertaken for the borough of Manhattan in New York City. This case study represents a common way in which spatial analysis is used for coastal vulnerability assessment.

Two datasets were used for this case study: a DEM that indicated the height above sea level for each pixel (0.5 m spatial resolution) and an independent set of 100 point elevations extracted from a LiDAR survey of Manhattan. Both datasets were sourced from the Department of Environmental Protection and accessed through the NYC Open Data portal. In the first step, values of the LiDAR point elevations were compared to the DEM in order to calculate the root mean square error and associated 95% confidence intervals for the DEM. The 95% confidence interval allowed the analyst to state a range of values within which they were 95% confident that their data point lay. This is a conventional way of expressing the uncertainty associated with DEMs.

In the second step, the land area of Manhattan that fell below an elevation of 6 m was calculated. Following application of the Bruun rule, this corresponded to the area that would be inundated by a 6 m tsunami (Bruun, 1983, 1988). Note that the Bruun rule has been widely criticised as an oversimplification of complex inundation dynamics, but is still widely used as a guide for coastal vulnerability modelling. In the third step, the uncertainty associated with the DEM was incorporated into the analysis. This was achieved by adding the vertical error corresponding to the upper 95% confidence limit to the height of the tsunami and re-evaluating the area inundated by an adjusted, larger tsunami height. Finally, a comparison of the land areas inundated in the first and second model runs was undertaken to illustrate the implications of uncertainty associated with DEMs for the overall results of the coastal vulnerability analysis (Figure 8.8).

Comparison of the LiDAR point elevations with the DEM yielded a RMSE of approximately 4.04 m for the DEM. The central limit theorem could be used to calculate the associated 95% confidence interval limit as $1.96 \times$ RMSE under the assumption that the error was random (i.e. equally likely to an over- or underestimation of the actual elevation) and normally distributed. This yielded a vertical linear error of 7.91 m. Thus, the analyst could make the statement for the elevation value associated with every pixel in the DEM that they could be 95% confident that the accepted elevation in real life of the ground area corresponding to that particular pixel fell within the range of the value specified by the DEM grid ± 7.91 m.

In the second step of the analysis, the area of the DEM that fell beneath an elevation of 6 m was calculated to be 25.36 km^2. In the third step of the analysis, the uncertainty associated with the DEM was incorporated into the assessment of the area inundated. For this coastal vulnerability exercise, which involved projecting the height of a hypothetical tsunami onto a DEM, this uncertainty range could be transferred to the height of the tsunami because vulnerability was assessed as a function of the vertical relationship between the tsunami height and the DEM height. The confidence limit of 7.91 m was therefore added to the tsunami, to project a tsunami of 13.91 m in height. This step was taken because it was possible, within the 95% confidence limit, that the DEM could actually lie 7.91 m lower than indicated by the values quoted. This would have the same effect on the area inundated as raising the height of the tsunami. The area of the DEM falling beneath an elevation of 13.91 m was found to be 35.30 km^2. When the uncertainty of the DEM dataset was accounted for, this therefore equated to an additional land area of 9.94 km^2.

In practical terms, this difference in area related to some of the world's highest-density and most expensive land. At real estate prices of $1363 per square foot (http://www.dnainfo.com/new-york/20140401/upper-east-side/manhattans-real-estate-prices-reach-record-1363-foot), the difference in value of the land inundated equates to US$1400 billion. This figure assumed single-storey properties, which is clearly not the case in the high-rise suburb of Manhattan. Accounting for skyscrapers would likely make this number much higher. It would therefore be of interest to an insurance company calculating the vulnerability of Manhattan properties to incorporate uncertainty into their calculations!

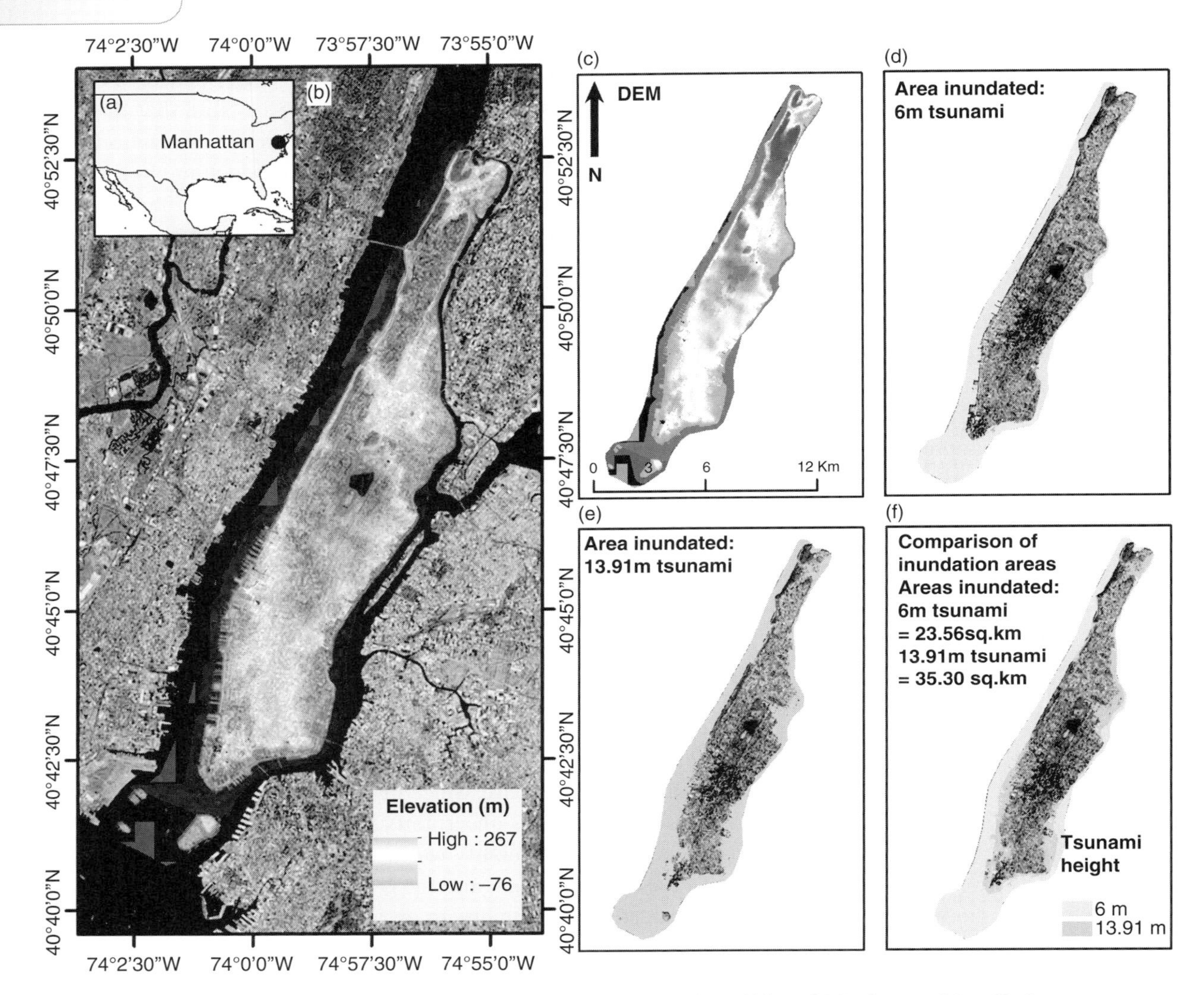

Figure 8.8 Uncertainty associated with an assessment of the vulnerability of Manhattan, New York, to a 6 m tsunami: (a) location of Manhattan on the North American continent; (b) DEM overlaid onto a satellite image of Manhattan; (c) DEM of Manhattan; (d) area inundated by a 6 m tsunami indicated in light blue; (e) area inundated by a 13.91 m tsunami indicated in dark blue (the upper 95% confidence limit); (f) area inundated by a 6 m tsunami (light blue) overlaid onto the area inundated by a 13.91 m tsunami for comparison of assessment outputs, allowing for uncertainty associated with the DEM. (A black and white version of this figure will appear in some formats. For the colour version, please refer to the plate section.)

8.13 Summary of Key Points: Chapter 8

- All spatial analysis has uncertainty associated with it, which must be evaluated through an assessment of accuracy or error.
- Errors in spatial analysis may be stochastic (random) or systematic (caused by a consistent shift away from an accepted or true value). The root mean square error (RMSE) measure is commonly used to quantify errors.
- The error associated with analysis is linked to the scale of the analysis. In general, the error associated with larger-scale analysis is small relative to the error associated with smaller-scale analysis.
- Statistical methods for evaluating the uncertainty of analysis can be non-spatial (mean, standard deviation, standard error, probability modelling of confidence intervals, sensitivity analysis, error propagation assessment) or spatial (residual mapping, confidence interval mapping).
- Spatial error can be mapped and modelled using geographically weighted regression techniques. This can be used to increase the reliability of predictive models.
- A validation exercise can evaluate the uncertainty associated with digital maps by systematically comparing them to an independent ground referencing dataset by means of an error (or confusion) matrix. Comparisons yield metrics such as the producer, user and overall accuracy of the map.

9 Communicating and Incorporating the Results of Spatial Analysis Successfully

9.1 Can Science Solve the World's Problems?

I was recently asked whether science can solve the world's problems. I thought about the spatial analysis work I have done over the last 15 years, the methods I have employed (many of which are described in this book) and the results that have been generated. It occurred to me that scientists have established the best philosophical framework I can think of for producing knowledge. When faced by a research challenge, a scientist must pose a hypothesis, which can then be tested and subjected to scrutiny. A hypothesis might lie at the centre of a spatial analysis exercise that seeks to explain or predict something in relation to the research challenge. A hypothesis that resists attempts at falsification will be considered the best available account for the time being. The essence of science is therefore a growing body of theory that has resisted attempts at falsification, rather than a definitive or proven set of laws. By adopting this framework of logical reasoning for building knowledge, scientists position themselves well to solve the world's problems. In addition to this, they often have access to state-of-the-art fieldwork instruments and analytical equipment, which, when combined with a creative, informed mind and some careful critical thinking, produces an excellent foundation for problem solving.

Within such a problem-solving framework, spatial analysis is an exciting endeavour because it has been supported by continual developments in technology over the last 50 years. These have presented opportunities to analyse large and valuable datasets that are increasing in quality (accuracy, consistency, resolution and completeness). Some of these technological developments include:

- incremental increases in the spatial, spectral and temporal resolution of remote sensing instruments for acquiring observational datasets,
- improvements in the accuracy of the global positioning system (GPS) technology for locating the data on which spatial analysis is based,
- developments in readily available and widely used spatial applications for smartphones that increase community involvement in the collection and analysis of spatial data,
- availability of an increasing range of field instruments for collecting spatially referenced datasets (e.g. terrestrial LiDAR, ground-penetrating radar, drones, autonomous underwater vehicles), and
- more sophisticated methods for working with field samples and biota in laboratory environments (e.g. dating sediments or manipulating environmental conditions such as seawater temperature and acidity in aquaria).

These technological developments have been accompanied by analytical developments, such as the appearance and refinement of geographical information system (GIS) software as a platform for carrying out analysis, and the associated growth of processing and statistical techniques for analysing and interpreting large spatially referenced datasets. These include remote sensing image classification algorithms, time-series analysis of images, geographically weighted regressions, measurements of spatial autocorrelation, semivariograms and correlograms. When these important analytical developments are combined with the technological progress that has been made and placed within the philosophical context of scientific knowledge production, an exciting opportunity is presented to the spatial analyst. This is an opportunity to create new insight by carrying out analysis that makes a useful contribution towards solving the world's problems. Such an opportunity is accompanied by an important responsibility to carry out robust, unbiased and transparent enquiry because spatial analysis has a powerful influence within broader society.

9.2 The Power of Maps and Spatial Analysis

Several spatial analysts have commented on the power that maps and other outputs of spatial analysis have in broader society:

- Digital maps are powerful geospatial images that can influence people's conception of space (Kraak and Ormeling, 2011).
- Results that appear in numerical form and are derived from computers tend to carry innate authority (Goodchild et al., 2005).
- Maps play a fundamental role in linking knowledge with power (Crampton, 2011).
- Driven by changes in technology, data supply and users, the increasingly complex data handling and analysis supplied by GIS is being powerfully used in policy development, decision-making and research in a range of political, economic and academic settings (Burrough and McDonnell, 2011).
- Power is the ability to do work. Which is what maps do: *They work* (Wood and Fels, 1992).

To understand the power of maps and spatial analysis, it is useful to draw on interdisciplinary perspectives. A positivist viewpoint of traditional scientific method holds that the reality of objects in the world can be expressed in mathematical terms. Systematic observation and measurement offer a 'cartographic truth' that can be independently verified. Through the application of science, the world can be categorised into ever more precise and accurate units, aided by the advent of geospatial technologies. It follows from this viewpoint that spatial analysis will continue to gain power as it is ever more comprehensively applied, driven by technological and analytical developments.

A critical approach seeks to uncover what types of assumptions, of familiar notions, of established, unexamined ways of thinking, the accepted practices of a field of knowledge are based on (Foucault, 2000). A more critical stance may read a map as a cultural portrayal, a mechanism for defining social relationships, sustaining social rules and strengthening social values (Crampton, 2001). This lies at the heart of the discipline of critical cartography (Crampton, 2011).

A critical perspective views the results of spatial analysis, including maps and associated statistical diagnostics, with an awareness of the contexts in which they were derived. By positioning spatial analysis within the societal context in which it was carried out, a richer interpretation of its role in furthering our understanding of places can be provided. A dual framework suggested by Brian Harley, who was both a cartographer and a critical scholar of maps, provides a useful basis for examining the power of maps (Harley, 1989). This draws on the external and the internal dimensions of their power. External power is realised by the patrons of cartography and wielded through the use of cartographic products. Internal power is largely determined by the actions taken by cartographers themselves when making maps. Cartographic process is therefore the key to this internal power, which Harley (1989, page 13) defines as:

> The way maps are compiled and the categories of information selected; the way they are generalised, a set of rules for the abstraction of the landscape; the way the elements in the landscape are formed into hierarchies; and the way various rhetorical styles that also reproduce power are employed to represent the landscape. To catalog the world is to appropriate it, so that all these technical processes represent acts of control over its image which extend beyond the professed uses of cartography.

Such a definition can be applied broadly to anything that determines the knowledge of the world made available to people through maps, including the entire discourse of positivism that underpins cartography and the perspectives brought to bear through geospatial analysis.

The power that maps have to shape our understanding of place has been a central theme in the field of critical cartography, which seeks to determine how, and in what ways, maps have and exercise power (Wood and Fels, 1992; Harley and Laxton, 2002; Dodge et al., 2011). This is particularly pertinent because the development of widely available and free mapping applications such as Google Earth is taking the practice of map making out of the hands of experts and making it available to anyone with a smartphone, tablet, home computer and broadband Internet connection (Crampton, 2011). Many of the comments and observations made by critical cartographers are relevant to the broader field of spatial analysis. Just as the decisions made by the cartographer during the map production process should be subject to interrogation, so should the conceptual and analytical decisions made by the spatial analyst be examined.

9.3 Critical Thinking, Interpretation and Responsible Communication of Spatial Analysis Results

Critical thinking involves interrogating one's reasoning in a manner designed to clarify, raise the efficiency of, and recognise errors and biases in arguments. The purpose of critical thinking is to maximise the rationality of the thinker, rather than to solve problems (other than indirectly, by improving one's own process of thinking) (Paul and Binker, 1990). Section 8.4 presented a series of analytical approaches for critically evaluating the results and outputs of spatial analysis. These were set within a context of established statistical procedures for describing and quantifying uncertainty. Here, that critical evaluation is broadened to consider the wider interpretation and communication of the results of spatial analysis within society.

The outputs of spatial analysis often take the form of a map. It is the job of this map to communicate key messages associated with the output or findings of that analysis. The map becomes a communication tool for rendering spatial patterns observable. Some questions that may help with the interpretation of results in the form of a map include the following:

- Are any patterns observable in the phenomenon represented by the map across the study area?
- What do the statistical diagnostics illustrate? If diagnostics such as R^2 values or probability measures are calculated locally and vary across the map, what does their distribution suggest?
- Is there a pertinent environmental gradient (e.g. onshore to offshore) that provides a meaningful axis along which trends in the data values can be interrogated?
- If trends are observable, do they substantiate or contradict any pre-existing expectations that were guided by existing research on the phenomenon of interest?
- How do the values associated with the results compare to similar studies? For example, if the resulting map depicts 'wave energy', are the values of the map larger or smaller than published wave energy values for comparable locations elsewhere? Are values of the same order of magnitude, i.e. is the difference a substantive or negligible one?

Where the answers to these questions provide new, meaningful or insightful ways of understanding a phenomenon, these should be communicated in a manner that renders these fresh insights clearly visible. Visualisation techniques such as appropriate colour schemes, contrast stretches and inset maps may be useful. Enough ancillary information should be provided for a person to independently make their own judgement on the validity and importance of the results. It may be helpful to provide associated plots and tables listing pertinent statistical diagnostics. In addition, it is important to provide a clear and transparent description of how any maps provided have been assembled. Effective communication of spatial analysis results makes transparent the decisions that were made during the process of their generation, interpretation and communication. In doing so, it opens them up for interrogation.

9.4 Principles of Digital Map Design and Graphical Representation Techniques

The manner in which results are displayed as a map determines how the information will be understood. Conventional cartographic rules define the basic elements to be provided for interpreting maps. These basic elements provide essential information for interpreting map features:

- a title summarising key content,
- a north arrow,
- a statement of scale (e.g. a scale bar or quoted ratio),
- a legend,
- a grid, graticule or margin showing a coordinate system that indicates absolute location,
- a statement of the geographic or projected coordinates used, and associated datums and units, and
- an inset box to guide readers as to the regional location context in relation to widely recognizable features.

Although convention dictates what basic elements must be included, the analyst must decide what scale is best for presenting the information, what layout best suits the configuration of elements, and what design features are appropriate. These decisions are important because considerations such as what visual contrast to apply, or how to hierarchically organise the symbology of a particular variable, will ultimately determine the legibility and clarity of the message contained in the map. Table 9.1 summarises the primary and secondary visual variables that govern how prominently an

Table 9.1 The primary and secondary visual variables that influence the prominence of features on a map. Summarised from Robinson (1958).

Variable	Description
Primary visual variables	
Shape	The form of a graphic mark, which may be geometric (square, circle, triangle) or irregular (icon of a bridge or tree)
Size	The dimensions (height, area, length) of a symbol
Orientation	A directional frame of reference provided for lines and elongated symbols
Colour (hue)	The degree to which a colour stimulus can be described as similar to, or different from, red, green, blue and yellow
Colour (tone or value)	The relative lightness or darkness of a colour
Colour (chroma)	The degree to which a colour departs from a grey tone of the same value or lightness
Secondary visual variables	
Arrangement (pattern)	The shape and configuration of components that make up a pattern, which may be systematic or random
Texture (pattern)	The size and spacing of components that make up a pattern, which may be coarse or fine
Orientation (pattern)	The directional arrangement of marks positioned with respect to a frame of reference

individual feature appears on a map. In addition to these, labels can be used to draw attention to and assist the users' understanding of selected features. In most cases, these decisions of presentation will be guided by the intended audience of the map and what message it carries.

The manipulation of colour has a substantive influence on how information is perceived. If a raster grid presents continuous data values, the scale along which a colour scheme is graduated determines which data values are visible. Transitions from dark to light are often associated with corresponding transitions from low to high data values. Similarly the 'traffic light' colour scheme that transitions from red, through amber to green is often used to depict graduated variations in phenomena. For example, this is commonly used to depict levels of risk, with red conventionally indicating high risk and green indicating low risk.

Manipulation of contrast also has a critical influence on the amount of detail that can be perceived in a dataset. For a raster data layer, a contrast stretch will apply a consistent multiplier to each value to ensure that the range of data values is optimally displayed on the computer screen. For example, if a computer screen is able to display 255 different colour tones, but a dataset has only 25 values, it may be useful to stretch the contrast range applied to the colour by a multiplier of 10.2, so that a greater proportion of the screen display capacity is utilised. In this way, more detail of variability within the dataset will be discernible from the visualisation. This can usually be achieved by adjusting the image display settings. This does not change the values of the underlying dataset itself. Rather, it transforms the data for the purpose of displaying it.

Figure 9.1 illustrates the different levels of detail visible on a digital elevation model (DEM) of a section of the New South Wales coastline. This length of coastline coincides with the city of Wollongong and is characterised by substantial topographic relief across a north–south aligned escarpment ridge that is approximately 450 m above sea level. The majority of the development associated with the city occurs on the flatter and lower coastal plain that extends approximately 2 km back from the shoreline. A visual comparison of Figure 9.1(c) and (d) reveals that much more detail of the topographic relief from the DEM can be seen in Figure 9.1(d). This is because of the application of a different colour scheme that graduates through a sequence of several tones and a 'histogram equalise' contrast stretch. Note also the utility of the inset maps on the left-hand side. This provides useful geographical context for anyone who is not familiar with the New South Wales coastline, but who may recognise the shape of continental Australia.

The type of map design selected may emphasise data features, such as choropleth maps, proportioned or graduated symbols, isolines or contours, cartograms and flow maps. Maps may display multiple different types of information by overlaying datasets or by assigning graduated symbols to illustrate some form of hierarchical information associated with features (e.g. defining the size of circles depicting cities on the basis of their population size). Maps employed in a multivariate analysis can be effectively presented alongside one another as repeated maps of the same geographic area. Such repetition allows easy visual comparison of datasets. Developments in Web mapping (see Figure 4.17) have made it easier to customise the appearance of a

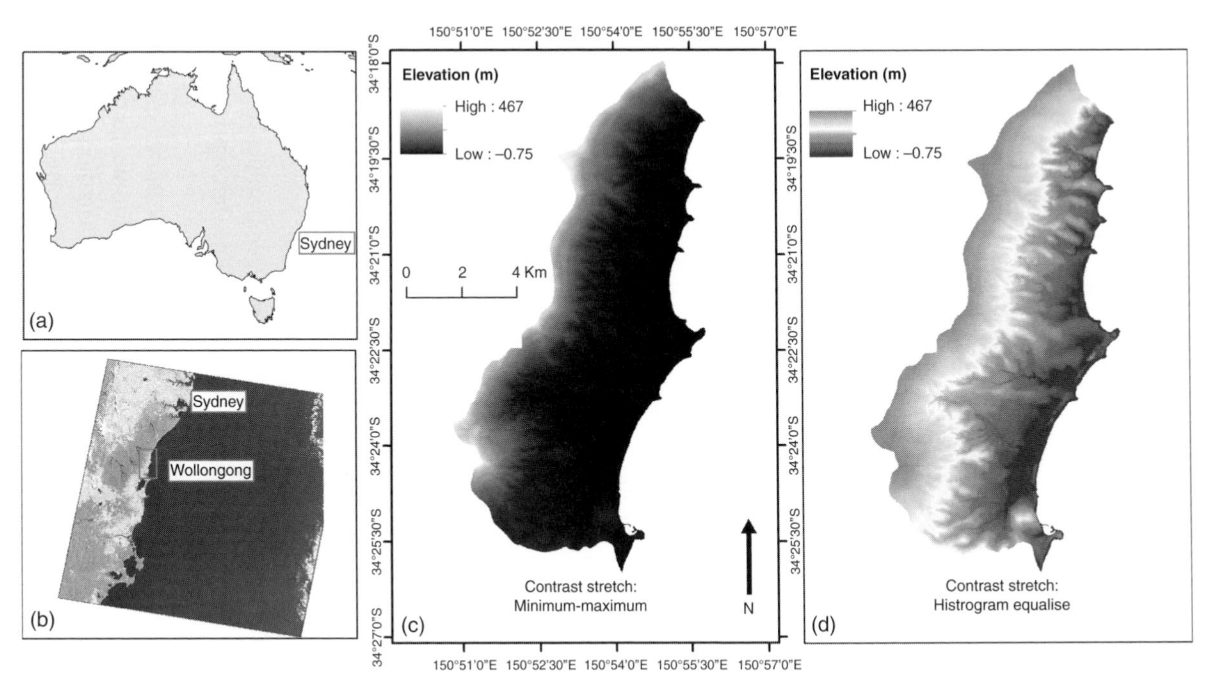

Figure 9.1 (a) and (b) The locations of Sydney and Wollongong on the eastern coast of Australia and New South Wales, respectively. (c) A DEM of the coast of New South Wales, where the city of Wollongong is located. (d) The same DEM with a 'histogram equalise' contrast stretch and a different colour scheme applied. Panels (c) and (d) illustrate the different ways in which information can be presented and, by extension, perceived as a result of applying different colour schemes and image contrasts. (A black and white version of this figure will appear in some formats. For the colour version, please refer to the plate section.)

map. Often an interactive map viewer embedded into a Web page will provide a user with the ability to interact with datasets, turning on and off different layers of interest and, in some cases, to format and analyse data layers.

9.5 Incorporating the Results of Spatial Analysis Into Organisational Practice

The commitment of an organisation to the conduct of spatial analysis and the uptake of results depends largely on the extent to which this analysis can have a practical impact on the work, tasks and jobs of employees. Are the technologies proven to be effective within the organisation's sphere of activity? Will spatial analysis bring benefits to the organisation and how can these be assessed? Table 1.1 helps to answer these questions by reviewing different kinds of spatial analysis by coastal environment type, while Section 1.11 reviews coastal management questions for which geospatial analysis offers insight.

Bartlett (2000) summarises the ways in which GIS can contribute to coastal management as follows:

- by providing the capability to handle, integrate and synthesise large databases, expanding management across longer lengths of coastline and across wider administrative divisions;
- by encouraging the development and use of coastal data standards;
- by employing a network to facilitate data currency and sharing, storage and retrieval; and
- by providing the ability to model, test and compare alternative management strategies before implementation.

Incorporating the results of spatial analysis into organisational practice will also depend on the extent to which an organisation has already adopted allied technologies, such as investments in GIS software and datasets or associated technology for data acquisition. When considering wider, internal adoption of spatial analysis, some pertinent hardware, software and human resources questions for organisations might include the following.

- Hardware:
  - What hardware is necessary (number of computers, processing speed, storage capacity, memory)?
  - What peripherals are necessary (e.g. graphic display devices, printers, plotters, scanners, data loggers, field equipment, remote sensing instruments)?
  - What about reliability, maintenance, expansion and upgrades of hardware?
  - Will new hardware be compatible with existing organisational hardware?

Table 9.2 Costs and benefits associated with increased organisational uptake of GIS as a primary tool for spatial analysis (Carver et al., 2006).

Cost	Benefit
Direct costs and benefits	
• Hardware/software purchase, development, customisation, upgrades	• Reduced costs of information production
• Consumables (printer, supplies, etc.)	• Fewer staff required
• Maintenance and support contracts	• Time savings for routine and repetitive tasks (e.g. in data processing)
• Networks	• Greater range of information more readily available
• Data acquisition, maintenance, storage, updating	• Faster provision of information
• Staff training, insurance, administration	• New range of outputs: maps, analysis results, etc.
Indirect costs and benefits	
• Risk and vulnerability associated with increased reliance on computers	• Improved information sharing and flows
• More skilled workforce required	• Better-informed decision-making
	• Stronger competitive ability
	• Motivated workforce
	• Greater analysis and understanding of problems
	• Analytical justification underpinning decision-making
	• Improved data visualisation

- Software:
  - What functions are necessary to meet the requirements of the organisation?
  - Is the package user-friendly?
  - Is the software interoperable? i.e. can data be exchanged with other packages used by the organisation?
  - Which operating system will be required?
- Human resources:
  - What level of skills do staff of the organisation already have?
  - What level of training is available?
  - What kinds of skills are necessary (e.g. GIS analyst vs computer technician vs managers using results)?

Cost will also be an overriding factor determining the extent to which investment can be made across these three areas. Table 9.2 summarises some of the costs and benefits associated with increased organisational uptake of GIS.

9.6 Driving Innovation in Web-Based GIS and Online Mapping: Google Earth

The largest Web-based GIS and online mapping service was developed by a series of computer graphics companies in California's Silicon Valley. Collectively they drove the technical innovations underpinning Google Earth. Intrinsic Graphics and Silicon Graphics, both founded in the 1980s, designed applications to render graphics at high speed and resolution, particularly in three-dimensional graphic display. Later, Keyhole Inc. was founded in 2001 to explore commercial applications of rapidly rendering spatially referenced graphic information on a globe. This latter company was acquired by Google in 2004, whose mission statement was *'To organize the world's information and make it universally accessible and useful'*. Google would develop this technological framework into Google Earth, the largest repository of geographic information and most widely used online tool for undertaking spatial analysis.

These companies brought about significant innovations that facilitated the development of contemporary Web-based mapping applications. One of these was to develop seamless rendering to support high-speed display of graphic Earth observation imagery. Displays needed to support large geographic areas, multiple spatial scales and three dimensions. High-speed graphic display and rendering would enable users of Google Earth to view a digital image of the whole Earth taken from space and zoom in while viewing a crisp, flicker-free image refreshing at up to 50 frames per second. This display could be tilted, rotated and panned by the viewer and quickly rendered on the screen at different resolutions. Clip-mapping technology was developed to support seamless rendering through an arrangement of correlated images of increasingly reduced resolution arranged in a pyramid fashion, known as a *MipMap* (Tanner et al., 1998). The development of 'ray tracing' allowed images to be rapidly rendered in three dimensions by anticipating the direction in which the viewer was looking, and filling that part of the screen before accessing surrounding 'peripheral vision' data. Such topographic visualisation relied on elevation datasets such as the 30 m global grid provided by the Shuttle Radar Topography Mission (SRTM), with higher-resolution datasets supplying more detail where available.

In parallel with these innovations, large volumes of satellite imagery and aerial photography were acquired from geoinformation companies. These were made available for viewing online for free. The company image archive in 2010 represented a huge repository of approximately 10 PB of data, the equivalent of 500 billion pages of printed text (Brotton, 2013). Although the imagery available to view was not real-time, the archive was updated every 30 days, such that the youngest images were six months old. Particular attention was paid to urban areas of high interest, where more detailed datasets were regularly updated. Standard criteria for new images included Sun angle (ideally a high angle to reduce shadow), cloud cover, haze and pollution, an evaluation of whether the image represented an improvement on the previous one and verification of spatial referencing information. Having passed initial quality checks, images could then be processed to the standard format and coordinate system of Google Earth and distributed to the live Google Earth database servers. Archives of multiple images for a given geographical area could be accessed through the historical imagery function and additional feature attribute information including place names, borders, road networks and *in situ* photographs could be overlaid.

9.7 Web Atlases and Tools for Online Mapping

A coastal Web atlas is a collection of digital maps, datasets, supplementary tables, illustrations and information that systematically illustrates the coast, alongside cartographic analysis and decision support tools, all of which are accessible via the Internet (Wright et al., 2010). Web atlases often take the form of spatial data infrastructures. These bring together spatially referenced information such as maps derived from remotely sensed imagery for use in management decision-making. Information is often hosted by an online data portal. This will include a metadata catalogue with descriptions of available datasets and imagery and links to networked servers. Users may visually browse and query data or build online maps within the portal (Wright, 2009).

The overriding objective of spatial data infrastructures is to facilitate access to geographic information that is held by a wide range of stakeholders, with a view to maximising overall usage of the spatial information they host (Hamylton and Prosper, 2012). Digital atlases are commonly being developed by government mapping and coastal management authorities, industry sectors, academic institutions and non-governmental organisations. Table 9.3 summarises some examples of currently available Web atlases alongside the institutions that have developed them. The International Coastal Atlas network encourages global interoperability between aspects of coastal Web atlases, including user and developer guides, handbooks and articles on best practice, information on standards and Web services, expertise and technical support directories, education and outreach (Wright and Bartlett, 2000).

Table 9.3 A summary of Web atlases currently available online.

Web atlas	Developer
Oregon Coastal Atlas	Oregon State University
The Marine Irish Digital Atlas	University College Cork
The Chesapeake Bay Coastal Atlas	National Oceanic and Atmospheric Administration's Office of Coastal Resource Management
The African Marine Atlas	Intergovernmental Oceanographic Commission of UNESCO
The Amirante Islands Atlas of the Seychelles	The University of Cambridge
The Caribbean Marine Atlas	UNESCO-supported joint initiative between Barbados, Cuba, Dominica, Guyana, Jamaica, St Lucia, Trinidad & Tobago and Turks & Caicos
The Sistema de Informacion Geografica del Litoral Andaluz	Regional Ministry for the Environment and Regional Government of Andalusia

Table 9.4 A summary of open-source applications for collecting, georeferencing, visualising and delivering geographic information online.

Tool	Description
ArcServer	Esri server GIS software for creating and managing GIS Web services, applications and data
ArcGIS Online	A cloud-based mapping platform for making and sharing maps
ArcGIS Pro	Software for creating, managing and sharing geographic maps, data and analytical models, including sharing results of spatial analysis online
Open Data Kit	Designed to support mobile data collection and management, including metadata field design, storage on a server and aggregation of fields for data management
OpenStreetMap	An editable map of the world to which users are able to add spatially referenced information
OpenLayers	An application programming interface (API) that consists of a Javascript library for displaying map data in Web browsers
Leaflet Maps Marker Pro	Javascript API that supports both mobile and desktop platforms
Google Earth Engine	A free, editable map of the world to which users are able to add spatially referenced information
My Maps	An application that lets users position editable markers, polylines and polygons onto a map and embed created maps into a Web page or blog

Table 9.4 summarises some emerging tools for online mapping developed by Google, Esri, Leaflet and OpenLayers. These are designed to:

- collect *in situ* information and add spatial referencing and metadata to these records,
- deliver and visualise geographic information, and
- host and customise interactive Web viewers.

It should be noted that online mapping technologies, particularly open-source approaches offered by large corporate entities, are often subject to criticism. Controlling the information supplied through a map interface has a high corporate value. The basemap that underpins interfaces is often supplied with a selective range of content that has been predetermined by the provider. For example, the online advertising service *Adwords* determines the content displayed through a search query entered into Google's maps application. This includes the placement of advertising copy alongside the search results. Advertisement placement is based on bids tendered by businesses for such a marketing service. In 2012, the revenue from selectively displaying spatially referenced information through Adwords was US$42.5 billion (Brotton, 2013).

9.8 Concluding Comments

This book has presented a range of techniques for applying spatial analysis towards a better understanding of coastal environments. It began with a vision of how spatial analysis can productively be applied to coastal environments. The nature and characteristics of spatial data were then explored, alongside some basic analytical operations that can be carried out with spatial data. Procedures for mapping coastal environments have been described that draw on remote sensing technology, an important source of data for spatial analysis. Digital maps provide the data that can be used as a foundation for monitoring and modelling exercises. Information sources and methodological frameworks for monitoring historical changes in coastal environments have been explored. Similarly, models have been described that enable the spatial analyst to predict potential future changes to the structure and form of coastal environments. As well as predictive models, explanatory models have been described that seek to further understand spatial relationships that exist along environmental gradients that traverse coastlines. The analysis underpinning these models has employed geostatistical techniques that draw explicitly on the spatial referencing, or coordinates, associated with the data values that are being manipulated. Collectively, these elements provide a powerful basis for spatial analysis that is targeted towards a practical purpose. This is supported by evaluation of the inherent uncertainty associated with results.

Since the 1990s, scholars have critically evaluated spatial analysis from philosophical, psychological, epistemological, ethical and social perspectives. For example, from an epistemological perspective, it is difficult to utilise spatial analysis technology without adopting a positivist, reductionist stance that privileges certain views of the world over others. A related ethical concern relates to the digital divide, across which the costs associated with conducting spatial analysis perpetuate inequities among developed and developing countries, socioeconomic groups, individuals, public agencies and private firms. This is particularly important because of the power that is often wielded by the results of spatial analysis. From a social perspective, spatial analysis and maps cannot be disconnected from the social dynamics that have driven their construction. It is beyond the scope of this book to provide a comprehensive review of the different critiques that have been applied to spatial analysis and allied disciplines. Nevertheless, an awareness and understanding of the range of critical perspectives often applied to the disciplinary field is likely to promote reflective practice among spatial analysts. In turn, this will increase the likelihood of spatial analysis fulfilling its potential benefits to all stakeholders.

It is clear that changes in users, data and technology have increased the sophistication and volume of meaningful spatial analysis that can be carried out in coastal environments. Associated developments of explicitly spatial techniques and software that enable the manipulation and interrogation of data have opened up a wide array of potential analytical applications. The constraints on how spatial analysis is applied in coastal environments are not so much to do with knowledge of mathematics or statistics, as to do with one's imagination. The challenge for managing and understanding coastal environments is to think about space and spatial processes in new and creative ways. It is an exciting time to be a spatial analyst.

9.9 Summary of Key Points: Chapter 9

- Digital maps and the outputs of spatial analysis have a powerful influence in society. They should be presented, communicated and interpreted responsibly.
- Decisions made by an analyst when producing a map have important implications for the message conveyed by that map, e.g. in the cartographic design principles, colours and symbols selected to represent features.
- Organisational or institutional uptake of spatial analysis depends on costs associated with hardware, software and human resources, alongside the extent to which spatial analysis has a proven practical impact on the work, tasks and jobs of employees.
- Spatial information and the results of spatial analysis are increasingly available through spatial data infrastructures, Web atlases, Web viewers and online repositories such as Google Earth. This increases their influence in society.

GLOSSARY

Accuracy The degree to which a measured or estimated value conforms to the 'correct' or 'true' value

Acoustic The study of mechanical waves in gases, liquids and solids, including vibrations and sound waves

Active remote sensing A remote sensing system that emits a source of radiation and analyses the return component (e.g. LiDAR or radar)

Aircraft pitch, roll and yaw Rotations of a fixed-wing aircraft during the flight of an airborne survey around the side-to-side (lateral), front-to-back (longitudinal) and vertical axes, respectively

Algorithm A procedure or formula for solving a mathematical problem, often implemented by a computer as a self-contained set of operations

Analysis of variance (ANOVA) A collection of statistical techniques for analysing the statistical significance of the variation among grouped values of a given phenomenon

Anisotropy The property of a given phenomenon exhibiting different values when measured along axes in different directions

Atmospheric correction The correction of aircraft- or satellite-received remote sensing data to allow for the effects on the radiation as it passes through the atmosphere (e.g. scattering and absorption of light photons)

Attenuation Reduction in the intensity of a beam of electromagnetic radiation as a result of both scattering and absorption

Attribute a property associated with a field in a geographic information system (GIS) database

Attribute join A type of table join operation in which fields from one layer's attribute table are appended to another layer's attribute table based on a common attribute of the features in the two layers

Autocorrelation The similarity between observations of a value as a function of the distance between them

Autocovariance A function that quantifies the covariance of a process or property with itself across pairs of points distributed in time or space

Azimuthal projection A map projection in which a region of the Earth is projected on a plane tangential to the surface, usually at a pole or the equator

Bathymetry The study and mapping of underwater depths

Benthic Relating to the lowest level of a body of water, i.e. the seafloor

Biogeography The study of how species and ecosystems are distributed in space and time

Biogeomorphological The study of the interaction between organisms and landforms

Boolean logic A form of algebra, often used in computer science, in which values representing logical concepts are reduced to 'true' or 'false', which correspond to binary numbers of 1 and 0, respectively

Brushing A dynamic statistical exploration technique for interrogating values of a phenomenon simultaneously in both mapped and tabular form

Buffer A zone defined by a boundary of consistent distance around a mapped object

Calibrate The process of establishing a relationship between two quantities, usually through an empirical comparison of repeated observations or samples

Cartesian coordinates Coordinates derived from a Cartesian system that specifies the unique location of each point on a plane; a pair of numerical coordinates specifies location in relation to two or three axes

Change matrix A matrix in which two digital maps generated for different time periods are compared by populating the matrix with the number of pixels or mapping units relating to each map class

Change vector analysis A change detection technique based on multi-temporal satellite images that employs spectral information to characterise the magnitude and direction of change between two image dates

Charge-coupled device (CCD) array An array of detectors established to collect photons of light, usually arranged in a grid circuit and etched onto a silicon surface; photons incident on this surface generate an electronic charge that can be turned into a digital record of the light patterns falling on the device

Choropleth mapping A thematic map in which features are visualised, usually through shading or symbols, in proportion to the values of a given variable measured for each location

Choynowski test A test for detecting underlying clusters in point datasets

Coefficient A number to be multiplied by a term within an algebraic expression

Coefficient of determination (R^2) An indication of how well a dataset fits a statistical model; specifically, the percentage of variation in the dataset explained by the model; usually quoted as a number between zero and one

Conceptual model A schematic arrangement of ideas, usually in the form of a diagram of linked boxes, to help people understand the subject that a model, or intended model, represents

Confidence interval A range of values within which it can be stated with a given level of certainty (e.g. 95%) that the value of an estimated parameter falls

Confirmatory spatial data analysis (confirmatory modelling) A family of statistical techniques for testing how well a series of measured variables represent predefined constructs

Confusion matrix A matrix in which a digital map is compared to a point dataset of independently collected field records representing locations visited where the ground surface was recorded using the same land-cover classification scheme as the digital map; the matrix is populated with the points depending on how they relate to land-cover class labels specified in the field dataset and digital map;

can be used to derive measures of producer's, user's and overall accuracy

Conical projection A map projection on which the Earth is shown as projected onto a cone with the centre of the map representing the cone apex and other areas radiating out from this point

Contextual operator A statistical operator designed to work with continuous raster datasets in which new values are derived based on a given pixel and its surrounding pixels (i.e. some definition of the localised context)

Contour A line on a map along which equal values of a given phenomenon (e.g. elevation) can be specified

Control base station A global positioning system (GPS) receiver placed at a fixed location that continually collects position information that is subsequently used to correct errors in position information that has been simultaneously collected by a nearby portable GPS receiver unit

Correlation A statistical technique that can reveal the existence and strength of a relationship between two variables

Correlogram A tool for detecting spatial autocorrelation from a plot of the Moran's I statistic as a function of distance

Covariate A variable that displays a relationship to a variable of interest, which may or may not be of direct interest to the study

Critical cartography A body of critical theory that links geographic knowledge with power

Cylindrical projection A map projection on which the Earth is shown as projected onto a cylinder that can be unfolded into a flat plane without tearing, stretching or shrinking; the cylinder is typically positioned with the globe centred horizontally, to minimise distortion in equatorial regions and maximise distortion at the poles

Datum A fixed reference point in a vertical or horizontal plane, with reference to which all other features are mapped

Degrees of freedom The number of values that are free to vary in the estimation of a given statistical value, typically specified as the number of parameters minus one ($N - 1$)

Dependent variable A variable measured during an experiment that is affected by inputs

Descriptive resolution The number of distinct descriptive classes (i.e. labels) incorporated into a map legend

Digital elevation model (DEM) The continuous representation of elevation values over a topographic surface, referenced to a common datum in a regular array

Digital number A generic term for pixel radiance values that have not yet been converted to physically meaningful units (e.g. radiance)

Digital Shoreline Analysis System (DSAS) Computer software established to calculate rates of shoreline change via comparison of multiple historic shoreline positions depicted on aerial photographs within a GIS

Digitising The act of converting geographic features on a map into digital format by visually interpreting and manually tracing over them using a digitising tablet or digitising tool within GIS software on a computer

Echosounder A device that uses sound waves to measure underwater depths and seafloor texture

Edge effects Changes or transitions that occur to the population or community structure of a given ecosystem at the boundary of two habitats

Electromagnetic radiation Radiation made up of oscillating electric and magnetic fields carrying energy from one point to another

Electromagnetic spectrum The range of frequencies over which electromagnetic radiation can be propagated

Ellipsoid A surface that approximates the geoid, a 'truer' representation of the three-dimensional globe than a sphere

Empirical An approach to analysis that is based on, or verified by, observation and experience as opposed to theory or logic

Endogenous Originating from an internal source

Error matrix *See* 'Confusion matrix'

Error propagation The effects of the uncertainty associated with multiple variables when bringing them together into a combined function

Exogenous Originating from an external source

Explanatory model A model that serves to explain a given observation in the natural world

Explanatory variable A variable thought to explain a phenomenon or observation of a phenomenon (usually described as the dependent variable)

Exploratory spatial data analysis (ESDA) A collection of statistical techniques for describing and visualising spatially referenced information

Falsification The act of proving a statement or theory to be wrong

Feature space The space created by a scatterplot of the pixel values of two or more wavebands of remote sensing imagery

Fetch The distance over which wind is able to blow uninterrupted across the sea surface

Field of view The total area viewed by a remote sensing system as the instantaneous sensor is scanned over the land surface

First-order effects Relating to the simplest or most fundamental level of organisation; in the context of mathematical equations, terms incorporated will only relate to the first power of an independent variable or function

Focal operator A spatial function applied to a raster dataset that computes an output value for each cell on the basis of the cell value, plus values of neighbouring cells (e.g. a convolution, kernel or moving window)

Gaussian curve A statistical distribution of data that takes the form of a characteristic bell shape, with the height of the curve's peak at the centre; also known as a normal curve

Geary statistic A measure of spatial autocorrelation that is particularly sensitive to local autocorrelation effects

Geodesy The mathematical discipline dealing with the shape and area of the Earth

Geographical coordinates A pair of numbers relating to positions along the two axes of a coordinate grid that enables every location on Earth to be specified; the two axes are conventionally north–south and east–west aligned

Geographical projection A system that systematically transforms the locations on the surface of a sphere or ellipsoid to a flat plane, usually resulting in some distortion to angles or areas

Geographical scale *See* 'Scale'

Geographically weighted regression A method for analysing spatially varying relationships that allows a regression equation to be fitted to incorporate fluctuating values for independent variables, depending on their location in space

Geometric correction The process of mapping a remotely sensed image onto a chosen projection, such as latitude/longitude, or a coordinate grid

Geomorphology The study of the physical features of the Earth's surface

Georectify The process of digitally aligning a satellite image or aerial photograph with a map of the same land area using a number of corresponding reference (ground control) points

Georeference The process of allocating a coordinate system to data so that it can be viewed and analysed alongside other geographic data in a common frame of reference

Geostatistics A family of statistical techniques for analysing and predicting the values associated with spatial phenomena

Glint Light reflecting off the sea surface as a result of specular reflection

Global operator A spatial function applied to a raster dataset that computes an output value for each cell as a function of all the cells in the input raster

Global positioning system (GPS) US satellite navigation system designed to enable users to position themselves on the Earth by determining their x, y and z coordinates; the satellite network is composed of 21 operational NAVSTAR satellites in near-circular orbits across six orbital planes; orbits are oriented to ensure that at least four satellites are visible above the horizon at any time for any observer

Ground control point (GCP) A distinctive feature of the landscape recognisable both in an image and on the ground (or on a map), used as a reference point for geometric correction of an image

Ground referencing The collection of independent information in the field for cross-referencing with remotely sensed or mathematically calculated data for the purposes of calibrating or validating a model or map

Habitat suitability model A model that attempts to relate habitat suitability to raster-based information, such as land cover and elevation

Hierarchical database A database structure in which information is stored in a tree-like fashion, whereby records can be traced back to parent records

Hydrodynamics The discipline concerned with forces acting on or exerted by fluids (e.g. waves and tides)

Hyperspectral imager A remote sensing instrument that samples the electromagnetic spectrum in many (> 100) narrow wavebands in order to record a near-continuous representation of the spectral signature of a target

Hypothesis A proposed idea or explanation to be tested through experimentation

Image classification The conversion of a remote sensing image from radiance data to thematic, classed information in the form of a map depicting features of interest

Independent variable A variable measured during an experiment that is unaffected by inputs

Inferential statistics/model Statistical analysis or a model designed to reach a conclusion on the basis of evidence and reasoning

Interpolation The process of estimating values at unsampled locations based on known values at sampled locations; well-known techniques include inverse distance weighting, spline and kriging

Intersect A geometric technique for combining two or more spatial datasets that preserves features that fall within areas common to all input datasets

Inverse distance weighting (IDW) An interpolation technique in which the value at an unsampled location is estimated from a set of sample points that have been weighted in accordance with their distance from the estimation location, such that points that are farther away are accorded less weight than those that are close by

Kernel density estimation A statistical estimation of the probability density function of a given variable in space

Kriging An interpolation technique in which the value at an unsampled location is estimated from a set of sample points that have been weighted based on a geostatistical function that takes into account the distance between the measured points, the prediction locations and the overall spatial arrangement of the points

Kulldorff's scan test A spatial statistical test that identifies excesses in the number of cases present in either space or time using a circular or cylindrical sample window

Latitude A geographical coordinate that specifies the north–south position of a point on the Earth's surface

LiDAR (light detection and ranging) An active remote sensing technique in which an instrument emits pulses of laser light to measure the distance to reflective surfaces

Littoral Relating to, or situated at, the shore of the sea or a lake

Local operator A spatial function applied to a raster dataset that computes an output value for each cell as a function of the corresponding cell in the input raster at the same location

Longitude A geographical coordinate that specifies the east–west position of a point on the Earth's surface

Map algebra A GIS analysis technique in which multiple maps are combined through the application of mathematical operations and algebraic functions to create new map themes

Map scale *See* 'Scale'

Margin of error The amount of error in a model or experiment's results, expressed as a range of values within which there is a stated probability that the real value falls

Marked point pattern analysis Analysis of the spatial distribution of the values associated with a series of point locations.

Masking A technique used to cover or hide features on a map or remote sensing image for exclusion from analysis

MAUP (modifiable areal unit problem) A problem encountered during the spatial analysis of aggregated data in which the results differ depending on the way in which an aggregation scheme is applied to a single dataset

Maximum likelihood classification A type of supervised classification that uses a probability function to discriminate the most likely class for each pixel of a remote sensing image to be classified

Measurement The act of assigning a quantity to some phenomenon in accordance with a conventionally accepted system of quantification; measurement levels include nominal, ordinal, interval and ratio

Model specification The process of determining which independent variables should be included in a model, and how they should be manipulated

Monte Carlo randomisation A technique in which a large quantity of randomly generated numbers are studied using a probabilistic method to find an approximate solution to a numerical problem

Moran's *I* statistic A measure of spatial autocorrelation that is based on the cross-product of pairs of data values in space; the statistic yields a measure that is large and positive in the presence of spatial dependence

Moving window analysis A moving window is a rectangular arrangement of cells that applies an operation to each cell in a raster dataset while shifting in position entirely, often iteratively centring the window on every pixel in the grid

Multi-criteria decision analysis In GIS, multi-criteria decision analysis is concerned with the identification of locations that meet a range of criteria specifying attributes that the selected locations should possess; such analysis supports the process of decision-making, particularly selecting the best location for a given application

Multispectral imager A type of electro-optical sensing instrument that samples radiance data for several spectral bands for each pixel of an image

Multi-temporal analysis Analysis that incorporates datasets derived from multiple points in time

Multivariate analysis A statistical method that evaluates the relationship between two or more variables

Nadir Vertically downwards, i.e. towards the centre of the Earth

Nearest-neighbour methods A spatial function applied to a raster dataset that computes an output value for each cell as a function of that cell and/or the nearest cell in an input raster

Negative feedback A reaction that occurs within a system that minimises the effect of the initial stimulus

Network database A database structured as a collection of topologically connected network elements that are derived from network sources and associated with network attributes

Non-isotropic/anisotropic A property that varies systematically along a given orientation, i.e. it is not uniform in all directions

Normalised difference vegetation index (NDVI) A normalised mathematical operation performed on the reflectances measured in two spectral bands of an optical/near-infrared remotely sensed image to yield a parameter that is correlated with the amount of vegetation present in an image pixel

Nugget The independent measurement or microscale error at the origin of either the covariance or semivariogram model that appears as a discontinuity at a scale that is too fine to detect

Null hypothesis The hypothesis that there is no significant difference between specified, sampled populations

Object-based image analysis (OBIA) A form of raster-based image analysis that begins with image segmentation, such that the units of organisation in subsequent analysis are 'objects', or segments, as opposed to grid pixels

Oblique An angle at which an aerial photograph is taken with the axis of the camera held between the horizontal plane of the ground and the vertical plane perpendicular to the ground

Oceanography The branch of science that studies the physical and biological properties of the sea

Optical rangefinder A device that measures the distance between an observer and a target

Optimisation The selection of the best mathematical function, with regard to some specified criteria, from a set of potential alternatives

Orthophotograph An aerial photograph that has been corrected for the effects of distortions relating to camera tilt, lens geometry and ground relief

Orthorectification The process of correcting the geometry of an image to remove distortion related to relief of the terrain over which the image was taken

Overlay The superimposition of two or more maps or image layers registered to a common coordinate system

Parameter A variable that determines the outcome of a function or operation

Parametric statistics Parametric statistics assume the value of a parameter for a given analysis

Parsimony The principle that, with respect to modelling relationships in the natural world, things are usually connected in the simplest or most economical way

Passive remote sensing system A remote sensing instrument that detects naturally occurring radiation, either in the form of reflected sunlight or thermal radiation

Photogrammetry The technique of obtaining quantitative measurements of the geometric properties of a feature of interest from one or more photographs of it

Pixel The discrete units (picture elements) of which digital images are composed; these commonly take the form of squares in a grid

Poisson distribution A discrete frequency distribution plot that expresses the probability of obtaining a given number of results within a given time frame

Polygon A closed shape on a vector-based map defined by a connected sequence of x, y coordinate pairs.

Polynomial A mathematical expression consisting of more than two algebraic terms that have been raised to a power of 2 or more

Portal A data portal is a repository for storing, viewing and interrogating spatially referenced data

Positive feedback A reaction that occurs within a system that acts in such a way as to accentuate the effect of the initial stimulus

Positivism A philosophical system that recognises only types of knowledge that can be scientifically verified or that which is capable of logical or mathematical proof

Precision The quality of being exact, as represented by the number of digits (decimal places) provided in a number

Predictive model A model that serves to predict the value of a phenomenon of interest in the natural world, either for unsampled locations in space or through time

Principal components analysis (PCA) A form of multivariate statistical analysis in which variation within a dataset of multiple dimensions is reduced to the dimensions along which the most variation is observable (i.e. the principal components)

Projection The geometric transformation between the geographical coordinates of a point on the Earth's surface and the Cartesian coordinates on a map or remotely sensed image

Propagation of error *See* 'Error propagation'

Quantitative revolution A major paradigmatic turning point in the history of geography characterised by increased use of statistical techniques, particularly multivariate analysis, in geographical research

R^2 *See* 'Coefficient of determination'

Radiance The electromagnetic power falling on a plane surface of a given area, from a given solid angle (units $W\ m^{-2}\ sr^{-1}$)

Random error Stochastic or inconsistent errors present in measured or modelled values of a constant attribute or quantity

Range The distance at which the characteristic levelling out of the Geary statistic (or some measure of semivariance) occurs for paired point samples in a dataset; sample locations separated by distances closer than the range are spatially autocorrelated, whereas locations farther apart than the range are not

Raster format Representation of spatial data, such as a map or image, as a grid of pixels

Regression A statistical technique for exploring the relationship between the values of one output variable (the dependent variable) and the corresponding values of one or more input variables (the independent variable(s))

Relational database A database structure in which collections of tables, structured as spatial data matrices, are logically associated with each other by shared fields

Remote sensing The collection and interpretation of information about the natural environment and the surface of the Earth from a distance (i.e. remotely), primarily by sensing radiation that is naturally emitted or reflected by the Earth's surface, or by detecting signals transmitted from a device and reflected back to it; examples of remote sensing methods include aerial photography, radar and satellite imaging

Representation A method of expressing data about a given phenomenon so that it can be viewed, stored and transferred in digital format

Resampling The process of assigning digital numbers to the pixels in an image that has been spatially transformed by geometric correction (e.g. the spatial resolution has been adjusted), using the digital numbers of the original (untransformed) image as input

Residual The difference between an observed value and an expected or modelled value of a dependent variable in a regression model

Residual mapping The act of plotting residual values in their geographic location in order to explore their spatial distribution

Resolution The fineness, or level of precision, with which an instrument can distinguish between different values of some measured property

Root mean square error (RMSE) The square root of the mean of the square of the sum total of all the error; a frequently used numerical summary of the difference between values predicted by a model and those observed

Scale The ratio of the length (linear dimension) of an object as it is represented on a map or image to the real length of the object

Semivariogram A summary measure of spatial autocorrelation for a series of paired point samples or observations of values for a given phenomenon; it is represented as a plot of distance between pairs of sample locations (x axis) vs semivariance of measured values for those locations, commonly represented as the Geary's C statistic (y axis); such a plot helps to explore, visualise and empirically define a mathematical model to describe the variability of values for that phenomenon with location

Sensitivity analysis Analysis designed to test the stability or robustness of a model by assessing the effects of making small changes in model parameters or data structure on the output model results

Sessile An immobile organism that is fixed in one place

Sextant An instrument with a graduated arc of 60 degrees and a sighting mechanism, used for measuring the angular distances between objects, especially for sighting the angle between celestial objects (such as the Sun) and the horizon to determine altitudes in navigation and surveying

Sill point The semivariance value (usually calculated as the Geary's C statistic) that the semivariogram model attains at the range, i.e. the point where paired point samples are no longer autocorrelated

Sonar (sound navigation and ranging) A system for the detection of objects underwater by emitting sound pulses and detecting or measuring their return after being reflected

Spatial autocorrelation The property of random variables taking values, at pairs of locations a certain distance apart, that are more similar (positive autocorrelation) or less similar (negative autocorrelation) than expected for randomly associated pairs of random observations

Spatial data infrastructure (SDI) An interconnected framework of geographic data, metadata, users and tools that come together as a functional unit for making efficient use of spatial data

Spatial data matrix A common structure for storing, querying and analysing geographically referenced data in a matrix format in which rows relate to individual features or objects and columns relate to attributes associated with them; every row is associated with a coordinate pair that locates it in space

Spatial dependence The tendency for point samples in a dataset to adhere to Tobler's first law of geography

Spatial error model A regression model that incorporates a spatially explicit term based on the modelled error of the dependent variable at a neighbouring location

Spatial join A type of table join operation in which fields from one layer's attribute table are appended to another layer's attribute table based on the relative locations of the features in the two layers

Spatial lag A spatial lag is a distance, or lag, introduced into a model, usually to assist with the introduction of an autoregressive parameter (i.e. a term in which the dependent variable itself is introduced into the equation, but at a location subject to a spatial lag)

Spatial resolution A measure of the smallest distance between two objects that can be distinguished by a sensor; in raster datasets, this refers to the length of the side of a pixel square

Spatially lagged autoregressive parameter A spatial lag model expresses the notion that the value of a dependent variable at a given location is related to the values of the same variable measured at nearby locations, reflecting some kind of interaction effect; the spatially lagged autoregressive parameter represents the term in the model that corresponds to the dependent variable at a nearby location, i.e. after a spatial lag has been applied (*See* 'Spatial lag')

Species distribution model A type of model developed to explore how the occurrence of a species is related to environmental characteristics and, in turn, how a species might respond to changes in its environment

Spectral resolution The ability of a sensor to distinguish between different wavelengths (or frequencies) in an electromagnetic signal

Spectral signature The unique pattern of reflected electromagnetic radiation that identifies a target or feature object from the portions of the spectrum they reflect and absorb

Spectrometer An instrument that measures the intensity of electromagnetic radiation as a function of its frequency, once it has been reflected off a target; spectrometers are usually used for measuring the visible portion of the spectrum

Spline interpolation An interpolation technique in which the value at an unsampled location is estimated from a set of sample points using a polynomial function to approximate surface values in a manner that minimises overall curvature, resulting in a smooth surface that passes exactly through input data values at sampled locations

Standard deviation A statistical measure of the spread of values from their mean, calculated as the square root of the sum of the squared deviations from the mean value, divided by the number of elements; the standard deviation for a distribution is the square root of the variance

Standard error The standard deviation of the sampling distribution of a statistic, commonly the mean of a sample

Stationarity A characteristic of a spatial process in which statistical properties such as the mean, variance and autocorrelation do not change over space or time

Stochastic error *See* 'Random error'

Structured Query Language (SQL) A commonly used programming syntax for retrieving and manipulating data from a relational database hosted in a GIS

Supervised image classification The process of assigning each pixel in an image to one of a number of classes by comparing the properties of the pixel with the properties of those pixels known to belong to various classes (training data)

Surrogate variable A variable that acts as substitute for another variable to fulfil a given role within a mathematical model

Swath The strip of the Earth's surface from which data are collected with a remote sensor

Systematic error A reproducible inaccuracy introduced into a measurement dataset that is applied consistently, either as an offset or as a scaling factor

Tacheometry A system of surveying in which the relative horizontal and vertical positioning of a collection of points is established through a series of distances and angles

Temporal resolution The time interval between successive opportunities to acquire a remotely sensed image of a give location

Terrain variable A variable, usually derived from a remotely sensed dataset, that describes some characteristic of the terrain, e.g. a digital elevation model, slope or rugosity layer depicting the land surface

Thematic resolution *See* 'Descriptive resolution'

Theodolite triangulation A system of surveying in which a theodolite is used to characterise land features by constructing a geodetic framework based on triangles

Time-series analysis Analysis designed to elucidate the longer-term trend, seasonal and irregular fluctuations in a collection of observations that represent repeated measurements over time

Tobler's first law of geography An expression of the idea of spatial autocorrelation by Waldo Tobler (1930): *'Everything is related to everything else, but near things are more related than distant things'*

Topographic character Relating to the arrangement or accurate representation of the physical features of an area

Topology The study of geometrical properties and spatial relationships of a set of constituent parts that are related in some way or another; in relation to GIS, topological information defines how point, line and polygon features are related and how information can be shared between them

Transducer A device that converts variations in a physical quantity, such as pressure or light intensity, into an electrical signal

Trend surface analysis An analysis method to characterise general tendencies within a sampled data surface by fitting a polynomial surface by least squares regression through the sample data points; this method results in a surface that minimises the variance of the surface in relation to the input values

Triangulated irregular network (TIN) A digital data structure used to represent a surface composed of vectors that partition geographic space into contiguous, non-overlapping triangles; the vertices of each triangle are sample data points with corresponding x, y and z values, connected by lines to form Delaunay triangles

Triangulation The tracing and measurement of a network of triangles in order to determine the distances and relative positions of features in a given area, specifically by measuring the length of one side of each triangle (the baseline) and by establishing the angles and lengths of the other two sides through observation and deduction

Trilateration Determining the position of a point on the Earth's surface with respect to two other points by measuring the distances between all three points

Type I error A false positive; the incorrect rejection of the null hypothesis through statistical testing

Type II error A false negative; the failure to reject a false null hypothesis through statistical testing

Uncertainty The state of something, e.g. a parameter within a model, not being known

Universal Transverse Mercator (UTM) grid A widely used projected, conical coordinate system that divides the world into 60 north and south zones, each six degrees wide

Unsupervised image classification The process of assigning each pixel in an image to one of a number of classes based on the reflectance properties of pixels; statistically discrete groupings are called 'clusters'; the user identifies the number of clusters to generate and which bands to use, and with this information, the image classification software generates clusters using different image clustering algorithms such as K-means and ISODATA

Variable An element, feature or factor that can be changed within a mathematical model; an independent variable stands alone and is not liable to change by other variables, while a dependent variable is dependent on other factors and is likely to change as a function of the independent variables

Variance In general terms, the quality of being different; in statistical terms, variance can be calculated in relation to a population mean as the square of the standard deviation

Vector data A representation of the world using points, lines and polygons; vector models are useful for storing data that have discrete boundaries, such as country borders and shorelines

Water column correction The correction of aircraft- or satellite-received remote sensing data to allow for the effects on the radiation as it passes through the water column, e.g. the differential scattering and absorption experienced by different wavelengths of light, in particular the tendency of longer-wavelength (e.g. red) light to be attenuated more than shorter-wavelength (e.g. blue) light

Wavebands A range of wavelengths or frequencies of electromagnetic radiation

Waveform analysis The analysis of a waveform, created through the variable response across frequencies or wavelengths of electromagnetic radiation

Wavelength The distance between adjacent peaks or troughs, measured in the direction of propagation, in a harmonic wave of energy (e.g. light)

Weights matrix A matrix composed of numbers that indicate the importance of a variable for a particular calculation; the matrix elements correspond to geographically referenced data (e.g. the pixels of a raster grid) and can be used to formally specify the spatial relationships within the dataset; the larger the weight assigned to the variable, the more that variable will influence the outcome of the operation

REFERENCES

ABUODHA, P. A. & WOODROFFE, C. D. 2010. Assessing vulnerability to sea-level rise using a coastal sensitivity index: a case study from southeast Australia. *Journal of Coastal Conservation*, 14, 189–205.

ADLER-GOLDEN, S. M., ACHARYA, P. K., BERK, A., MATTHEW, M. W. & GORODETZKY, D. 2005. Remote bathymetry of the littoral zone from AVIRIS, LASH, and QuickBird imagery. *IEEE Transactions on Geoscience and Remote Sensing*, 43, 337–347.

AGISOFT. 2012. *Agisoft PhotoScan user manual*. Professional edition, version 0.9.0. AgiSoft LLC.

ALESHEIKH, A. A., GHORBANALI, A. & NOURI, N. 2007. Coastline change detection using remote sensing. *International Journal of Environmental Science and Technology*, 4, 61–66.

ALLEN, Y. C., WILSON, C. A., ROBERTS, H. H. & SUPAN, J. 2005. High resolution mapping and classification of oyster habitats in nearshore Louisiana using sidescan sonar. *Estuaries*, 28, 435–446.

ALMUNIA, J., BASTERRETXEA, G., ARÍ, J. & ULANOWICZ, R. 1999. Benthic–pelagic switching in a coastal subtropical lagoon. *Estuarine, Coastal and Shelf Science*, 49, 363–384.

ANDERS, F. J. & BYRNES, M. R. 1991. Accuracy of shoreline change rates as determined from maps and aerial photographs. *Shore and Beach*, 59, 17–26.

ANDRÉFOUËT, S., CLAEREBOUDT, M., MATSAKIS, P., PAGES, J. & DUFOUR, P. 2001. Typology of atoll rims in Tuamotu Archipelago (French Polynesia) at landscape scale using SPOT HRV images. *International Journal of Remote Sensing*, 22, 987–1004.

ANGELOUDIS, A., AHMADIAN, R., FALCONER, R. A. & BOCKELMANN-EVANS, B. 2016. Numerical model simulations for optimisation of tidal lagoon schemes. *Applied Energy*, 165, 522–536.

ANSELIN, L. 1995. Local indicators of spatial association – LISA. *Geographical Analysis*, 27, 93–115.

ANSELIN, L. & BERA, A. K. 1998. Spatial dependence in linear regression models with an introduction to spatial econometrics. In *Handbook of applied economic statistics*, Eds A. Ullah & D. E. A. Giles, Statistics: Textbooks and Monographs, vol. 155, pp. 237–290. Marcel Dekker.

ANSELIN, L., SYABRI, I. & KHO, Y. 2006. GeoDa: an introduction to spatial data analysis. *Geographical Analysis*, 38, 5–22.

ARBIA, G., GRIFFITH, D. & HAINING, R. 1998. Error propagation modelling in raster GIS: overlay operations. *International Journal of Geographical Information Science*, 12, 145–167.

ARIAS-GONZÁLEZ, J. E., ACOSTA-GONZÁLEZ, G., MEMBRILLO, N., GARZA-PÉREZ, J. R. & CASTRO-PÉREZ, J. M. 2012. Predicting spatially explicit coral reef fish abundance, richness and Shannon–Weaver index from habitat characteristics. *Biodiversity and Conservation*, 21, 115–130.

ATKINSON, P. M. & FOODY, G. M. 2002. Uncertainty in remote sensing and GIS: fundamentals. In *Uncertainty in remote sensing and GIS*, Eds G. M. Foody & P. M. Atkinson, pp. 1–18. John Wiley.

BAAS, A. C. 2002. Chaos, fractals and self-organization in coastal geomorphology: simulating dune landscapes in vegetated environments. *Geomorphology*, 48, 309–328.

BAILEY, H. & THOMPSON, P. M. 2009. Using marine mammal habitat modelling to identify priority conservation zones within a marine protected area. *Marine Ecology Progress Series*, 378, 279–287.

BAILEY, T. C. & GATRELL, A. C. 1995. *Interactive spatial data analysis*. Longman Scientific & Technical.

BARNARD, P. L., RUBIN, D. M., HARNEY, J. & MUSTAIN, N. 2007. Field test comparison of an autocorrelation technique for determining grain size using a digital 'beachball' camera versus traditional methods. *Sedimentary Geology*, 201, 180–195.

BARNES, R. 2010. Regional and latitudinal variation in the diversity, dominance and abundance of microphagous microgastropods and other benthos in intertidal beds of dwarf eelgrass, *Nanozostera* spp. *Marine Biodiversity*, 40, 95–106.

BARNES, R. & HAMYLTON, S. 2013. Abrupt transitions between macrobenthic faunal assemblages across seagrass bed margins. *Estuarine, Coastal and Shelf Science*, 131, 213–223.

BARNES, R. & HAMYLTON, S. 2016. On the very edge: faunal and functional responses to the interface between benthic seagrass and unvegetated sand assemblages. *Marine Ecology Progress Series*, 553, 33–48.

BARNES, R. S. K. & HUGHES, R. N. 2009. *An introduction to marine ecology*. John Wiley.

BARTHOLOMEW, A., BOHNSACK, J. A., SMITH, S. G., AULT, J. S., HARPER, D. E. & MCCLELLAN, D. B. 2008. Influence of marine reserve size and boundary length on the initial response of exploited reef fishes in the Florida Keys National Marine Sanctuary, USA. *Landscape Ecology*, 23, 55–65.

BARTLETT, D. J. 2000. Working on the frontiers of science: applying GIS to the coastal zone. In *Marine and coastal geographical information systems*, Eds D. J. Wright & D. J. Bartlett, pp. 11–24. Taylor & Francis.

BAVAUD, F. 1998. Models for spatial weights: a systematic look. *Geographical Analysis*, 30, 153–171.

BEACH PROTECTION AUTHORITY, QUEENSLAND. 1989. *Green Island data report*. Government Printer, Queensland.

BEGON, M., TOWNSEND, C. & HARPER, J. 2006. *Ecology: from individuals to ecosystems*. Blackwell Publishing.

BELL, S. S., FONSECA, M. S. & KENWORTHY, W. J. 2008. Dynamics of a subtropical seagrass landscape: links between disturbance and mobile seed banks. *Landscape Ecology*, 23, 67–74.

BELLUCO, E., CAMUFFO, M., FERRARI, S., MODENESE, L., SILVESTRI, S., MARANI, A. & MARANI, M. 2006. Mapping salt-marsh vegetation by multispectral and hyperspectral remote sensing. *Remote Sensing of Environment*, 105, 54–67.

BENENSON, W. & STÖCKER, H. 2002. *Handbook of physics*. Springer.

BENFIELD, S., GUZMAN, H., MAIR, J. & YOUNG, J. 2007. Mapping the distribution of coral reefs and associated sublittoral habitats in Pacific Panama: a comparison of optical satellite sensors and classification methodologies. *International Journal of Remote Sensing*, 28, 5047–5070.

BERBEROGLU, S. & AKIN, A. 2009. Assessing different remote sensing techniques to detect land use/cover changes in the eastern Mediterranean. *International Journal of Applied Earth Observation and Geoinformation*, 11, 46–53.

BERRY, B. J. 1964. Approaches to regional analysis: a synthesis. *Annals of the Association of American Geographers*, 54, 2–11.

BIE, S. & BECKETT, P. 1973. Comparison of four independent soil surveys by air-photo interpretation, Paphos area (Cyprus). *Photogrammetria*, 29, 189–202.

BILIO, M. & NIERMANN, U. 2004. Is the comb jelly really to blame for it all? *Mnemiopsis leidyi* and the ecological concerns about the Caspian Sea. *Marine Ecology Progress Series*, 269, 173–183.

BINDING, C., BOWERS, D. & MITCHELSON-JACOB, E. 2005. Estimating suspended sediment concentrations from ocean colour measurements in moderately turbid waters; the impact of variable particle scattering properties. *Remote Sensing of Environment*, 94, 373–383.

BIRIBO, N. 2012. Morphological development of reef islands on Tarawa Atoll, Kiribati. Thesis, University of Wollongong, Australia.

BISHOP, C. T., DONELAN, M. A. & KAHMA, K. K. 1992. Shore Protection Manual's wave prediction reviewed. *Coastal Engineering*, 17, 25–48.

BIVAND, R. S., PEBESMA, E. & GÓMEZ-RUBIO, V. 2013. *Applied spatial data analysis with R*. Springer.

BOAK, E. H. & TURNER, I. L. 2005. Shoreline definition and detection: a review. *Journal of Coastal Research*, 21, 688–703.

BOARD, C. 1967. *Maps as models*. Methuen.

BOLSTAD, P. V., GESSLER, P. & LILLESAND, T. M. 1990. Positional uncertainty in manually digitized map data. *International Journal of Geographical Information Systems*, 4, 399–412.

BOLSTAD, W. M. 2013. *Introduction to Bayesian statistics*. John Wiley.

BOOLE, G. 1854. *The laws of thought*. Walton and Maberly, London.

BRANCH, G. & PRINGLE, A. 1987. The impact of the sand prawn *Callianassa kraussi* Stebbing on sediment turnover and on bacteria, meiofauna, and benthic microflora. *Journal of Experimental Marine Biology and Ecology*, 107, 219–235.

BREY, L. 2009. Acclimation of kelp photosynthesis to seasonal changes in the underwater radiation regime of an Arctic fjord system. Dissertation, University of Bremen.

BROTTON, J. 2013. *A history of the world in 12 Maps*. Penguin.

BROWN, C. J., SMITH, S. J., LAWTON, P. & ANDERSON, J. T. 2011. Benthic habitat mapping: a review of progress towards improved understanding of the spatial ecology of the seafloor using acoustic techniques. *Estuarine, Coastal and Shelf Science*, 92, 502–520.

BRUNETTE, W., SODT, R., CHAUDHRI, R., GOEL, M., FALCONE, M., VAN ORDEN, J. & BORRIELLO, G. 2012. Open data kit sensors: a sensor integration framework for android at the application level. In *MobiSys'12: Proc. 10th Int. Conf. on Mobile Systems, Applications, and Services*, pp. 351–364. ACM.

BRUUN, P. 1983. Review of conditions for uses of the Bruun rule of erosion. *Coastal Engineering*, 7, 77–89.

BRUUN, P. 1988. The Bruun rule of erosion by sea-level rise: a discussion on large-scale two- and three-dimensional usages. *Journal of Coastal Research*, 4, 627–648.

BRYSON, M., JOHNSON-ROBERSON, M., MURPHY, R. J. & BONGIORNO, D. 2013. Kite aerial photography for low-cost, ultra-high spatial resolution multi-spectral mapping of intertidal landscapes. *PloS one*, 8, e73550.

BUJA, A., COOK, D. & SWAYNE, D. F. 1996. Interactive high-dimensional data visualization. *Journal of Computational and Graphical Statistics*, 5, 78–99.

BURROUGH, P. A. & McDONNELL, R. A. 2011. *Principles of geographical information systems*. Oxford University Press.

CALMANT, S. & BAUDRY, N. 1996. Modelling bathymetry by inverting satellite altimetry data: a review. *Marine Geophysical Research*, 18, 123–134.

CAMPBELL, J. B. 1996. *Introduction to remote sensing*. Taylor & Francis.

CAMPBELL, J. B. 2002. *Introduction to remote sensing*. CRC Press.

CARR, M. H., NEIGEL, J. E., ESTES, J. A., ANDELMAN, S., WARNER, R. R. & LARGIER, J. L. 2003. Comparing marine and terrestrial ecosystems: implications for the design of coastal marine reserves. *Ecological Applications*, 13, 90–107.

CARVER, S. J., CORNELIUS, S. C. & HEYWOOD, D. I. 2006. *An introduction to geographical information systems*. Pearson Education.

CASAGRANDE, C., FABRE, P., RAPHAEL, E. & VEYSSIE, M. 1989. Water/oil interfaces. *Europhysics Letters*, 9, 251–255.

CASELLA, E., ROVERE, A., PEDRONCINI, A., MUCERINO, L., CASELLA, M., CUSATI, L. A., VACCHI, M., FERRARI, M. & FIRPO, M. 2014. Study of wave runup using numerical models and low-altitude aerial photogrammetry: a tool for coastal management. *Estuarine, Coastal and Shelf Science*, 149, 160–167.

CECCHERELLI, G., PIAZZI, L. & CINELLI, F. 2000. Response of the non-indigenous *Caulerpa racemosa* (Forsskål) J. Agardh to the native seagrass *Posidonia oceanica* (L.) Delile: effect of density of shoots and orientation of edges of meadows. *Journal of Experimental Marine Biology and Ecology*, 243, 227–240.

CHAPMAN, V. & TREVARTHEN, C. 1953. General schemes of classification in relation to marine coastal zonation. *Journal of Ecology*, 41, 198–204.

CHAPPUIS, E., TERRADAS, M., CEFALÌ, M. E., MARIANI, S. & BALLESTEROS, E. 2014. Vertical zonation is the main distribution pattern of littoral assemblages on rocky shores at a regional scale. *Estuarine, Coastal and Shelf Science*, 147, 113–122.

CHOLLETT, I. & MUMBY, P. 2012. Predicting the distribution of *Montastraea* reefs using wave exposure. *Coral Reefs*, 31, 493–503.

CHORLEY, R. J. 1964. Geography and analogue theory. *Annals of the Association of American Geographers*, 54, 127–137.

CHORLEY, R. J. 1967. Models in geomorphology. In *Models in geography*, Eds R. J. Chorley & P. Haggett, pp. 59–96. Methuen.

CHORLEY, R. J. & KENNEDY, B. A. 1971. *Physical geography: a systems approach*. Prentice-Hall.

CHOU, Y.-H. 1997. *Exploring spatial analysis in geographic information systems*. OnWord Press.

CHRISMAN, N. 1982. A theory of cartographic error and its measurement in digital data bases. In *Proc. Fifth Int. Symp. on Computer-Assisted Cartography and Int. Soc. for Photogrammetry and Remote Sensing. Commission IV*, pp. 159–168. Crystal City.

CHRISTIAN, R. R., FORÉS, E., COMIN, F., VIAROLI, P., NALDI, M. & FERRARI, I. 1996. Nitrogen cycling networks of coastal ecosystems: influence of trophic status and primary producer form. *Ecological Modelling*, 87, 111–129.

CHURCH, J. A., WHITE, N. J., COLEMAN, R., LAMBECK, K. & MITROVICA, J. X. 2004. Estimates of the regional distribution of sea level rise over the 1950–2000 period. *Journal of Climate*, 17, 2609–2625.

CLARK, C. 1993. Satellite remote sensing of marine pollution. *International Journal of Remote Sensing*, 14, 2985–3004.

CLARKE, J. I. 1966. Morphometry from maps. *Essays in Geomorphology*, 252, 235–274.

CLIFF, A. D. & ORD, J. K. 1981. *Spatial processes: models and applications*. Pion.

COHEN, K. M. 2005. 3D geostatistical interpolation and geological interpretation of paleo-groundwater rise in the Holocene coastal prism in the Netherlands. In *River deltas – concepts, models, and examples*, Eds L. Giosan & J. P. Bhattacharya, SEPM Publ. No. 83, pp. 341–364. Society for Sedimentary Geology.

COLLEN, J., GARTON, D. & GARDNER, J. 2009. Shoreline changes and sediment redistribution at Palmyra Atoll (equatorial Pacific Ocean): 1874–present. *Journal of Coastal Research*, 25, 711–722.

CONGALTON, R. G. 1991. A review of assessing the accuracy of classifications of remotely sensed data. *Remote Sensing of Environment*, 37, 35–46.

CONGALTON, R. G. & GREEN, K. 2008. *Assessing the accuracy of remotely sensed data: principles and practices*. CRC Press.

COOK, D., MAJURE, J. J., SYMANZIK, J. & CRESSIE, N. 1996. Dynamic graphics in a GIS: exploring and analyzing multivariate spatial data using linked software. *Computational Statistics*, 11, 467–480.

COOK, D., SYMANZIK, J., MAJURE, J. J. & CRESSIE, N. 1997. Dynamic graphics in a GIS: more examples using linked software. *Computers & Geosciences*, 23, 371–385.

COOLEY, T., ANDERSON, G., FELDE, G., HOKE, M., RATKOWSKI, A., CHETWYND, J., GARDNER, J., ADLER-GOLDEN, S., MATTHEW, M. & BERK, A. 2002. FLAASH, a MODTRAN4-based atmospheric correction algorithm, its application and validation. In *2002 IEEE Int. Geoscience and Remote Sensing Symp., IGARSS'02*, pp. 1414–1418. IEEE.

COOPER, J. & PILKEY, O. 2004. Longshore drift: trapped in an expected universe. *Journal of Sedimentary Research*, 74, 599–606.

COWELL, P. & THOM, B. 1994. *Morphodynamics of coastal evolution*, pp. 33–86. Cambridge University Press.

COWEN, R. K. & SPONAUGLE, S. 2009. Larval dispersal and marine population connectivity. *Annual Review of Marine Science*, 1, 443–466.

CRAIG, R. G. & LABOVITZ, M. L. 1980. Sources of variation in Landsat autocorrelation. In *Proc. 14th Int. Symp. on Remote Sensing of Environment*, San José, Costa Rica, pp. 1755–1767.

CRAMPTON, J. W. 2001. Maps as social constructions: power, communication and visualization. *Progress in Human Geography*, 25, 235–252.

CRAMPTON, J. W. 2011. *Mapping: a critical introduction to cartography and GIS*. John Wiley.

CRESSIE, N. A. 1993. *Statistics for spatial data*. John Wiley.

CRESSIE, N. A. & LIU, D. 2012. Spatial statistics in geographic information systems (GISs). In *Encyclopedia of environmetrics*, 2nd edn, Eds. A. H. El-Shaarawi & W. W. Piegorsch, pp. 1–5. John Wiley.

CRONE, G. R. 1966. *Maps and their makers: an introduction to the history of cartography*. Hutchinson.

CROWELL, M., LEATHERMAN, S. P. & BUCKLEY, M. K. 1991. Historical shoreline change: error analysis and mapping accuracy. *Journal of Coastal Research*, 7, 839–852.

CRUZ, J. 2007. *Ocean wave energy: current status and future perspectives*. Springer.

CURRAN, P. & NOVO, E. 1988. The relationship between suspended sediment concentration and remotely sensed spectral radiance: a review. *Journal of Coastal Research*, 4, 351–368.

CURRIER, K. 2015. Mapping with strings attached: kite aerial photography of Durai Island, Anambas Islands, Indonesia. *Journal of Maps*, 11, 589–597.

DALE, M. R. & FORTIN, M.-J. 2014. *Spatial analysis: a guide for ecologists*. Cambridge University Press.

DARWISH, A., LEUKERT, K. & REINHARDT, W. 2003. Image segmentation for the purpose of object-based classification. In *Int. Geoscience and Remote Sensing Symp., III*, pp. 2039–2041.

DAUBER, J., HIRSCH, M., SIMMERING, D., WALDHARDT, R., OTTE, A. & WOLTERS, V. 2003. Landscape structure as an indicator of biodiversity: matrix effects on species richness. *Agriculture, Ecosystems & Environment*, 98, 321–329.

DAVIES, A. J. & GUINOTTE, J. M. 2011. Global habitat suitability for framework-forming cold-water corals. *PloS one*, 6, e18483.

DAVIES, J. 1972. *Geographical variation in coastline development*, p. 204. Oliver and Boyd.

DAWSON, J. L. & SMITHERS, S. G. 2010. Shoreline and beach volume change between 1967 and 2007 at Raine Island, Great Barrier Reef, Australia. *Global and Planetary Change*, 72, 141–154.

DAY, J. C. 2002. Zoning – lessons from the Great Barrier Reef Marine Park. *Ocean & Coastal Management*, 45, 139–156.

DEBENHAM, F. 1936. Cheaper plane tables. *Geography*, 21, 222–224.

DEBENHAM, F. 1937. *Exercises in cartography*. Blackie.

DEBENHAM, F. 1956. *Map making*. Blackie.

DECKER, D. 2001. *GIS data sources*. John Wiley.

DEKKER, A. G., PHINN, S. R., ANSTEE, J., BISSETT, P., BRANDO, V. E., CASEY, B., FEARNS, P., HEDLEY, J., KLONOWSKI, W. & LEE, Z. P. 2011. Intercomparison of shallow water bathymetry, hydro-optics, and benthos mapping techniques in Australian and Caribbean coastal environments. *Limnology and Oceanography: Methods*, 9, 396–425.

DESA. 2013. *World population prospects: the 2012 revision*. Population Division, Department of Economic and Social Affairs, United Nations Secretariat.

DETHIER, M. N. & SCHOCH, G. C. 2005. The consequences of scale: assessing the distribution of benthic populations in a complex estuarine fjord. *Estuarine, Coastal and Shelf Science*, 62, 253–270.

DIGGLE, P. J. 2013. *Statistical analysis of spatial and spatio-temporal point patterns*. CRC Press.

DODGE, M., KITCHIN, R. & PERKINS, C. 2011. *The map reader: theories of mapping practice and cartographic representation*. John Wiley.

DOO, S. S., HAMYLTON, S. & BYRNE, M. 2012. Reef-scale assessment of intertidal large benthic foraminifera populations on One Tree Island, Great Barrier Reef and their future carbonate production potential in a warming ocean. *Zoological Studies*, 51, 1298–1307.

DOWD, M., GRANT, J. & LU, L. 2014. Predictive modeling of marine benthic macrofauna and its use to inform spatial monitoring design. *Ecological Applications*, 24, 862–876.

DUARTE, C. M. 1991. Seagrass depth limits. *Aquatic Botany*, 40, 363–377.

DUCE, S., VILA-CONCEJO, A., HAMYLTON, S., WEBSTER, J., BRUCE, E. & BEAMAN, R. 2016. A morphometric assessment and classification of coral reef spur and groove morphology. *Geomorphology*, 265, 68–83.

DUNN, R. & HARRISON, A. 1993. Two-dimensional systematic sampling of land use. *Journal of the Royal Statistical Society, Series C, Applied Statistics*, 42, 585–601.

EASTMAN, J. R. 2003. *IDRISI Kilimanjaro: guide to GIS and image processing*. Clark Labs, Clark University, Worcester, MA.

EKEBOM, J., LAIHONEN, P. & SUOMINEN, T. 2003. A GIS-based step-wise procedure for assessing physical exposure in fragmented archipelagos. *Estuarine, Coastal and Shelf Science*, 57, 887–898.

EL-ASHRY, M. & WANLESS, H. R. 1967. Shoreline features and their changes. *Photogrammetric Engineering*, 33, 184–189.
ELITH, J. & LEATHWICK, J. R. 2009. Species distribution models: ecological explanation and prediction across space and time. *Annual Review of Ecology, Evolution, and Systematics*, 40, 677.
ELLIS, M. Y. 1978. *Coastal mapping handbook*. US Government Printing Office.
ELVIDGE, C. D., DIETZ, J. B., BERKELMANS, R., ANDREFOUET, S., SKIRVING, W., STRONG, A. E. & TUTTLE, B. T. 2004. Satellite observation of Keppel Islands (Great Barrier Reef) 2002 coral bleaching using IKONOS data. *Coral Reefs*, 23, 123–132.
EVANS, R. D., MURRAY, K. L., FIELD, S. N., MOORE, J. A., SHEDRAWI, G., HUNTLEY, B. G., FEARNS, P., BROOMHALL, M., McKINNA, L. I. & MARRABLE, D. 2012. Digitise this! A quick and easy remote sensing method to monitor the daily extent of dredge plumes. *PloS one*, 7, e51668.
FABBRI, K. P. 1998. A methodology for supporting decision making in integrated coastal zone management. *Ocean & Coastal Management*, 39, 51–62.
FAIRBRIDGE, R. W. & TEICHERT, C. 1948. The Low Isles of the Great Barrier Reef: a new analysis. *Geographical Journal*, 111, 67–88.
FARINA, A. 2008. *Principles and methods in landscape ecology: towards a science of the landscape*. Springer.
FEAGIN, R. A., SHERMAN, D. J. & GRANT, W. E. 2005. Coastal erosion, global sea-level rise, and the loss of sand dune plant habitats. *Frontiers in Ecology and the Environment*, 3, 359–364.
FIGUEIREDO JR, A., SWIFT, D. & CLARKE, T. 1981. Sand ridges on the inner Atlantic shelf of North America: morphometric comparisons with Huthnance stability model. *Geo-Marine Letters*, 1, 187–191.
FIORENTINO, D., CARUSO, T. & TERLIZZI, A. 2012. Spatial autocorrelation in the response of soft-bottom marine benthos to gas extraction activities: the case of amphipods in the Ionian Sea. *Marine Environmental Research*, 79, 79–85.
FISCHER, M. & GETIS, A. 1997. *Recent developments in spatial analysis*. Springer.
FLEECER, J., PALMER, M. & MOSER, E. 1990. On the scale of aggregation of meio-benthic copepods on a tidal mudflat. *Marine Ecology*, 11, 227–237.
FLETCHER, C., ROONEY, J., BARBEE, M., LIM, S.-C. & RICHMOND, B. 2003. Mapping shoreline change using digital orthophotogrammetry on Maui, Hawaii. *Journal of Coastal Research*, 19, Special Issue 38, 106–124.
FLOOD, P. G. 1974. Sand movements on Heron Island – a vegetated sand cay Great Barrier Reef Province, Australia. In *Proc. Second Int. Symp. on Coral Reefs*, pp. 387–394.
FLOOD, P. G. 1979. Heron Island erosion problems. *Reeflections*. Great Barrier Reef Marine Park Authority.
FLOOD, P. G. & HEATWOLE, H. 1986. Coral cay instability and species-turnover of plants at Swain Reefs, Southern Great Barrier Reef, Australia. *Journal of Coastal Research*, 2, 479–496.
FONSECA, M. S. & BELL, S. S. 1998. Influence of physical setting on seagrass landscapes near Beaufort, North Carolina, USA. *Marine Ecology Progress Series*, 171, 109.
FORD, F. A. 1999. *Modeling the environment: an introduction to system dynamics models of environmental systems*. Island Press.
FORD, M. 2011. Shoreline changes on an urban atoll in the central Pacific Ocean: Majuro Atoll, Marshall Islands. *Journal of Coastal Research*, 28, 11–22.

FORMAN, R. T. T. 1995. *Land mosaics: the ecology of landscapes and regions*. Cambridge University Press.

FORMAN, R. T. T. & GODRON, M. 1981. Patches and structural components for a landscape ecology. *BioScience*, 31, 733–740.

FORMAN, R. T. T. & GODRON, M. 1986. *Landscape ecology*. John Wiley.

FOTHERINGHAM, A. S. & ROGERSON, P. 2013. *Spatial analysis and GIS*. CRC Press.

FOTHERINGHAM, A. S., BRUNSDON, C. & CHARLTON, M. 2000. *Quantitative geography: perspectives on spatial data analysis*. Sage.

FOUCAULT, M. 2000. So is it important to think? In *Power: the essential works of Michel Foucault 1954–1984*, Ed. J. B. Faubion, vol. 3, pp. 454–458. The New Press, New York.

FOWLER, S. & OREGIONI, B. 1976. Trace metals in mussels from the NW Mediterranean. *Marine Pollution Bulletin*, 7, 26–29.

FRANK, T. D. & JELL, J. S. 2006. Recent developments on a nearshore, terrigenous-influenced reef: Low Isles Reef, Australia. *Journal of Coastal Research*, 22, 474–486.

FRANKLIN, J. 2010. *Mapping species distributions: spatial inference and prediction*. Cambridge University Press.

FROHN, R. C. 1997. *Remote sensing for landscape ecology: new metric indicators for monitoring, modeling, and assessment of ecosystems*. CRC Press.

GAINES, S., GAYLORD, B., GERBER, L., HASTINGS, A. & KINLAN, B. 2007. Connecting places. The ecological consequences of dispersal in the sea. *Oceanography*, 20, 90–99.

GANCI, G. & SHORTIS, M. R. 1996. A comparison of the utility and efficiency of digital photogrammetry and industrial theodolite systems. *International Archives of Photogrammetry and Remote Sensing*, 31, 182–187.

GELMAN, A. 2004. Exploratory data analysis for complex models. *Journal of Computational and Graphical Statistics*, 13, 755–779.

GEOSCIENCE AUSTRALIA. 2008. *OzCoasts: Australian online coastal information*. Marine & Coastal Environment Group, Geoscience Australia.

GIARDINO, C., CANDIANI, G., BRESCIANI, M., LEE, Z., GAGLIANO, S. & PEPE, M. 2012. BOMBER: a tool for estimating water quality and bottom properties from remote sensing images. *Computers & Geosciences*, 45, 313–318.

GIBSON, C. & PHILLIPSON, J. 1983. The vegetation of Aldabra Atoll: preliminary analysis and explanation of the vegetation map. *Philosophical Transactions of the Royal Society of London B: Biological Sciences*, 302, 201–235.

GILBERT, C. L. 1986. Practitioner's corner: Professor Hendry's econometric methodology. *Oxford Bulletin of Economics and Statistics*, 48, 283–307.

GILHOUSEN, D. B. 1987. A field evaluation of NDBC moored buoy winds. *Journal of Atmospheric and Oceanic Technology*, 4, 94–104.

GLASGOW, H. B., BURKHOLDER, J. M., REED, R. E., LEWITUS, A. J. & KLEINMAN, J. E. 2004. Real-time remote monitoring of water quality: a review of current applications, and advancements in sensor, telemetry, and computing technologies. *Journal of Experimental Marine Biology and Ecology*, 300, 409–448.

GOODCHILD, M. F., LONGLEY, P., MAGUIRE, D. J. & RHIND, D. W. 2005. *Geographic information systems and science*. John Wiley.

GOODMAN, J. & USTIN, S. L. 2007. Classification of benthic composition in a coral reef environment using spectral unmixing. *Journal of Applied Remote Sensing*, 1, 011501.

GOODMAN, J. A., PURKIS, S. J. & PHINN, S. R. 2013. *Coral reef remote sensing: a guide for mapping, monitoring and management*. Springer.

GORNITZ, V. 1991. Global coastal hazards from future sea level rise. *Palaeogeography, Palaeoclimatology, Palaeoecology*, 89, 379–398.

GORNITZ, V. M., DANIELS, R. C., WHITE, T. W. & BIRDWELL, K. R. 1994. The development of a coastal risk assessment database: vulnerability to sea-level rise in the US Southeast. *Journal of Coastal Research*, 10, Special Issue 12, 327–338.

GOURLAY, M. 1988. Coral cays: products of wave action and geological processes in a biogenic environment. In *Proc. 6th Int. Coral Reef Symp.*, Townsville, Australia, pp. 491–496.

GRAF, W. L. 2006. Downstream hydrologic and geomorphic effects of large dams on American rivers. *Geomorphology*, 79, 336–360.

GREEN, D. & HARTLEY, S. 2000. Integrating photointerpretation and GIS for vegetation mapping: some issues of error. In *Vegetation mapping: from patch to planet*, In R. Alexander & A. C. Millington, pp. 103–134. John Wiley.

GREEN, E., MUMBY, P., EDWARDS, A. & CLARK, C. 1996. A review of remote sensing for the assessment and management of tropical coastal resources. *Coastal Management*, 24, 1–40.

GREEN, E. P., MUMBY, P. J., EDWARDS, A. J. & CLARK, C. D. 2005. *Remote sensing handbook for tropical coastal management*. Unesco.

GREIG-SMITH, P. 1961. Data on pattern within plant communities: I. The analysis of pattern. *Journal of Ecology*, 49, 695–702.

GRIFFITH, D. A. & LAYNE, L. J. 1999. *A casebook for spatial statistical data analysis: a compilation of analyses of different thematic data sets*. Oxford University Press.

GROBER-DUNSMORE, R., FRAZER, T. K., BEETS, J. P., LINDBERG, W. J., ZWICK, P. & FUNICELLI, N. A. 2008. Influence of landscape structure on reef fish assemblages. *Landscape Ecology*, 23, 37–53.

GUILLOT, B. & POUGET, F. 2015. UAV application in coastal environment, example of the Oleron Island for dunes and dikes survey. *International Archives of Photogrammetry, Remote Sensing and Spatial Information Sciences*, 40, 321.

GUINAN, J., GREHAN, A. J., DOLAN, M. F. & BROWN, C. 2009. Quantifying relationships between video observations of cold-water coral cover and seafloor features in Rockall Trough, west of Ireland. *Marine Ecology Progress Series*, 375, 125–138.

GUISAN, A. & ZIMMERMANN, N. E. 2000. Predictive habitat distribution models in ecology. *Ecological Modelling*, 135, 147–186.

HAGGETT, P. & CHORLEY, R. J. 1967. Models, paradigms and the new geography. *Models in Geography*, 19, 41.

HAINING, R. 1993. *Spatial data analysis in the social and environmental sciences*. Cambridge University Press.

HAINING, R. P. 2003. *Spatial data analysis*. Cambridge University Press.

HAMMERSCHLAG, N., GALLAGHER, A. J., WESTER, J., LUO, J. & AULT, J. S. 2012. Don't bite the hand that feeds: assessing ecological impacts of provisioning ecotourism on an apex marine predator. *Functional Ecology*, 26, 567–576.

HAMYLTON, S. 2011. Estimating the coverage of coral reef benthic communities from airborne hyperspectral remote sensing data: multiple discriminant function analysis and linear spectral unmixing. *International Journal of Remote Sensing*, 32, 9673–9690.

HAMYLTON, S. 2012. A comparison of spatially explicit and classic regression modelling of live coral cover using hyperspectral remote-sensing data in the Al Wajh Lagoon, Red Sea. *International Journal of Geographical Information Science*, 26, 2161–2175.

HAMYLTON, S. & EAST, H. 2012. A geospatial appraisal of ecological and geomorphic change on Diego Garcia Atoll, Chagos Islands (British Indian Ocean Territory). *Remote Sensing*, 4, 3444–3461.

HAMYLTON, S. M. & PROSPER, J. 2012. Development of a spatial data infrastructure for coastal management in the Amirante Islands, Seychelles. *International Journal of Applied Earth Observation and Geoinformation*, 19, 24–30.

HAMYLTON, S. M. & PUOTINEN, M. 2015. A meta-analysis of reef island response to environmental change on the Great Barrier Reef. *Earth Surface Processes and Landforms*, 40, 1006–1016.

HAMYLTON, S. & SPENCER, T. 2007. Classification of seagrass habitat structure as a response to wave exposure at Etoile Cay, Seychelles. *EARSeL eProceedings*, 6(2), 94–100.

HAMYLTON, S. & SPENCER, T. 2011. Geomorphological modelling of tropical marine landscapes: optical remote sensing, patches and spatial statistics. *Continental Shelf Research*, 31, S151–S161.

HAMYLTON, S., SPENCER, T. & HAGAN, A. 2012a. Spatial modelling of benthic cover using remote sensing data in the Aldabra lagoon, western Indian Ocean. *Marine Ecology Progress Series*, 460, 35–47.

HAMYLTON, S. M., HAGAN, A. B. & DOAK, N. 2012b. Observations of dugongs at Aldabra Atoll, western Indian Ocean: lagoon habitat mapping and spatial analysis of sighting records. *International Journal of Geographical Information Science*, 26, 839–853.

HAMYLTON, S., PESCUD, A., LEON, J. & CALLAGHAN, D. 2013. A geospatial assessment of the relationship between reef flat community calcium carbonate production and wave energy. *Coral Reefs*, 32, 1025–1039.

HAMYLTON, S., LEON, J. X., SAUNDERS, M. I. & WOODROFFE, C. 2014. Simulating reef response to sea-level rise at Lizard Island: a geospatial approach. *Geomorphology*, 222, 151–161.

HAMYLTON, S., CARVALHO, R., DUCE, S., ROELFSEMA, C. & VILA-CONCEJO, A. 2016. Linking pattern to process in reef sediment dynamics at Lady Musgrave Island, southern Great Barrier Reef. *Sedimentology*, 63, 1634–1650.

HANNA, E., NAVARRO, F. J., PATTYN, F., DOMINGUES, C. M., FETTWEIS, X., IVINS, E. R., NICHOLLS, R. J., RITZ, C., SMITH, B. & TULACZYK, S. 2013. Ice-sheet mass balance and climate change. *Nature*, 498, 51–59.

HARBORNE, A. R., MUMBY, P. J., ŻYCHALUK, K., HEDLEY, J. D. & BLACKWELL, P. G. 2006. Modeling the beta diversity of coral reefs. *Ecology*, 87, 2871–2881.

HARBORNE, A. R., MUMBY, P. J. & FERRARI, R. 2012. The effectiveness of different meso-scale rugosity metrics for predicting intra-habitat variation in coral-reef fish assemblages. *Environmental Biology of Fishes*, 94, 431–442.

HARISS, M., PURKIS, S., ELLIS, J., SWART, P. & REIJMER, J. 2013. Mapping depositional facies on Great Bahama Bank: an integration of groundtruthing and remote sensing methods. In *AGU Fall Meeting Abstracts*, p. 1717.

HARLEY, C. D. & HELMUTH, B. S. 2003. Local- and regional-scale effects of wave exposure, thermal stress, and absolute versus effective shore level on patterns of intertidal zonation. *Limnology and Oceanography*, 48, 1498–1508.

HARLEY, J. B. 1989. Deconstructing the map. *Cartographica: International Journal for Geographic Information and Geovisualization*, 26, 1–20.

HARLEY, J. B. & LAXTON, P. 2002. *The new nature of maps: essays in the history of cartography*. Johns Hopkins University Press.

HARRIS, R. & JARVIS, C. 2014. *Statistics for geography and environmental science*. Routledge.

HASEGAWA, H. 1990. Shoreline changes of the Hatenohama coral sand cays at the eastern end of Kume Island, the Ryukyus, Japan. *Geographical Review of Japan A*, 63, 676–692.

HEDLEY, J. D. 2013. Hyperspectral applications. In *Coral reef remote sensing*, Eds J. A. Goodman, S. J. Purkis & S. R. Phinn, pp. 79–108. Springer.

HEDLEY, J. D. & MUMBY, P. J. 2003. A remote sensing method for resolving depth and subpixel composition of aquatic benthos. *Limnology and Oceanography*, 48, 480–488.

HEDLEY, J., HARBORNE, A. & MUMBY, P. 2005. Technical note: simple and robust removal of sun glint for mapping shallow-water benthos. *International Journal of Remote Sensing*, 26, 2107–2112.

HEDLEY, J., ROELFSEMA, C., KOETZ, B. & PHINN, S. 2012. Capability of the Sentinel 2 mission for tropical coral reef mapping and coral bleaching detection. *Remote Sensing of Environment*, 120, 145–155.

HEEZEN, B. C. & THARP, M. 1977. World Ocean Floor Panorama, for *National Geographic* magazine. Full Colour, Painted by H. Berann. Mercator Projection, Scale 1:23,230,300.

HEUMANN, B. W. 2011. Satellite remote sensing of mangrove forests: recent advances and future opportunities. *Progress in Physical Geography*, 35, 87–108.

HIJMANS, R. J., CAMERON, S. E., PARRA, J. L., JONES, P. G. & JARVIS, A. 2005. Very high resolution interpolated climate surfaces for global land areas. *International Journal of Climatology*, 25, 1965–1978.

HINCHEY, E. K., NICHOLSON, M. C., ZAJAC, R. N. & IRLANDI, E. A. 2008. Preface: marine and coastal applications in landscape ecology. *Landscape Ecology*, 23, Supplement 1, 1–5.

HINKEL, J. & KLEIN, R. J. 2009. Integrating knowledge to assess coastal vulnerability to sea-level rise: the development of the DIVA tool. *Global Environmental Change*, 19, 384–395.

HIRST, B. & TODD, P. 2003. Defining boundaries within the tidal interface. In *Proc. Coastal GIS Conf.*, University of Wollongong, pp. 7–8. University of Wollongong Press.

HOCHBERG, E. J. & ATKINSON, M. 2000. Spectral discrimination of coral reef benthic communities. *Coral Reefs*, 19, 164–171.

HOCHBERG, E. J., ATKINSON, M. J. & ANDRÉFOUËT, S. 2003. Spectral reflectance of coral reef bottom-types worldwide and implications for coral reef remote sensing. *Remote Sensing of Environment*, 85, 159–173.

HOCK, K., WOLFF, N. H., CONDIE, S. A., ANTHONY, K. & MUMBY, P. J. 2014. Connectivity networks reveal the risks of crown-of-thorns starfish outbreaks on the Great Barrier Reef. *Journal of Applied Ecology*, 51, 1188–1196.

HOLDEN, H. & LEDREW, E. 1998. Spectral discrimination of healthy and non-healthy corals based on cluster analysis, principal components analysis, and derivative spectroscopy. *Remote Sensing of Environment*, 65, 217–224.

HOLDEN, H. & LEDREW, E. 1999. Hyperspectral identification of coral reef features. *International Journal of Remote Sensing*, 20, 2545–2563.

HOLDEN, H. & LEDREW, E. 2000. Accuracy assessment of hyperspectral classification of coral reef features. *Geocarto International*, 15, 7–14.

HOSSAIN, M. S., CHOWDHURY, S. R., DAS, N. G., SHARIFUZZAMAN, S. & SULTANA, A. 2009. Integration of GIS and multicriteria decision analysis for urban aquaculture development in Bangladesh. *Landscape and Urban Planning*, 90, 119–133.

HUBBARD, D. 1997. Reefs as dynamic systems. In *Life and death of coral reefs*, Ed C. Birkeland, pp. 43–67. Chapman & Hall.

HUGHES, I. & HASE, T. 2010. *Measurements and their uncertainties: a practical guide to modern error analysis*. Oxford University Press.

HYLAND, J. L., VAN DOLAH, R. F. & SNOOTS, T. R. 1999. Predicting stress in benthic communities of southeastern US estuaries in relation to chemical contamination of sediments. *Environmental Toxicology and Chemistry*, 18, 2557–2564.

IERODIACONOU, D., MONK, J., RATTRAY, A., LAURENSON, L. & VERSACE, V. 2011. Comparison of automated classification techniques for predicting benthic biological communities using hydroacoustics and video observations. *Continental Shelf Research*, 31, S28–S38.

INKPEN, R. & WILSON, G. 2013. *Science, philosophy and physical geography*. Routledge.

INMAN, D. L. & NORDSTROM, C. 1971. On the tectonic and morphologic classification of coasts. *Journal of Geology*, 79, 1–21.

JACKSON, C. W. 2010. *Basic user guide for the AMBUR package for R*, version 1.0a.

JARMALAVIČIUS, D., SATKŪNAS, J., ŽILINSKAS, G. & PUPIENIS, D. 2012. Dynamics of beaches of the Lithuanian coast (the Baltic Sea) for the period 1993–2008 based on morphometric indicators. *Environmental Earth Sciences*, 65, 1727–1736.

JOHNSON, G. C., SLOYAN, B. M., KESSLER, W. S. & McTAGGART, K. E. 2002. Direct measurements of upper ocean currents and water properties across the tropical Pacific during the 1990s. *Progress in Oceanography*, 52, 31–61.

JOHNSTON, M. W. & PURKIS, S. J. 2011. Spatial analysis of the invasion of lionfish in the western Atlantic and Caribbean. *Marine Pollution Bulletin*, 62, 1218–1226.

JONES, E. J. W. 1999. *Marine geophysics*. John Wiley.

JORGENSEN, S. J., REEB, C. A., CHAPPLE, T. K., ANDERSON, S., PERLE, C., VAN SOMMERAN, S. R., FRITZ-COPE, C., BROWN, A. C., KLIMLEY, A. P. & BLOCK, B. A. 2010. Philopatry and migration of Pacific white sharks. *Proceedings of the Royal Society B: Biological Sciences*, 277, 679–688.

JOYCE, K. E. 2005. A method for mapping live coral cover using remote sensing. Unpublished PhD thesis, University of Queensland, Australia.

JOYCE, K. E. & PHINN, S. R. 2013. Spectral index development for mapping live coral cover. *Journal of Applied Remote Sensing*, 7, 073590.

JUPP, D. 1988. Background and extensions to depth of penetration (DOP) mapping in shallow coastal waters. In *Proc. Symp. on Remote Sensing of the Coastal Zone, IV*, pp. 2.1–2.19.

KARPOUZLI, E. & MALTHUS, T. 2003. The empirical line method for the atmospheric correction of IKONOS imagery. *International Journal of Remote Sensing*, 24, 1143–1150.

KERSHAW, K. A. 1957. The use of cover and frequency in the detection of pattern in plant communities. *Ecology*, 38, 291–299.

KINLAN, B. P. & GAINES, S. D. 2003. Propagule dispersal in marine and terrestrial environments: a community perspective. *Ecology*, 84, 2007–2020.

KIRKMAN, H. 1990. Seagrass distribution and mapping. Seagrass research methods. *Monographs on Oceanographic Methodology*, 9, 19–25.

KITSIOU, D., COCCOSSIS, H. & KARYDIS, M. 2002. Multi-dimensional evaluation and ranking of coastal areas using GIS and multiple criteria choice methods. *Science of the Total Environment*, 284, 1–17.

KLEISNER, K. M. 2010. Modeling the spatial autocorrelation of pelagic fish abundance. *Marine Ecology Progress Series*, 411, 203–213.

KLEMAS, V. 2012. Airborne remote sensing of coastal features and processes: an overview. *Journal of Coastal Research*, 29, 239–255.

KLEMAS, V. V. 2015. Coastal and environmental remote sensing from unmanned aerial vehicles: an overview. *Journal of Coastal Research*, 31, 1260–1267.

KLONOWSKI, W. M., FEARNS, P. R. & LYNCH, M. J. 2007. Retrieving key benthic cover types and bathymetry from hyperspectral imagery. *Journal of Applied Remote Sensing*, 1, 011505.

KOMAR, P. D. 1998. *Beach processes and sedimentation*. Prentice-Hall.

KRAAK, M.-J. & ORMELING, F. 2011. *Cartography: visualization of spatial data*. Guilford Press.

KRAAN, C., VAN DER MEER, J., DEKINGA, A. & PIERSMA, T. 2009. Patchiness of macrobenthic invertebrates in homogenized intertidal habitats: hidden spatial structure at a landscape scale. *Marine Ecology Progress Series*, 383, 211–224.

KRAAN, C., AARTS, G., VAN DER MEER, J. & PIERSMA, T. 2010. The role of environmental variables in structuring landscape-scale species distributions in seafloor habitats. *Ecology*, 91, 1583–1590.

KRITZER, J. P. & SALE, P. F. 2010. *Marine metapopulations*. Academic Press.

KULLDORFF, M. 1997. A spatial scan statistic. *Communications in Statistics – Theory and Methods*, 26, 1481–1496.

KULLDORFF, M. & NAGARWALLA, N. 1995. Spatial disease clusters: detection and inference. *Statistics in Medicine*, 14, 799–810.

KUTSER, T., PARSLOW, J., CLEMENTSON, L., SKIRVING, W., DONE, T., WAKEFORD, M. & MILLER, I. 2000a. Hyperspectral detection of coral reef bottom types. In *Proc. of Ocean Optics XV*, Monaco, 13pp., CD-ROM.

KUTSER, T., SKIRVING, W., PARSLOW, J., DONE, T., CLEMENTSON, L., WAKEFORD, M. & MILLER, I. 2000b. Hyperspectral detection of coral reef health. In *Proc. 10th Australasian Remote Sensing Conf.*, Adelaide, pp. 931–948.

KUTSER, T., MILLER, I. & JUPP, D. L. 2006. Mapping coral reef benthic substrates using hyperspectral space-borne images and spectral libraries. *Estuarine, Coastal and Shelf Science*, 70, 449–460.

LANDGREBE, D. 1999. Information extraction principles and methods for multispectral and hyperspectral image data. *Information Processing for Remote Sensing*, 82, 3–38.

LEAMER, E. E. 1978. *Specification searches: ad hoc inference with nonexperimental data*. John Wiley.

LEATHWICK, J., MOILANEN, A., FRANCIS, M., ELITH, J., TAYLOR, P., JULIAN, K., HASTIE, T. & DUFFY, C. 2008. Novel methods for the design and evaluation of marine protected areas in offshore waters. *Conservation Letters*, 1, 91–102.

LE CORRE, N. J., JOHNSON, L. E., SMITH, G. K. & GUICHARD, F. 2015. Patterns and scales of connectivity: temporal stability and variation within a marine metapopulation. *Ecology*, 96, 2245–2256.

LEE, J. H., HUANG, Y., DICKMAN, M. & JAYAWARDENA, A. 2003. Neural network modelling of coastal algal blooms. *Ecological Modelling*, 159, 179–201.

LEE, Z., CARDER, K. L., MOBLEY, C. D., STEWARD, R. G. & PATCH, J. S. 1998. Hyperspectral remote sensing for shallow waters. 1. A semianalytical model. *Applied Optics*, 37, 6329–6338.

LEE, Z., CARDER, K. L., MOBLEY, C. D., STEWARD, R. G. & PATCH, J. S. 1999. Hyperspectral remote sensing for shallow waters: 2. Deriving bottom depths and water properties by optimization. *Applied Optics*, 38, 3831–3843.

LEFSKY, M. A., COHEN, W. B., PARKER, G. G. & HARDING, D. J. 2002. Lidar remote sensing for ecosystem studies. *BioScience*, 52, 19–30.

LEGENDRE, P. 1993. Spatial autocorrelation: trouble or new paradigm? *Ecology*, 74, 1659–1673.

LEGENDRE, P. & FORTIN, M. J. 1989. Spatial pattern and ecological analysis. *Vegetatio*, 80, 107–138.

LEGENDRE, P. & TROUSSELLIER, M. 1988. Aquatic heterotrophic bacteria: modeling in the presence of spatial autocorrelation. *Limnology and Oceanography*, 33, 1055–1067.

LEGENDRE, P., DALE, M. R., FORTIN, M. J., GUREVITCH, J., HOHN, M. & MYERS, D. 2002. The consequences of spatial structure for the design and analysis of ecological field surveys. *Ecography*, 25, 601–615.

LEON, J. 2010. Torres Strait reefs and carbonate production: a geospatial approach. Thesis, School of Earth and Environmental Sciences, University of Wollongong, Australia.

LEON, J. & WOODROFFE, C. D. 2011. Improving the synoptic mapping of coral reef geomorphology using object-based image analysis. *International Journal of Geographical Information Science*, 25, 949–969.

LESSER, M. & MOBLEY, C. 2007. Bathymetry, water optical properties, and benthic classification of coral reefs using hyperspectral remote sensing imagery. *Coral Reefs*, 26, 819–829.

LINTZ, J. & SIMONETT, D. 1976. *Remote sensing of environment*. Addison Wesley.

LIU, H., SHERMAN, D. & GU, S. 2007. Automated extraction of shorelines from airborne light detection and ranging data and accuracy assessment based on Monte Carlo simulation. *Journal of Coastal Research*, 23, 1359–1369.

LIVINGSTONE, D. 1993. *The geographical tradition: episodes in the history of a contested enterprise*. Wiley-Blackwell.

LLOYD, C. 2010. *Spatial data analysis: an introduction for GIS users*. Oxford University Press.

LONGHURST, A. R. 2010. *Ecological geography of the sea*. Academic Press.

LOUCHARD, E. M., REID, R. P., STEPHENS, F. C., DAVIS, C. O., LEATHERS, R. A. & DOWNES, T. V. 2003. Optical remote sensing of benthic habitats and bathymetry in coastal environments at Lee Stocking Island, Bahamas: a comparative spectral classification approach. *Limnology and Oceanography*, 48, 511–521.

LOYA, Y. 1972. Community structure and species diversity of hermatypic corals at Eilat, Red Sea. *Marine Biology*, 13, 100–123.

LU, D., MAUSEL, P., BATISTELLA, M. & MORAN, E. 2005. Land-cover binary change detection methods for use in the moist tropical region of the Amazon: a comparative study. *International Journal of Remote Sensing*, 26, 101–114.

LUCZKOVICH, J. J., WAGNER, T., MICHALEK, J. & STOFFLE, R. 1993. Discrimination of coral reefs, seagrass meadows, and sand bottom types from space: a Dominican Republic case study. *Photogrammetric Engineering and Remote Sensing*, 59, 385–389.

LUNDBLAD, E., WRIGHT, D., MILLER, J., LARKIN, E., RINEHART, R., BATTISTA, T., ANDERSON, S., NAAR, D. & DONAHUE, B. 2006. Classifying benthic terrains with multibeam bathymetry, bathymetric position and rugosity: Tutuila, American Samoa. *Marine Geodesy*, 29, 89–111.

LUŠIĆ, Z. 2013. The use of horizontal and vertical angles in terrestrial navigation. *Transactions on Maritime Science*, 2, 5–14.

LYZENGA, D. R. 1978. Passive remote sensing techniques for mapping water depth and bottom features. *Applied Optics*, 17, 379–383.

LYZENGA, D. R. 1981. Remote sensing of bottom reflectance and water attenuation parameters in shallow water using aircraft and Landsat data. *International Journal of Remote Sensing*, 2, 71–82.

MA, Z. & REDMOND, R. L. 1995. Tau coefficients for accuracy assessment of classification of remote sensing data. *Photogrammetric Engineering and Remote Sensing*, 61, 435–439.

MACARTHUR, R. H. & WILSON, E. O. 1967. *The theory of island biogeography*. Princeton University Press.

MACLEOD, R. D. & CONGALTON, R. G. 1998. A quantitative comparison of change-detection algorithms for monitoring eelgrass from remotely sensed data. *Photogrammetric Engineering and Remote Sensing*, 64, 207–216.

MALTHUS, T. J. & MUMBY, P. J. 2003. Remote sensing of the coastal zone: an overview and priorities for future research. *International Journal of Remote Sensing*, 24, 2805–2815.

MANDELBROT, B. B. 1967. How long is the coast of Britain. *Science*, 156, 636–638.

MARTINEZ, P. A. & HARBAUGH, J. W. 1989. Computer simulation of wave and fluvial-dominated nearshore environments. In *Applications in coastal modeling*, Eds V. C. Lakhan & A. S. Trenhaile, Elsevier Oceanography Series, vol. 49, pp. 297–340. Elsevier.

MARTINO, R. L. & SANDERSON, D. D. 1993. Fourier and autocorrelation analysis of estuarine tidal rhythmites, lower Breathitt Formation (Pennsylvanian), eastern Kentucky, USA. *Journal of Sedimentary Research*, 63, 105–119.

MATHER, P. M. 1979. Theory and quantitative methods in geomorphology. *Progress in Physical Geography*, 3, 471–487.

MATHER, P. M. & KOCH, M. 2011. *Computer processing of remotely-sensed images: an introduction*. John Wiley.

MAXWELL, D., STELZENMÜLLER, V., EASTWOOD, P. & ROGERS, S. 2009. Modelling the spatial distribution of plaice (*Pleuronectes platessa*), sole (*Solea solea*) and thornback ray (*Raja clavata*) in UK waters for marine management and planning. *Journal of Sea Research*, 61, 258–267.

MAY, V. 1993. Coastal tourism, geomorphology and geological conservation: the example of south central England. In *Tourism vs environment: the case for coastal areas*, Ed P. P. Wong. Springer.

McBRATNEY, A. B. & WEBSTER, R. 1986. Choosing functions for semi-variograms of soil properties and fitting them to sampling estimates. *Journal of Soil Science*, 37, 617–639.

McCUTCHEON, S. & McCUTCHEON, B. 2003. *The facts on file marine science handbook*. Infobase.

McGARVEY, R., FEENSTRA, J. E., MAYFIELD, S. & SAUTTER, E. V. 2010. A diver survey method to quantify the clustering of sedentary invertebrates by the scale of spatial autocorrelation. *Marine and Freshwater Research*, 61, 153–162.

McGRANAHAN, G., BALK, D. & ANDERSON, B. 2007. The rising tide: assessing the risks of climate change and human settlements in low elevation coastal zones. *Environment and Urbanization*, 19, 17–37.

McINTYRE, M. L., NAAR, D. F., CARDER, K. L., DONAHUE, B. T. & MALLINSON, D. J. 2006. Coastal bathymetry from hyperspectral remote sensing data: comparisons with high resolution multibeam bathymetry. *Marine Geophysical Researches*, 27, 129–136.

McKERGOW, L. A., PROSSER, I. P., HUGHES, A. O. & BRODIE, J. 2005. Sources of sediment to the Great Barrier Reef world heritage area. *Marine Pollution Bulletin*, 51, 200–211.

McLAUGHLIN, S., McKENNA, J. & COOPER, J. 2002. Socio-economic data in coastal vulnerability indices: constraints and opportunities. *Journal of Coastal Research*, 36, 487–497.

McLEAN, R., STODDART, D., HOPLEY, D. & POLACH, H. 1978. Sea level change in the Holocene on the northern Great Barrier Reef. *Philosophical Transactions of the Royal Society A: Mathematical, Physical and Engineering Sciences*, 291, 167–186.

MEEKAN, M., STEVEN, A. & FORTIN, M. 1995. Spatial patterns in the distribution of damselfishes on a fringing coral reef. *Coral Reefs*, 14, 151–161.

MELLIN, C., RUSSELL, B. D., CONNELL, S. D., BROOK, B. W. & FORDHAM, D. A. 2012. Geographic range determinants of two commercially important marine molluscs. *Diversity and Distributions*, 18, 133–146.

MELTON, M. A. 1957. *An analysis of the relations among elements of climate, surface properties, and geomorphology*. Technical Report No. 11, Office of Naval Research.

MERRITTS, D. & VINCENT, K. R. 1989. Geomorphic response of coastal streams to low, intermediate, and high rates of uplift, Medocino triple junction region, northern California. *Geological Society of America Bulletin*, 101, 1373–1388.

MERTON, L., BOURN, D. & HNATIUK, R. 1976. Giant tortoise and vegetation interactions on Aldabra Atoll – Part 1: Inland. *Biological Conservation*, 9, 293–304.

MICHALEK, J. L., WAGNER, T. W., LUCZKOVICH, J. J. & STOFFLE, R. W. 1993. Multispectral change vector analysis for monitoring coastal marine environments. *Photogrammetric Engineering and Remote Sensing*, 59, 381–384.

MILLER, J. 2010. Species distribution modeling. *Geography Compass*, 4, 490–509.

MILNE, P. 1972. *In situ* underwater surveying by plane table and alidade. *Underwater Journal and Information Bulletin*, 4, 59–63.

MILTON, E. 1987. Review article principles of field spectroscopy. *Remote Sensing*, 8, 1807–1827.

MOBLEY, C. D. 1994. *Light and water: radiative transfer in natural waters*. Academic Press.

MOORE, L. J. 2000. Shoreline mapping techniques. *Journal of Coastal Research*, 16, 111–124.

MORTON, R. A., GIBEAUT, J. C. & PAINE, J. G. 1995. Meso-scale transfer of sand during and after storms: implications for prediction of shoreline movement. *Marine Geology*, 126, 161–179.

MUMBY, P. & GREEN, E. 2000. Mapping coral reefs and macroalgae. In *Remote sensing handbook for tropical coastal management*, Eds E. P. Green, P. J. Mumby, A. J. Edwards & C. D. Clark, pp. 155–174. Unesco.

MUMBY, P. J., SKIRVING, W., STRONG, A. E., HARDY, J. T., LeDREW, E. F., HOCHBERG, E. J., STUMPF, R. P. & DAVID, L. T. 2004. Remote sensing of coral reefs and their physical environment. *Marine Pollution Bulletin*, 48, 219–228.

MUSSETTER, W. 1956. Tacheometric surveying. *Surveying and Mapping*, 16, 137–156.

MUTTITANON, W. & TRIPATHI, N. 2005. Land use/land cover changes in the coastal zone of Ban Don Bay, Thailand using Landsat 5 TM data. *International Journal of Remote Sensing*, 26, 2311–2323.

MYERS, M., HARDY, J., MAZEL, C. & DUSTAN, P. 1999. Optical spectra and pigmentation of Caribbean reef corals and macroalgae. *Coral Reefs*, 18, 179–186.

NASON, J. 1975. Reconnaissance and plot mapping of coral atolls: a simplified rangefinder method. *Atoll Research Bulletin*, 185, 13–20.

NAVEH, Z. 1987. Biocybernetic and thermodynamic perspectives of landscape functions and land use patterns. *Landscape Ecology*, 1, 75–83.

NAVIDI, W., MURPHY, W. S. & HEREMAN, W. 1998. Statistical methods in surveying by trilateration. *Computational Statistics & Data Analysis*, 27, 209–227.

NELLEMANN, C. & CAMERON, R. D. 1996. Effects of petroleum development on terrain preferences of calving caribou. *Arctic*, 49, 23–28.

NELSON, J. 1993. Conservation and use of the Mai Po Marshes, Hong Kong. *Natural Areas Journal*, 13, 215–219.

NEREM, R. S., LEULIETTE, E. & CAZENAVE, A. 2006. Present-day sea-level change: a review. *Comptes Rendus Geoscience*, 338, 1077–1083.

NEUMANN, B., VAFEIDIS, A. T., ZIMMERMANN, J. & NICHOLLS, R. J. 2015. Future coastal population growth and exposure to sea-level rise and coastal flooding – a global assessment. *PloS one*, 10, e0118571.

NOETZLI, J., HUGGEL, C., HOELZLE, M. & HAEBERLI, W. 2006. GIS-based modelling of rock–ice avalanches from Alpine permafrost areas. *Computational Geosciences*, 10, 161–178.

ODEN, S. R. 1978. Spatial autocorrelation in biology. 1. Methodology. *Biological Journal of the Linnean Society*, 10, 199–228.

O'DRISCOLL, R. L. 1998. Description of spatial pattern in seabird distributions along line transects using neighbour *K* statistics. *Marine Ecology Progress Series*, 165, 81–94.

ODUM, E. P., ODUM, H. T. & ANDREWS, J. 1971. *Fundamentals of ecology*. W. B. Saunders.

OGDEN, J. 1997. Ecosystem interactions in the tropical coastal seascape. In *Life and death of coral reefs*, Ed C. Birkeland, pp. 288–297. Chapman and Hall.

OLIVER, T. S., ROGERS, K., CHAFER, C. J. & WOODROFFE, C. D. 2012. Measuring, mapping and modelling: an integrated approach to the management of mangrove and saltmarsh in the Minnamurra River estuary, southeast Australia. *Wetlands Ecology and Management*, 20, 353–371.

OLSEN, M. J., YOUNG, A. P. & ASHFORD, S. A. 2012. TopCAT – topographical compartment analysis tool to analyze seacliff and beach change in GIS. *Computers & Geosciences*, 45, 284–292.

O'NEILL, R. V., RIITTERS, K. H., WICKHAM, J. & JONES, K. B. 1999. Landscape pattern metrics and regional assessment. *Ecosystem Health*, 5, 225–233.

PAINE, R. T. & LEVIN, S. A. 1981. Intertidal landscapes: disturbance and the dynamics of pattern. *Ecological Monographs*, 51, 145–178.

PANIGADA, S., ZANARDELLI, M., MACKENZIE, M., DONOVAN, C., MÉLIN, F. & HAMMOND, P. S. 2008. Modelling habitat preferences for fin whales and striped dolphins in the Pelagos Sanctuary (western Mediterranean Sea) with physiographic and remote sensing variables. *Remote Sensing of Environment*, 112, 3400–3412.

PATTON, D. R. 1975. A diversity index for quantifying habitat 'edge'. *Wildlife Society Bulletin*, 3, 171–173.

PAUL, R. W. & BINKER, A. 1990. *Critical thinking: what every person needs to survive in a rapidly changing world.* Center for Critical Thinking and Moral Critique.

PEREIRA, E., BENCATEL, R., CORREIA, J., FÉLIX, L., GONÇALVES, G., MORGADO, J. & SOUSA, J. 2009. Unmanned air vehicles for coastal and environmental research. *Journal of Coastal Research*, 25, Special Issue 56, 1557–1561.

PÉREZ-ALBERTI, A. & TRENHAILE, A. S. 2015. An initial evaluation of drone-based monitoring of boulder beaches in Galicia, north-western Spain. *Earth Surface Processes and Landforms*, 40, 105–111.

PETERSON, D. L., BRASS, J. A., SMITH, W. H., LANGFORD, G., WEGENER, S., DUNAGAN, S., HAMMER, P. & SNOOK, K. 2003. Platform options of free-flying satellites, UAVs or the International Space Station for remote sensing assessment of the littoral zone. *International Journal of Remote Sensing*, 24, 2785–2804.

PHILPOT, W. D. 1989. Bathymetric mapping with passive multispectral imagery. *Applied Optics*, 28, 1569–1578.

PHINN, S. R., ROELFSEMA, C. M. & MUMBY, P. J. 2012. Multi-scale, object-based image analysis for mapping geomorphic and ecological zones on coral reefs. *International Journal of Remote Sensing*, 33, 3768–3797.

PINCKNEY, J. & SANDULLI, R. 1990. Spatial autocorrelation analysis of meiofaunal and microalgal populations on an intertidal sandflat: scale linkage between consumers and resources. *Estuarine, Coastal and Shelf Science*, 30, 341–353.

POLOCZANSKA, E. S., HAWKINS, S. J., SOUTHWARD, A. J. & BURROWS, M. T. 2008. Modeling the response of populations of competing species to climate change. *Ecology*, 89, 3138–3149.

POPPER, K. 2014. *Conjectures and refutations: the growth of scientific knowledge*. Routledge.

PRICE, W. A. 1954. Dynamic environments: reconnaissance mapping, geologic and geomorphic, of continental shelf of Gulf of Mexico. *Transactions of the Gulf Coast Association of Geological Societies*, 4, 75–107.

PRICE, W. A. 1955. *Correlation of shoreline type with offshore bottom conditions*. Department of Oceanography, Agricultural and Mechanical College of Texas.

PUOTINEN, M. 2007. Modelling the risk of cyclone wave damage to coral reefs using GIS: a case study of the Great Barrier Reef, 1969–2003. *International Journal of Geographical Information Science*, 21, 97–120.

PURKIS, S. J. 2005. A 'reef-up' approach to classifying coral habitats from IKONOS imagery. *IEEE Transactions on Geoscience and Remote Sensing*, 43, 1375–1390.

PURKIS, S. J. & BROCK, J. C. 2013. Lidar overview. In *Coral reef remote sensing* , Eds J. A. Goodman, S. J. Purkis & S. R. Phinn, pp. 115–138. Springer.

RAAIJMAKERS, R., KRYWKOW, J. & VAN DER VEEN, A. 2008. Flood risk perceptions and spatial multi-criteria analysis: an exploratory research for hazard mitigation. *Natural Hazards*, 46, 307–322.

RADIARTA, I. N., SAITOH, S.-I. & MIYAZONO, A. 2008. GIS-based multi-criteria evaluation models for identifying suitable sites for Japanese scallop (*Mizuhopecten yessoensis*) aquaculture in Funka Bay, southwestern Hokkaido, Japan. *Aquaculture*, 284, 127–135.

RALPH, P. J. & SHORT, F. T. 2002. Impact of the wasting disease pathogen, *Labyrinthula zosterae*, on the photobiology of eelgrass *Zostera marina*. *Marine Ecology Progress Series*, 226, e271.

REDFERN, J., FERGUSON, M., BECKER, E., HYRENBACH, K., GOOD, C. P., BARLOW, J., KASCHNER, K., BAUMGARTNER, M. F., FORNEY, K. & BALLANCE, L. 2006. Techniques for cetacean-habitat modeling. *Marine Ecology Progress Series*, 310, 271–295.

REES, G. 1999. *The remote sensing data book*. Cambridge University Press.

REISS, H., CUNZE, S., KÖNIG, K., NEUMANN, H. & KRÖNCKE, I. 2011. Species distribution modelling of marine benthos: a North Sea case study. *Marine Ecology Progress Series*, 442, 71–86.

REY, S. J. & ANSELIN, L. 2010. PySAL: a Python library of spatial analytical methods. In *Handbook of applied spatial analysis*. Springer.

RICHARDS, K. 2009. Geography and the physical sciences tradition. *Key Concepts in Geography*, 2, 21–45.

RICHARDSON, P. L. & WALSH, D. 1986. Mapping climatological seasonal variations of surface currents in the tropical Atlantic using ship drifts. *Journal of Geophysical Research: Oceans*, 91, 10537–10550.

RIEGL, B. & GUARIN, H. 2013. Acoustic methods overview. In *Coral reef remote sensing*. Springer.

RIITTERS, K. H., O'NEILL, R., HUNSAKER, C., WICKHAM, J. D., YANKEE, D., TIMMINS, S., JONES, K. & JACKSON, B. 1995. A factor analysis of landscape pattern and structure metrics. *Landscape Ecology*, 10, 23–39.

RIPLEY, B. D. 1991. *Statistical inference for spatial processes*. Cambridge University Press.

RITZ, D. A. 1994. Social aggregation in pelagic invertebrates. *Advances in Marine Biology*, 30, 155–216.

ROBBINS, B. D. & BELL, S. S. 1994. Seagrass landscapes: a terrestrial approach to the marine subtidal environment. *Trends in Ecology & Evolution*, 9(8), 301–304.

ROBINSON, A. H. 1958. *Elements of cartography*. John Wiley.

ROBINSON, I. S. 2004. *Measuring the oceans from space: the principles and methods of satellite oceanography*. Springer.

ROBINSON, L., ELITH, J., HOBDAY, A., PEARSON, R., KENDALL, B., POSSINGHAM, H. & RICHARDSON, A. 2011. Pushing the limits in marine species distribution modelling: lessons from the land present challenges and opportunities. *Global Ecology and Biogeography*, 20, 789–802.

ROCA, E., GAMBOA, G. & TÀBARA, J. D. 2008. Assessing the multidimensionality of coastal erosion risks: public participation and multicriteria analysis in a Mediterranean coastal system. *Risk Analysis*, 28, 399–412.

RODGERS III, J. C., MURRAH, A. W. & COOKE, W. H. 2009. The impact of Hurricane Katrina on the coastal vegetation of the Weeks Bay Reserve, Alabama from NDVI data. *Estuaries and Coasts*, 32, 496–507.

ROELFSEMA, C. & PHINN, S. 2006. *Spectral reflectance library of selected biotic and abiotic coral reef features in Heron Reef*. Centre for Remote Sensing & Spatial Information Science, University of Queensland.

ROGERS, K., SAINTILAN, N. & COPELAND, C. 2012. Modelling wetland surface elevation dynamics and its application to forecasting the effects of sea-level rise on estuarine wetlands. *Ecological Modelling*, 244, 148–157.

ROSSER, N. J., PETLEY, D. N., LIM, M., DUNNING, S. & ALLISON, R. J. 2005. Terrestrial laser scanning for monitoring the process of hard rock coastal cliff erosion. *Quarterly Journal of Engineering Geology and Hydrogeology*, 38, 363–375.

ROVERE, A., CASELLA, E., VACCHI, M., MUCERINO, L., PEDRONCINI, A., FERRARI, M. & FIRPO, M. 2014. Monitoring beach evolution using low-altitude aerial photogrammetry and UAV drones. In *EGU General Assembly Conf. Abstracts*, 8341. Copernicus Publications.

SALA, E. & GRAHAM, M. H. 2002. Community-wide distribution of predator–prey interaction strength in kelp forests. *Proceedings of the National Academy of Sciences*, 99, 3678–3683.

SALLENGER JR, A., KRABILL, W., SWIFT, R., BROCK, J., LIST, J., HANSEN, M., HOLMAN, R., MANIZADE, S., SONTAG, J. & MEREDITH, A. 2003. Evaluation of airborne topographic lidar for quantifying beach changes. *Journal of Coastal Research*, 19, 125–133.

SANDULLI, R. & PINCKNEY, J. 1999. Patch sizes and spatial patterns of meiobenthic copepods and benthic microalgae in sandy sediments: a microscale approach. *Journal of Sea Research*, 41, 179–187.

SAUNDERS, M. I., LEON, J. X., CALLAGHAN, D. P., ROELFSEMA, C. M., HAMYLTON, S., BROWN, C. J., BALDOCK, T., GOLSHANI, A., PHINN, S. R. & LOVELOCK, C. E. 2014. Interdependency of tropical marine ecosystems in response to climate change. *Nature Climate Change*, 4, 724–729.

SCHNEIDER, D. C. 1990. Spatial autocorrelation in marine birds. *Polar Research*, 8, 89–97.

SCOFFIN, T. 1982. Reef aerial photography from a kite. *Coral Reefs*, 1, 67–69.

SCOFFIN, T. 1993. The geological effects of hurricanes on coral reefs and the interpretation of storm deposits. *Coral Reefs*, 12, 203–221.

SHALABY, A. & TATEISHI, R. 2007. Remote sensing and GIS for mapping and monitoring land cover and land-use changes in the northwestern coastal zone of Egypt. *Applied Geography*, 27, 28–41.

SHARPLES, C., MOUNT, R. & PEDERSEN, T. 2009. *The Australian coastal smartline geomorphic and stability map, Version 1: Manual and data dictionary*. Report of the University of Tasmania, for the Department of Climate Change and Geoscience, Australia.

SHEPPARD, C. R. 1998. Biodiversity patterns in Indian Ocean corals, and effects of taxonomic error in data. *Biodiversity & Conservation*, 7, 847–868.

SHI, W., FISHER, P. & GOODCHILD, M. F. 2003. *Spatial data quality*. CRC Press.

SHIH, S.-M. & KOMAR, P. D. 1994. Sediments, beach morphology and sea cliff erosion within an Oregon coast littoral cell. *Journal of Coastal Research*, 10, 144–157.

SIEBERT, T. & BRANCH, G. M. 2007. Influences of biological interactions on community structure within seagrass beds and sandprawn-dominated sandflats. *Journal of Experimental Marine Biology and Ecology*, 340, 11–24.

SINGH, A. 1989. Digital change detection techniques using remotely-sensed data. *International Journal of Remote Sensing*, 10, 989–1003.

SKLAR, F. H., COSTANZA, R. & DAY, J. W. 1985. Dynamic spatial simulation modeling of coastal wetland habitat succession. *Ecological Modelling*, 29, 261–281.

SKOP, E. & SØRENSEN, P. B. 1998. GIS-based modelling of solute fluxes at the catchment scale: a case study of the agricultural contribution to the riverine nitrogen loading in the Vejle Fjord catchment, Denmark. *Ecological Modelling*, 106, 291–310.

SLAYMAKER, O. & EMBLETON-HAMANN, C. 2009. *Geomorphology and global environmental change*. Cambridge University Press.

SMITH, T. M., HINDELL, J. S., JENKINS, G. P., CONNOLLY, R. M. & KEOUGH, M. J. 2011. Edge effects in patchy seagrass landscapes: the role of predation in determining fish distribution. *Journal of Experimental Marine Biology and Ecology*, 399, 8–16.

SOBEL, D. 2007. *Longitude: the true story of a lone genius who solved the greatest scientific problem of his time*. Bloomsbury.

SPALDING, M., RAVILIOUS, C. & GREEN, E. P. 2001. *World atlas of coral reefs*. University of California Press.

SPEAK, P. 2008. *Deb: geographer, scientist, Antarctic explorer; a biography of Frank Debenham*. Polar.

SPENCER, T., HAGAN, A. & HAMYLTON, S. 2009. *Atlas of the Amirantes*. Department of Geography, University of Cambridge.

STEELE, J. H. 1991. Can ecological theory cross the land–sea boundary? *Journal of Theoretical Biology*, 153, 425–436.

STEERS, J. A. 1929. The Queensland coast and the Great Barrier Reefs. *Geographical Journal*, 74, 232–257.

STEERS, J. A. 1945. Coral reefs and air photography. *Geographical Journal*, 106, 232–235.

STEHMAN, S. 1996. Estimating the kappa coefficient and its variance under stratified random sampling. *Photogrammetric Engineering and Remote Sensing*, 62, 401–407.

STELZENMÜLLER, V., ROGERS, S. I. & MILLS, C. M. 2008. Spatio-temporal patterns of fishing pressure on UK marine landscapes, and their implications for spatial planning and management. *ICES Journal of Marine Science: Journal du Conseil*, 65, 1081–1091.

STEPHENSON, T. & STEPHENSON, A. 1949. The universal features of zonation between tide-marks on rocky coasts. *Journal of Ecology*, 37, 289–305.

STEVENS, S. S. 1946. On the theory of scales of measurement. *Science*, 103, 677–680.

STIEGELER, S. E. 1977. *Dictionary of Earth sciences*. Pica Press.

STOCKDON, H. F., SALLENGER JR, A. H., LIST, J. H. & HOLMAN, R. A. 2002. Estimation of shoreline position and change using airborne topographic lidar data. *Journal of Coastal Research*, 18, 502–513.

STOCKER, T. F. 2014. *Climate change 2013: the physical science basis. Working Group I contribution to the Fifth Assessment Report of the Intergovernmental Panel on Climate Change*. Cambridge University Press.

STODDART, D. R. 1962. *Three Caribbean atolls: Turneffe Islands, Lighthouse Reef, and Glover's Reef, British Honduras*. Atoll Research Bulletin 87. Pacific Science Board, National Academy of Sciences.

STODDART, D. R. 1978. Mapping reefs and islands. In *Coral reefs: research methods*, Eds D. R. Stoddart & R. E. Johannes, Monographs on Oceanographic Methodology, pp. 17–22. Unesco.

STODDART, D. R. & STEERS, J. 1977. The nature and origin of coral reef islands. *Biology and Geology of Coral Reefs*, 4, 59–105.

STODDART, D. R. & TAYLOR, J. 1971. *Geography and ecology of Diego Garcia Atoll, Chagos*

Archipelago. Smithsonian Institution.

STODDART, D. R., McLEAN, R., SCOFFIN, T. & GIBBS, P. 1978. Forty-five years of change on low wooded islands, Great Barrier Reef. *Philosophical Transactions of the Royal Society B: Biological Sciences*, 284, 63–80.

STOW, D., TINNEY, L. & ESTES, J. 1980. Deriving land use/land cover change statistics from LANDSAT: a study of prime agricultural land. In *Proc. Int. Symp. on Remote Sensing of Environment*, pp. 1–11.

STRONG, A. E. & McCLAIN, E. P. 1984. Improved ocean surface temperatures from space – comparisons with drifting buoys. *Bulletin of the American Meteorological Society*, 65, 138–142.

STUMPF, R. P., HOLDERIED, K. & SINCLAIR, M. 2003. Determination of water depth with high-resolution satellite imagery over variable bottom types. *Limnology and Oceanography*, 48, 547–556.

SUMMERFIELD, M. A. 2014. *Global geomorphology*. Routledge.

SWALES, A., BENTLEY, S. J., LOVELOCK, C. & BELL, R. G. 2007. Sediment processes and mangrove-habitat expansion on a rapidly-prograding muddy coast, New Zealand. In *Proc. Coastal Sediments '07*, pp. 1441–1454. ASCE.

TAIT, R. V. & DIPPER, F. 1998. *Elements of marine ecology*. Butterworth-Heinemann.

TANNER, C. C., MIGDAL, C. J. & JONES, M. T. 1998. The clipmap: a virtual mip-map. In *Proc. 25th Annual Conf. on Computer Graphics and Interactive Techniques*, pp. 151–158. ACM.

TANNER, J. E. 2005. Edge effects on fauna in fragmented seagrass meadows. *Austral Ecology*, 30, 210–218.

TEICHERT, C. 1995. Photointerpretation of coral reefs in Australia during World War II. *Carbonates and Evaporites*, 10, 102–104.

THARP, M. 1999. Connect the dots: mapping the seafloor and discovering the mid-ocean ridge. In *Lamont-Doherty Earth Observatory: twelve perspectives on the first fifty years, 1949–1999*, Ed. L. Lippsett. Lamont-Doherty Earth Observatory of Columbia University. See Woods Hole Oceanographic Institution. Mary Sears Woman Pioneer in Oceanography Award. *Marie Tharp bio.* Available at http://www.whoi.edu/sbl/liteSite.do?litesiteid=9092&articleId=13407.

THIELER, E. R. & DANFORTH, W. W. 1994. Historical shoreline mapping (I): improving techniques and reducing positioning errors. *Journal of Coastal Research*, 10, 549–563.

THIELER, E. R., SMITH, T. L., KNISEL, J. M. & SAMPSON, D. W. 2013. *Massachusetts Shoreline Change Mapping and Analysis Project, 2013 Update*. US Geological Survey.

THRUSH, S. F. 1991. Spatial patterns in soft-bottom communities. *Trends in Ecology & Evolution*, 6, 75–79.

TOBLER, W. R. 1970. A computer movie simulating urban growth in the Detroit region. *Economic Geography*, 46, Supplement, 234–240.

TOLSTOY, I. & CLAY, C. S. 1966. *Ocean acoustics*. McGraw-Hill.

TOMLIN, D. C. 1990. *Geographic information systems and cartographic modeling*. Prentice-Hall.

TORRES, L. G., ROSEL, P. E., D'AGROSA, C. & READ, A. J. 2003. Improving management of overlapping bottlenose dolphin ecotypes through spatial analysis and genetics. *Marine Mammal Science*, 19, 502–514.

TROLL, C. 1939. Luftbildplan und ökologische Bodenforschung: ihr zweckmäßiger Einsatz für die wissenschaftliche Erforschung und praktische Erschließung wenig bekannter Länder. *Zeitschrift der Gesellschaft für Erdkunde zu Berlin*, 1939(7/8), 241–298.

TUKEY, J. W. 1977. *Exploratory data analysis*. Addison-Wesley.

UNWIN, D. J. 1995. Geographical information systems and the problem of 'error and uncertainty'. *Progress in Human Geography*, 19, 549–558.

UPTON, G. & FINGLETON, B. 1985. *Spatial data analysis by example*, vol. 1, *Point pattern and quantitative data*. John Wiley.

VALAVANIS, V. D., PIERCE, G. J., ZUUR, A. F., PALIALEXIS, A., SAVELIEV, A., KATARA, I. & WANG, J. 2008. Modelling of essential fish habitat based on remote sensing, spatial analysis and GIS. *Hydrobiologia*, 612, 5–20.

VALDES, M. J. 1981. Heuristic models of inquiry. *New Literary History*, 12, 253–267.

VEIT, R. R., SILVERMAN, E. D. & EVERSON, I. 1993. Aggregation patterns of pelagic predators and their principal prey, Antarctic krill, near South Georgia. *Journal of Animal Ecology*, 62, 551–564.

VELIMIROV, B. & GRIFFITHS, C. 1979. Wave-induced kelp movement and its importance for community structure. *Botanica Marina*, 22, 169–172.

VERSTAPPEN, H. T. 1977. *Remote sensing in geomorphology*. Elsevier Scientific.

VILES, H. A. 1988. *Biogeomorphology*. Basil Blackwell.

VILES, H. A. & SPENCER, T. 2014. *Coastal problems: geomorphology, ecology and society at the coast*. Routledge.

VILLA, F., TUNESI, L. & AGARDY, T. 2002. Zoning marine protected areas through spatial multiple-criteria analysis: the case of the Asinara Island National Marine Reserve of Italy. *Conservation Biology*, 16, 515–526.

VINCENT, R. K. 1997. *Fundamentals of geological and environmental remote sensing*. Prentice-Hall.

VON BERTALANFFY, L. 1956. General system theory. *General Systems*, 1, 11–17.

WADELL, H. 1932. Volume, shape, and roundness of rock particles. *Journal of Geology*, 40, 443–451.

WAGNER, H. H. & FORTIN, M.-J. 2005. Spatial analysis of landscapes: concepts and statistics. *Ecology*, 86, 1975–1987.

WARD, T., VANDERKLIFT, M., NICHOLLS, A. & KENCHINGTON, R. 1999. Selecting marine reserves using habitats and species assemblages as surrogates for biological diversity. *Ecological Applications*, 9, 691–698.

WEBB, A. P. & KENCH, P. S. 2010. The dynamic response of reef islands to sea-level rise: evidence from multi-decadal analysis of island change in the Central Pacific. *Global and Planetary Change*, 72, 234–246.

WERNER, A. 1967. A calendar of the development of surveying. *Australian Surveyor*, 21, 265–289.

WETTLE, M., FERRIER, G., LAWRENCE, A. & ANDERSON, K. 2003. Fourth derivative analysis of Red Sea coral reflectance spectra. *International Journal of Remote Sensing*, 24, 3867–3872.

WILBER, P. & IOCCO, L. E. 2003. Using a GIS to examine changes in the bathymetry of borrow pits and in lower bay, New York Harbor, USA. *Marine Geodesy*, 26, 49–61.

WILLIAMS, S. J., PENLAND, S. & SALLENGER, A. H. 1992. *Atlas of shoreline changes in Louisiana from 1853 to 1989*. US Geological Survey.

WILLIAMS, W. & LANCE, G. 1977. Hierarchical classificatory methods. *Statistical Methods for Digital Computers*, 3, 269–295.

WILLIS, C. M. & GRIGGS, G. B. 2003. Reductions in fluvial sediment discharge by coastal dams in California and implications for beach sustainability. *Journal of Geology*, 111, 167–182.

WIRSING, A. J., HEITHAUS, M. R. & DILL, L. M. 2007. Can measures of prey availability improve our ability to predict the abundance of large marine predators? *Oecologia*, 153, 563–568.

WOLFRAM, S. 1994. *Cellular automata and complexity: collected papers*. Addison-Wesley.

WOOD, D. & FELS, J. 1992. *The power of maps*. Guilford Press.

WOODROFFE, C. D. 2002. *Coasts: form, process and evolution*. Cambridge University Press.

WOODY, C., SHIH, E., MILLER, J., ROYER, T., ATKINSON, L. P. & MOODY, R. S. 2000. Measurements of salinity in the coastal ocean: a review of requirements and technologies. *Marine Technology Society Journal*, 34, 26–33.

WRIGHT, D. J. 2009. Spatial data infrastructures for coastal environments. In *Remote sensing and geospatial technologies for coastal ecosystem assessment and management*, Ed X. Yang, pp. 91–112. Springer.

WRIGHT, D. J. & BARTLETT, D. J. 2000. *Marine and coastal geographical information systems*. CRC Press.

WRIGHT, D. J., SATALOFF, G., LAVOI, T., MEINER, A. & UHEL, R. 2010. Overview of coastal atlases. In *Coastal informatics: web atlas design and implementation*, Ed D. J. Wright, p. 80. Information Science Reference.

WRIGHT, L. D. & THOM, B. G. 1977. Coastal depositional landforms: a morphodynamic approach. *Progress in Physical Geography*, 1, 412–459.

WRIGHT, L. D., SHORT, A. & GREEN, M. 1985. Short-term changes in the morphodynamic states of beaches and surf zones: an empirical predictive model. *Marine Geology*, 62, 339–364.

YAMANO, H. & TAMURA, M. 2004. Detection limits of coral reef bleaching by satellite remote sensing: simulation and data analysis. *Remote Sensing of Environment*, 90, 86–103.

YAMANO, H., KAYANNE, H. & CHIKAMORI, M. 2005. An overview of the nature and dynamics of reef islands. *Global Environmental Research – English Edition*, 9, 9.

YAMANO, H., SHIMAZAKI, H., MURASE, T., ITOU, K., SANO, S., SUZUKI, Y., LEENDERS, N., FORSTREUTER, W. & KAYANNE, H. 2006. Construction of digital elevation models for atoll islands using digital photogrammetry. In *GIS for the Coastal Zone: a Selection of Papers from CoastGIS*, pp. 165–175. University of Wollongong.

YAMANO, H., KAYANNE, H., YAMAGUCHI, T., KUWAHARA, Y., YOKOKI, H., SHIMAZAKI, H. & CHIKAMORI, M. 2007. Atoll island vulnerability to flooding and inundation revealed by historical reconstruction: Fongafale Islet, Funafuti Atoll, Tuvalu. *Global and Planetary Change*, 57, 407–416.

YONGE, C. M. 1929. *Great Barrier Reef expedition, 1928–1929*. British Museum.

INDEX

Some index entries refer to figures and tables. These entries are identified by either an 'f' or a 't', so, for example, '226f. 8.6' is p. 226, Figure 8.6, and '187t. 7.1' is p. 187, Table 7.1.